T0122454

Big Data Analytics for Time-Critical Mobility Forecasting

George A. Vouros • Gennady Andrienko •
Christos Doulkeridis • Nikolaos Pelekis •
Alexander Artikis • Anne-Laure Jousselme •
Cyril Ray • Jose Manuel Cordero • David Scarlatti
Editors

Big Data Analytics for Time-Critical Mobility Forecasting

From Raw Data to Trajectory-Oriented
Mobility Analytics in the Aviation and
Maritime Domains

 Springer

Editors
George A. Vouros
Department of Digital Systems
University of Piraeus
Piraeus, Greece

Gennady Andrienko
Intelligent Analysis & Info Systems
Fraunhofer Institute IAIS
Sankt Augustin, Germany

Christos Doulkeridis
Department of Digital Systems
University of Piraeus
Piraeus, Greece

Nikolaos Pelekis
Department of Statistics and Insurance
Science
University of Piraeus
Piraeus, Greece

Alexander Artikis
Complex Event Recognition Group
NCSR "Demokritos"
Agia Paraskevi, Greece

Anne-Laure Jousselme
Centre for Maritime Research and
Experimentation (CMRE)
NATO Science and Technology
Organization (STO)
La Spezia, Italy

Cyril Ray
Naval Research Institute (IRENav)
Arts & Metiers-Paris Tech
Brest, France

Jose Manuel Cordero
Edificio Allende
CRIDA
Madrid, Spain

David Scarlatti
Boeing Research & Technology Europe
Madrid, Spain

ISBN 978-3-030-45166-0   ISBN 978-3-030-45164-6   (eBook)
https://doi.org/10.1007/978-3-030-45164-6

This Springer imprint is published by the registered company Springer Nature Switzerland AG.
The registered company address is: Gewerbestrasse 11, 6330 Cham, Switzerland

This book is dedicated to all those who struggle for a better own and others' trajectory with philotimo,[1] respect and with no obsession.

---

[1] Philotimo (or filotimo) is a Greek word, which is difficult to translate. One may start from here, among other references: https://en.wikipedia.org/wiki/Philotimo.

Preface

Spatiotemporal mobility data has a significant role and impact on the global economy and our everyday lives. The improvements along the last decades in terms of data management, planning of operations, security of operations, information provision to operators and end-users have been driven by location-centered information. While a shift of paradigm regarding mobility data towards trajectory-oriented tasks is emerging in several domains, the ever-increasing volume of data emphasizes the need for advanced methods supporting detection and prediction of events and trajectories, supplemented by advanced visual analytic methods, over multiple heterogeneous, voluminous, fluctuating, and noisy data streams of moving entities. This book provides a comprehensive and detailed description of Big Data solutions towards activity detection and forecasting in very large numbers of moving entities spread across large geographical areas. Specifically, following a trajectory-oriented approach, this book reports on the state-of-the-art methods for the detection and prediction of trajectories and important events related to moving entities, together with advanced visual analytics methods, over multiple heterogeneous, voluminous, fluctuating, and noisy data streams from moving entities, correlating them with data from archived data sources expressing, among others, entities' characteristics, geographical information, mobility patterns, regulations, and intentional data (e.g., planned routes), in a timely manner. Solutions provided are motivated, validated, and evaluated in user-defined challenges focusing on increasing the safety, efficiency, and economy of operations concerning moving entities in the air-traffic management and maritime domains.

The book contents have been structured into six parts:

The first part provides the motivating points and background for mobility forecasting supported by trajectory-oriented analytics. It presents specific problems and challenges in the aviation (air-traffic management) and the maritime domains and clarifies operational concerns and objectives in both domains. It presents domain-specific terminology used in examples and cases, in which technology is demonstrated, evaluated/validated, throughout the book. Equally important to the above is the presentation of the data sources exploited per domain, the big data challenges ahead in both domains, and of course, the requirements from technologies presented in subsequent parts of the book. These chapters present data exploited for operational purposes in the aviation and maritime domains and provide an initial understanding of spatiotemporal data through specific examples. They also present challenges and motivating points by means of operational scenarios where technology can help, putting the technologies presented in subsequent parts of the book in a unique frame: This helps us understand why technological achievements are necessary, what are the domain-specific requirements driving developments in analytics, data storage, and processing, and what are the data processing, data management, and data-driven analytics tools needed to advance operational goals towards trajectory-based operations.

The second part focuses on big data quality assessment and processing, as applied in the data sources and according to the requirements and objectives presented in the first part of the book. This, second part of the book, presents novel technologies, appropriate to serve mobility analytics components that are presented in subsequent sections. In doing so, workflows regarding data sources' quality assessment via visual analytics methods are considered to be essential to understand inherent features and imperfections of data, affecting the ways data should be processed and managed, as the first section of this part shows. In addition to this, methods for online construction of streamed data synopses are presented, towards addressing big data challenges presented by surveillance, mostly, data sources.

The third part of this book specifies solutions towards managing big spatiotemporal data: The first section specifies a generic ontology revolving around the notion of trajectory so as to model data and information that is necessary for analytics components. This ontology provides a generic model for constructing knowledge graphs integrating data from disparate data sources. In conjunction to this, this part describes novel methods for integrating data from archived and streamed data sources. Special emphasis is given to enriching data streams and integrating streamed and archival data to provide coherent views of mobility: This is addressed by real-time methods discovering topological and proximity relations among spatiotemporal entities. Finally, distributed storage of integrated dynamic and archived mobility data—i.e. large knowledge graphs constructed according to the generic model introduced—are within focus.

The next part focuses on mobility analytics methods exploiting (online) processed, synopsized, and enriched data streams as well as (offline) integrated, archived mobility data. Specifically, online future location prediction methods and trajectory prediction methods are presented, distinguishing between short-term and the challenging long-term predictions. Recognition of complex events in challenging cases for detecting complex events is thoroughly presented. In addition to this, an industry-strong maritime anomaly detection service capable of processing daily real-world data volumes is presented. This part focuses also on offline trajectory analytics, addressing trajectory clustering and detection of routes followed by mobile entities. Novel algorithms for subtrajectory clustering are proposed and evaluated.

The fifth part presents how methods addressing data management, data processing, and mobility analytics are integrated in a big data architecture that has distinctive characteristics when compared to known big data paradigmatic architectures. We call this architectural paradigm, which is based on well-defined principles for building analytics pipelines δ. This paradigm is instantiated to a specific architecture realizing the datAcron integrated system prototype. This part presents the software stack of the datAcron system, together with issues concerning individual, online, and offline components integration.

The last part focuses on important ethical issues that research on mobility analytics should address: This is deemed to be crucial, given the growth of interest in that topic in computer science and operational stakeholders, necessitating the sharing of data and distributing the processing among stakeholders.

All chapters present background information on the specific topics they address, detailed and rigorous specification of scientific and technological problems considered, and state-of-the-art methods addressing these problems, together with novel approaches that authors have developed, evaluated, and validated, mainly during the last 3 years of their involvement in the datAcron H2020 ICT Big Data Project. Evaluation and validation results per method are presented using data sets from both, maritime and aviation domain, showing the potential and the limitations of methods presented, also according to the requirements specified in the first part of the book. The chapters present also technical details about implementations of methods, aiming to address big data challenges, so as to achieve the latency and throughput requirements set in both domains.

In doing so, this book aims to present a reference book to all stakeholders in different domains with mobility detection and forecasting needs and computer science disciplines aiming to address data-driven mobility data exploration, processing, storage, and analysis problems.

I would like to take the opportunity to thank everybody who contributed to the exciting effort of developing mobility data processing, storage, analysis solutions in time-critical domains, whose state of the art is summarized in this book. These, as part of a much wider community, include all co-editors and chapter authors of this publication. This book is a concerted effort of many people who worked and continue to work together in different, but always exciting, lines of research for mobility analytics.

Piraeus, Greece   George A. Vouros
February 2020

Acknowledgements

The developments described in the chapters of this book have been developed in the course of work in several past and on-going research projects, whose support is also acknowledged by the authors of each chapter. However, ideas and large part of this work have been developed, evaluated, and validated, mainly during the 3 years of involvement in the datAcron H2020 ICT Big Data Project.

datAcron has been funded by the European Union's Horizon 2020 Programme under grant agreement No. 687591. datAcron is a research and innovation collaborative project whose aim was to introduce novel methods to detect threats and abnormal activity of very large numbers of moving entities in large geographic areas.

Towards this target, datAcron advanced the management and integrated exploitation of voluminous and heterogeneous data-at-rest (archival data) and data-in-motion (streaming data) sources, so as to significantly advance the capacities of systems to promote safety and effectiveness of critical operations for large numbers of moving entities in large geographical areas.

Technological developments in datAcron have been validated and evaluated in user-defined challenges that aim at increasing the safety, efficiency, and economy of operations concerning moving entities in the air-traffic management (ATM) and maritime domains.

The datAcron addressed the following core challenges:

- **Distributed management and querying of integrated spatiotemporal RDF data-at-rest and data-in-motion in integrated manners**: datAcron advanced RDF data processing and spatiotemporal query answering for very large numbers of real-world triples and spatiotemporal queries, providing also native support for trajectory data, handling (semantic) trajectories as first-class citizens in data processing. In situ data processing and link discovery for data integration are critical technologies to those targets.
- **Detection and prediction of trajectories of moving entities in the aviation and maritime domains**: datAcron developed novel methods for real-time trajectory reconstruction, aiming at efficient large-scale mobility data analytics. Real-time

trajectories forecasting for the aviation and maritime domains aim to a short forecasting horizon.

- **Recognition and forecasting of complex events in the aviation and maritime domains**: datAcron developed methods for event recognition under uncertainty in noisy settings, aiming at processing very large number of events/second with complex event definitions. In doing so, optimization of complex events patterns' structure and parameters by means of machine learning methods for constructing event patterns was within datAcron objectives.
- **Visual analytics in the aviation and maritime domains**: datAcron developed a general visual analytics infrastructure supporting all steps of analysis through appropriate interactive visualizations, including both generic components and components tailored for specific applications.

Contents

**Part II   Visual Analytics and Trajectory Detection and
            Summarization: Exploring Data and Constructing
            Trajectories**

3   Visual Analytics in the Aviation and Maritime Domains...............   59
Gennady Andrienko, Natalia Andrienko, Georg Fuchs, Stefan
Rüping, Jose Manuel Cordero, David Scarlatti, George A. Vouros,
Ricardo Herranz, and Rodrigo Marcos

**4   Trajectory Detection and Summarization over Surveillance
      Data Streams**...   85
Kostas Patroumpas, Eva Chondrodima, Nikos Pelekis,
and Yannis Theodoridis

Part III   Trajectory Oriented Data Management for Mobility Analytics

5   Modeling Mobility Data and Constructing Large Knowledge Graphs to Support Analytics: The datAcron Ontology
Georgios M. Santipantakis, George A. Vouros, Akrivi Vlachou, and Christos Doulkeridis

Editors and Contributors

About the Editors

Dr. Gennady Andrienko is a lead scientist responsible for the visual analytics research at Fraunhofer Institute IAIS (Sankt Augustin, Germany) and full professor at City University London. He co-authored monographs "Exploratory Analysis of Spatial and Temporal Data" (Springer, 2006) and "Visual Analytics of Movement" (Springer, 2013) and about 100 peer-reviewed journal papers. He is associate editor of three journals: "Information Visualization" (since 2012), "IEEE Transactions on Visualization and Computer Graphics" (2012–2016), and "International Journal of Cartography" (since 2014) and editorial board member of "Cartography and Geographic Information Science" and "Cartographica." He received Test of Time award at IEEE VAST 2018 and best paper awards at AGILE 2006, IEEE VAST 2011 and 2012 conferences, and EuroVA 2018 workshop.

Alexander Artikis is an Assistant Professor in the Department of Maritime Studies of the University of Piraeus. He is also a Research Associate in the Institute of Informatics and Telecommunications of NCSR Demokritos, the largest research center in Greece, leading the Complex Event Recognition lab. His research interests lie in the field of Artificial Intelligence. He has published over 100 papers in the top conferences and journals of the field, while, according to Google Scholar, his h-index is 31. He has participated in several EU-funded Big Data projects, being the scientific coordinator in some of them. He has been serving a member of the program committee of various conferences, such as AAAI, IJCAI, AAMAS, ECAI, KR, and DEBS.

Jose Manuel Cordero holds the Telecommunications Engineer degree from the Universidad de Sevilla, Spain (2002). He is currently a Principal Researcher at CRIDA (ATM R&D Reference Centre, depending on the Spanish ANSP, ENAIRE), with over 15 years of experience in the air-traffic management domain in the areas of performance monitoring and assessment, system simulation, and validation. In the last years, he focused his activity on Performance Management projects, including research activities in big data analytics, predictive models, and multi-objective optimization methods.

Christos Doulkeridis is an Assistant Professor at the Department of Digital Systems in the University of Piraeus. He has been awarded both a Marie-Curie fellowship and an ERCIM "Allain Bensoussan" fellowship for postdoctoral studies at the Norwegian University of Science and Technology in 2011 and 2009, respectively. He was the Principal Investigator of the research project "RoadRunner: Scalable and Efficient Analytics for Big Data" (2014–2015, funded by the General Secretariat for Research and Technology in Greece). He has participated in several H2020 EU research projects related to Big Data management and analytics ("Track&Know" 2018–2020, "BigDataStack," 2018–2020, and "datAcron," 2016–2018). He has been awarded the first position in the 2017 SemEval challenge on Sentiment Analysis in Twitter and the third position in the 2016 ACM SIGSPATIAL Cup on Hot Spot Analysis of Mobility Data. He has published in top international journals (including VLDB Journal, IEEE TKDE, IEEE JSAC, ACM TKDD, Data Mining and Knowledge Discovery, Information Systems, Distributed and Parallel Databases) and conferences (including ACM SIGMOD, VLDB, ICDE, EDBT, SSTD, PKDD, SIAM SDM) in the areas of data management, knowledge discovery, and distributed systems.

Dr. Anne-Laure Jousselme is with the NATO Centre for Maritime Research and Experimentation. She is member of the Boards of Director of the International Society of Information Fusion and Belief Functions and Applications Society. She is associate editor of the Perspectives on Information Fusion magazine and area editor of the International Journal of Approximate Reasoning. Her research interests include maritime anomaly detection, information fusion, reasoning under uncertainty, information quality.

Nikos Pelekis is Assistant Professor at the Department of Statistics and Insurance Science, University of Piraeus, Greece. His research interests include all topics of data science. He has been particularly working for almost 20 years in the field of Mobility Data Management and Mining. Nikos has co-authored one monograph and more than 80 refereed articles in scientific journals and conferences, receiving more than 1000 citations, while he has received 3 best paper awards and won the SemEval'17 competition and ranked 3rd in ACM SIGSPATIAL'16 data challenge. He has offered several invited lectures in Greece and abroad (including PhD/MSc/summer courses at Rhodes, Milano, KAUST, Aalborg, Trento, Ghent, JRC Ispra) on Mobility Data Management and Data Mining topics. He has been actively involved in more than 10 European and National R&D projects. Among them, he is or was principal researcher in GeoPKDD, MODAP, MOVE, DATASIM,

SEEK, DART and datAcron, Track & Know, MASTER. For more information: http://www.unipi.gr/faculty/npelekis/.

Cyril Ray is associate professor in computer science at Arts & Metiers—ParisTech, attached to Naval Academy Research Institute (IRENav) in France. His current research is oriented to the modeling and design of location-based services. His work mainly concerns theoretical aspects of the design of ubiquitous and adaptive location-based services applied to human mobility, maritime, and urban transportation systems. This research addresses the relationship between geographic information systems and the underlying computing architectures that support real-time tracking of mobile objects (pedestrian in indoor spaces, vehicles in urban areas, and ships at sea). This work includes, at different level, integration of location acquisition technologies, modeling of heterogeneous and large spatiotemporal datasets, movement data processing (cleaning, filtering, trajectory modeling, knowledge discovery), modeling of context-aware systems, and traffic simulation and prediction.

David Scarlatti works as Data Solutions Architect at Boeing Research & Technology Europe in the Aerospace Operational Efficiency group. He received Aeronautics Engineering degree in 1994 (Universidad Politecnica de Madrid), is Stanford Certified Project Manager (2008), Master of Science in Technology Management (2010, Open University- UK), and GIAC Certified Incident Handler (2011). In 1989, he started to work with computers at PHILIPS, then worked at INDRA (1993–2000) as Software Engineer, and at ORACLE as Technology Manager (2000–2005) where he led the European Professional Community on Data Warehousing. In May 2005, he joined Boeing where he has been applying data analytics technologies to a wide variety of aviation-related problems. His fields of expertise include big data, visualization, advanced computing in data analysis, cybersecurity, and human–machine interfaces. Is co-inventor in 17 patents (5 granted).

George A. Vouros holds a BSc in Mathematics (1986) and a PhD in Artificial Intelligence (1992) all from the University of Athens, Greece. Currently, he is a Professor in the Department of Digital Systems in the University of Piraeus and head of the AI-Lab in this Department. He has done research in the areas of Expert Systems, Knowledge Management, Collaborative Systems, Ontologies, and Agents and Multi-Agent Systems. He served/serves as program chair, chair, and member of organizing committees of national and international conferences (AAMAS, AAAI, IJCAI, ECAI, WI/IAT, AT, EUMAS, ICMLA, ESWC, CSCL, AIAI, ISWC) and as member of steering committees/boards of international conferences/workshops (EURAMAS, SETN, AT, COIN, OAEI). He has given keynote speeches in conferences and workshops (WoMo, ICTAI, CLIMA, IF&GIS) and he has organized several workshops (MATES@EDBT, Data-Driven ATM@WAC, Data-Enhanced Trajectory-Based Operations@ICRAT; the most recent ones). He served/ serves as guest editor in special issues in well-reputed journals (e.g., IJCIS,

AIR, GEOINFORMATICA, AICom, ISF). He is/was senior researcher in numerous EU-funded and National research projects (GSRT/ AMINESS, FP7/Grid4All, FP7/SEMAGROW, COST/Agreement Technologies the most recent ones). He recently coordinated the successful DART (SESARER) project on data-driven trajectory prediction in the aviation domain and coordinates the datAcron (H2020 ICT-16) Big Data project. For more information on recent activities and publications, see http://ai-group.ds.unipi.gr/georgev/.

Contributors

Gennady Andrienko Fraunhofer Institute IAIS, Sankt Augustin, Germany
City University of London, London, UK

Natalia Andrienko Fraunhofer Institute IAIS, Sankt Augustin, Germany
City University of London, London, UK

Alexander Artikis Department of Maritime Studies, University of Piraeus, Piraeus, Greece
Institute of Informatics & Telecommunications, NCSR Demokritos, Athens, Greece

Konstantina Bereta MarineTraffic, Athens, Greece

Elena Camossi NATO STO Centre for Maritime Research and Experimentation, La Spezia, Italy

Eva Chondrodima University of Piraeus, Piraeus, Greece

Gemma Galdon Clavell Eticas Research and Innovation, Barcelona, Spain

Jose Manuel Cordero CRIDA (Reference Center for Research, Development and Innovation in ATM), Madrid, Spain

Christos Doulkeridis University of Piraeus, Piraeus, Greece

Richard Dréo Arts et Metiers Institute of Technology, Ecole Navale, IRENav, Brest, France

Georg Fuchs Fraunhofer Institute IAIS, Sankt Augustin, Germany

Harris Georgiou University of Piraeus, Piraeus, Greece

Apostolis Glenis University of Piraeus, Piraeus, Greece

Ricardo Herranz Nommon Solutions and Technologies, Madrid, Spain

Clément Iphar NATO STO Centre for Maritime Research and Experimentation, La Spezia, Italy

Anne-Laure Jousselme NATO STO Centre for Maritime Research and Experimentation, La Spezia, Italy

Nikolaos Koutroumanis University of Piraeus, Piraeus, Greece

Rodrigo Marcos Nommon Solutions and Technologies, Madrid, Spain

Panagiotis Nikitopoulos University of Piraeus, Piraeus, Greece

Kostas Patroumpas Athena Research Center, Marousi, Greece

Nikos Pelekis University of Piraeus, Piraeus, Greece

Petros Petrou University of Piraeus, Piraeus, Greece

Victoria Peuvrelle Eticas Research and Innovation, Barcelona, Spain

Manolis Pitsikalis Institute of Informatics & Telecommunications, NCSR Demokritos, Athens, Greece

Cyril Ray Arts et Metiers Institute of Technology, Ecole Navale, IRENav, Brest, France

Stefan Rüping Fraunhofer Institute IAIS, Sankt Augustin, Germany

Georgios M. Santipantakis University of Piraeus, Piraeus, Greece

David Scarlatti Boeing Research & Development Europe, Madrid, Spain

Stylianos Sideridis University of Piraeus, Piraeus, Greece

Panagiotis Tampakis University of Piraeus, Piraeus, Greece

Yannis Theodoridis University of Piraeus, Piraeus, Greece

Akrivi Vlachou University of Piraeus, Piraeus, Greece

Marios Vodas MarineTraffic, Athens, Greece

George A. Vouros University of Piraeus, Piraeus, Greece

Dimitris Zissis Department of Product & Systems Design Engineering, University of the Aegean, Lesbos, Greece
MarineTraffic, Athens, Greece

Maximilian Zocholl NATO STO Centre for Maritime Research and Experimentation, La Spezia, Italy

Acronyms

ADS-B	Automatic-dependent surveillance-broadcast
AI	Aircraft intent
AIS	Automatic identification system
ANFR	French Frequencies Agency
ANSP	Air Navigation Service Providers
API	Application programming interface
APM	Aircraft performance model
ATC	Air traffic control
ATCO	Air traffic controller
ATFM	Air traffic flow management
aTHS	approximate Trajectory Hot Spot
ATM	Air-traffic management
AtoN	Aids to navigation
BADA	Base of aircraft data
BGP	Basic graph pattern
CAS	Calibrated airspeed
CDO	Continuous descent operations
CEF	Complex event forecasting
CEP	Complex event processing
CER	Complex event recognition
CoG	Course over ground
COLREGs	Collision regulations
CP	Cutting point
CRS	Coordinate reference system
CSV	Comma separated values
CTOT	Calculated take off time
datAcron	Big data analytics for time-critical mobility forecasting
DBSCAN	Density-based spatial clustering of application with noise

DCB	Demand and capacity balancing
DG ENTR	European Commission's Directorate-General for Enterprise and Industry
DiStRDF	Distributed spatiotemporal RDF engine
DM	Data management/manager
DMP	Data management plan
DoF	Degrees of freedom
DSTs	Decision support tools
DTJ	Distributed subTrajectory Join
DTW	Dynamic time warping
EC	Entry count
EEA	European Environmental Agency
EEZ	Exclusive economic zone
EPIRB	Emergency Position Indicating Radio Beacon
ESRI	Environmental Systems Research Institute, Inc.
ETA	Estimated time of arrival
ETD	Estimated time of departure
ETOT	Estimated take off time
EUMSS	European Union Maritime Security Strategy
FAO	Food and Agriculture Organization
FDR	Flight recorded data
FLP	Future location predictor
FMS	Flight management system
FP	Flight plan
GFS	Global Forecast System
GLM	Generalized linear models
GPS	Global Positioning System
GSHHG	Global Self-consistent, Hierarchical, High-resolution Geography Database
HDFS	Hadoop distributed file system
HMM	Hidden Markov Model
ICAO	International Civil Aviation Organization
IFREMER	Institut français de recherche pour l'exploitation de la mer
IMO	International Maritime Organization
ISA	International Standard Atmosphere
ISR	Intelligence, surveillance, and reconnaissance
ITU	International Telecommunication Union
IUU	Illegal, unreported, and unregulated
IVA	Interactive visual analytics
JRC	European Commission Joint Research Centre
JSON	JavaScript Object Notation

KDE	Kernel density estimation
LCSS	Longest Common SubSequence
LD	Link discovery
LED	Low-level events detector
LNG	Liquid natural gas
LRIT	Long range tracking and identification
MBB	Minimum bounding box
MBR	Minimum bounding rectangle
METAR	Meteorological Terminal Aviation Routine Weather Report
METOC	Meteorological and oceanographic
MLlib	Machine learning library
MMSI	Maritime Mobile Service Identity
MOB	Man overboard
MSA	Maritime situation awareness
MSI	Maritime situational indicator
NGO	Non-governmental organization
NMEA	National Marine Electronics Association
NOAA	National Oceanic and Atmospheric Administration
ODC-By	Open Data Commons Attribution License
ODC-ODbL	Open Data Commons Open Database License
OGC	Open Geospatial Consortium
PbD	Privacy by design
PMM	Point-mass model
QAR	Quick access recorder
RBT	Reference business trajectory
RDF	Resource description framework
RFL	Requested flight level
RMSE	Root mean square error
ROC	Rate of climb
RoT	Rate of turn
RTEC	Event calculus for run-time reasoning
SAR	Search And Rescue
SBT	Shared business trajectory
SESAR	Single European Sky ATM Research
SI	Semantic integrator
SIGMET	Significant meteorological information
SoG	Speed over Ground
SSCR	Sum of similarity between cluster members and cluster representatives
STD	Semantic Trajectory Database
TAF	Terminal aerodrome forecast

TAS	True airspeed
TBO	Trajectory-based operations
TCL	Talis Community License
TDA	Trajectory data analytics
THS	Trajectory hot spot
TP	Trajectory predictors
TSA	Trajectory segmentation algorithm
TSS	Traffic separation scheme
TTL	Terse RDF Triple Language
UN	United Nations
VA	Visual analytics
VHF	Very high frequency
VMS	Vessel monitoring system
VTS	Vessel traffic system
WKT	Well-known text
WGS	World Geodetic System
WGS84	World Geodetic System 1984
WPI	World Port Index
YARN	Yet Another Resource Negotiator

Part I
Time Critical Mobility Operations and Data: A Perspective from the Maritime and Aviation Domains

The first part of this book provides the motivating points and background for mobility forecasting supported by trajectory-oriented analytics. It presents specific problems and challenges in the Aviation (Air Traffic Management (ATM)) and the Maritime domains, clarifies operational concerns and objectives in both domains, and explains domain-specific terminology. It presents challenging cases which motivate technology presented in subsequent chapters of this book and which drive evaluation and validation of solutions presented. Equally important to the above is the presentation of the data sources exploited, the big data challenges ahead in both domains and, of course, the requirements from technologies presented in subsequent parts of the book.

Chapter 1
Mobility Data: A Perspective from the Maritime Domain

Cyril Ray, Anne-Laure Jousselme, Clément Iphar, Maximilian Zocholl, Elena Camossi, and Richard Dréo

Abstract This chapter overviews maritime operational situations and underlying challenges that the automated processing of maritime mobility data would support with the detection of threats and abnormal activities. The maritime use cases and scenarios are geared on fishing activities monitoring, aligning with the European Union Maritime Security Strategy. Six scenarios falling under three use cases are presented together with maritime situational indicators expressing users' needs when conducting operational tasks. This chapter also presents relevant data sources to be exploited for operational purposes in the maritime domain, and discusses the related big data challenges to be addressed by algorithmic solutions. An integrated dataset of heterogeneous sources for maritime surveillance is finally described, gathering 13 sources. This chapter concludes on the generation of specific datasets to be used for algorithms evaluation and comparison purposes.

1.1 Maritime Operational Scenarios: Challenges and Requirements

Effective Maritime Situation Awareness (MSA) requires not only detecting, tracking, and classifying vessels but also detecting, classifying, and predicting their behavior. This challenging and crucial task is at the core of the compilation of a

---

The material presented in this chapter is a compilation of excerpts from [7, 8, 14, 17].

C. Ray · R. Dréo
Arts et Metiers Institute of Technology, Ecole Navale, IRENav, Brest, France
e-mail: cyril.ray@ecole-navale.fr; richard.dreo@ecole-navale.fr

A.-L. Jousselme (✉) · C. Iphar · M. Zocholl · E. Camossi
NATO STO Centre for Maritime Research and Experimentation, La Spezia, Italy
e-mail: anne-laure.jousselme@cmre.nato.int; clement.iphar@cmre.nato.int; maximilian.zocholl@cmre.nato.int; elena.camossi@cmre.nato.int

© Springer Nature Switzerland AG 2020
G. A. Vouros et al. (eds.), *Big Data Analytics for Time-Critical Mobility Forecasting*, https://doi.org/10.1007/978-3-030-45164-6_1

maritime picture,[1] which involves extracting relevant contextual information (for instance, maritime routes or loitering areas) but also monitoring the real-time maritime traffic. The use of a set of sensors mixing cooperative self-identification systems such as the Automatic Identification System (AIS) and non-cooperative systems such as coastal radars or satellite imagery provides complementary and redundancy in information, as necessary to overcome the quite common spoofing of AIS signals and increase the clarity and the accuracy of the maritime picture. In many cases, intelligence information can also be helpful in refining and guiding the search in the huge amount of data to be processed, filtered, and analyzed, as well as representing the contextual information for some MSA problems.

Ensuring *security* and *control* of fishing activities is one of the most important aspect of the European Union Maritime Security Strategy (EUMSS) Action Plan,[2] which defines several strategic interests for the European Union and the Member States. Europe is the world's biggest market for seafood and the aim of the EUMSS is to promote better international governance across the world's seas and oceans to keep them clean, safe, and secure. Since fishing is an activity that exploits common natural resources, it needs to be regulated to safeguard fair access, sustainability, and profitability for all.

The Big Data Analytics for Time Critical Mobility Forecasting (datAcron) project aimed to develop novel methods for threat and abnormal activity detection in very large fleets of moving entities. To motivate and support the development of vessel movement analytics algorithms within datAcron, we present below operational challenges linked to the monitoring of fishing activities in European waters, through six scenarios organized in three use cases. Relevant Maritime Situational Indicator (MSI)s express users' needs for appropriate awareness of the situation, targeting expected outcomes for algorithms.

1.1.1 Monitoring Fishing Activities

Fishing activity monitoring is a complex maritime surveillance mission that encompasses several maritime risks and environmental issues such as environmental destruction and degradation but also maritime accidents, Illegal, Unreported, and Unregulated (IUU) fishing and trafficking problems. In particular, IUU fishing is a global threat to the marine environment and honest fishermen alike, whose global cost is estimated in about 10 Billion Euros per year. The European Union, in collaboration with International organizations, is committed to fighting IUU fishing worldwide.

---

[1]The maritime picture is a geographic presentation of all contacts in the maritime environment arising from all available sources, commercial or military.
[2]EUMSS Action Plan: http://ec.europa.eu/maritimeaffairs/policy/maritime-security/doc/20141216 -action-plan_en.pdf (published December 2014), accessed January 2020.

Besides the detection of IUU fishing activities, safety is another core issue of the EUMSS. In times of peace, fishing is one of the deadliest occupations. Fishing vessels may fish in areas with dense traffic, like traffic lanes and waiting areas, and to keep the fishing place hidden from others, they sometimes intentionally switch off their AIS device while fishing, endangering themselves and the surrounding traffic. Therefore, preserving the maritime environment from illegal fishing and ensuring fishing safety requires live processing and prediction of fishing vessel trajectories, identifying movement patterns (e.g., *close-encounters, change in speed* and *course*), detecting vessel activities (e.g., *fishing, loitering, tugging, rendezvous*), forecasting potential collisions between surrounding ships within a typical time scale of 5–15 min.

The maritime use cases and associated scenarios[3] described herein have been developed in collaboration with operational experts to capture domain requirements. The scenarios highlight the needs for live tracking of fishing vessels and surrounding traffic, as well as of contextually enhanced offline data analytics, including, for instance, cluster and spatial analysis together with motion pattern detection.

These scenarios have been elaborated in order to (1) stress trajectory and event detection, prediction and visualization algorithms against big data challenges in terms of *velocity, veracity, variety,* and *volume* (see Sect. 1.2) and (2) provide operational relevance to their future use. Each scenario of the three use cases of *Secured fishing, Maritime sustainable development,* and *Maritime security* highlights different users' **goals** and possible **actions**, as well as **information needs** expressed in terms of Maritime Situational Indicators as illustrated in Table 1.1.

1.1.1.1 Secured Fishing

Fishing is known to be one of most dangerous activity. Detecting and preventing collisions between ships and by optimizing *rendezvous* between rescuing ships in proximity of a vessel in danger and emergency services certainly contributes to Secured fishing, as detailed in the two scenarios below.

Collision Avoidance

Collisions involving fishing vessels are frequent, not only while fishing, which makes collision avoidance crucial for secured fishing. Figure 1.1a shows a collision between a cargo and a fishing vessel, which occurred during the night between the fishing vessel Sokalique and the cargo Ocean Jasper. Ocean Jasper continued

---

[3]While "[a] use case is the story of how the business or system and the user interact," the "[s]cenarios tell the full story" [4].

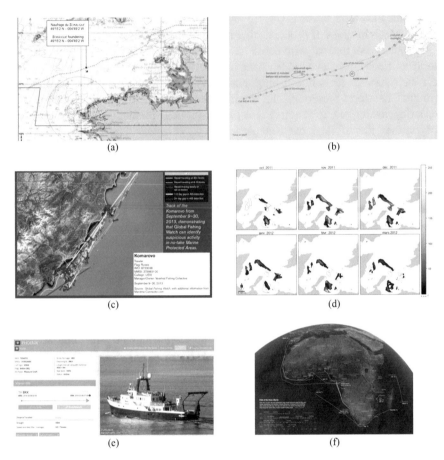

Fig. 1.1 Illustrations of operational scenarios for monitoring fishing activities. (**a**) Secured fishing: Collision between a cargo and a fishing vessel. (**b**) Secured fishing: Accident on a fishing ship. (**c**) Sustainable development: Track of the Komarovo in September 2013 (source *Global Fishing Watch*). (**d**) Sustainable development: Estimated fishing activity for scallop dredging vessels per month [9]. (**e**) Maritime security: The vessel Phoenix (screenshot from MarineTraffic.com). (**f**) Maritime security: Path of the Dona Liberta (Source SkyTruth)

her[4] route without assisting the sinking Sokalique, which asked the ships in her vicinity for help. While vessels maintain the responsibility to comply with Collision Regulations (COLREGs),[5] data analytics techniques may be used to enhance the situational awareness between vessels, specifically when it is anticipated that a

---

[4]In the remainder of the chapter vessels are referred to as "she" because it is commonplace in the maritime domain.

[5]COLREGs: www.imo.org/en/About/Conventions/ListOfConventions/Pages/COLREG.aspx (accessed January 2020).

vessel will be required to "give way" to a fishing vessel (sailing vessels and power-driven vessels).

Scenario: Secured fishing use case—collision avoidance.

Goal: In order to prevent a collision of fishing vessels with other ships, in the collision avoidance scenario the Vessel Traffic System (VTS) operator wants to predict which other vessels (such as cargos, tankers, ferries) will cross the areas where the fishing vessels are fishing.

Actions: Upon the prevention of possible collisions, the user can send a warning to the two vessels, so they can perform the appropriate action. The captains of the ships will base their decisions on the potential risk highlighted by the monitoring system.

Vessel in Distress/Man Overboard

Beside collisions, technical problems or accidents occurring on-board can put the fishing vessel in difficulty. Figure 1.1b displays the AIS data of a fishing vessel before and after an accident on-board. After leaving the port, the fishing vessel went towards a first fishing zone, then disappeared (i.e., likely switched the AIS off). After 35 min, the vessel re-appeared (the AIS unit was switched on), moved towards another fishing area, and switched off the AIS transmitter for another 35 min. Then, it re-appeared again, moved towards a third fishing area where it started fishing (AIS was off). The accident location, highlighted with a red ellipse in Fig. 1.1b, occurred in the morning, 7 h after the AIS unit was switched off. The AIS on-board was re-enabled immediately after the accident while the ship was heading back to the port. The rescue helicopter reached the boat 1 h later (red circle in the figure). A system that could alert vessels in the vicinity of a vessel in distress or in a Man Overboard (MOB) situation would be a valuable capability to optimize rescuing operations. Moreover, the information from Emergency Position Indicating Radio Beacon (EPIRB) and MOB equipped with AIS devices (fishermen sometimes wear small individual AIS beacon broadcasting in a range of 5–10 miles), if integrated in the system, could provide almost immediate event cueing.

Scenario: Secured fishing use case—vessel in distress.

Goal: In order to effectively intervene and help a vessel in distress, the VTS operator wants to identify the vessels in the vicinity of the vessel in distress, so they can provide early assistance while awaiting for the Search and Rescue team. In the case of a collision and escape of the responsible vessel, the VTS operator also wants to predict the trajectory of the fugitive.

Actions: Once the position of the vessel in distress/MOB is known or predicted, the VTS operator alerts the vessels in proximity, and the Navy or the Coast Guards launch the Search And Rescue (SAR) operation, involving helicopters and, if necessary, rescue vessels. In case of the collision and escape, the Navy or the Coast Guards start an interception of the fugitive.

1.1.1.2 Maritime Sustainable Development

Estimating the spatial distribution and the intensity of fishing activities is necessary for natural resources management, impact assessment, and maritime planning. However, the access to such information at high resolution is still a challenge. Since the European Union adopted the Vessel Monitoring System (VMS) to monitor fishing vessels, significant advances have been made [10]. Nevertheless, the results achievable by VMS analysis are limited by the restricted access and incompleteness of VMS data. Indeed, VMS data are confidential, and the use of the VMS is not mandatory for small vessels (i.e., less than 12 m long). In addition, VMS data analysis is usually conducted at a spatial resolution from 1 to 10 km, which can be too coarse for secure fishing operations.

Recently, the AIS data has been considered for the monitoring of fishing activities [11]. Characterizing the potential impacts of legal and illegal fishing activities on species and on the geographical areas and providing relevant information for European resources management based on AIS data remains however challenging. Two typical scenarios for maritime sustainable development can be distinguished: Protection of marine areas from fishing and fishing pressure on area.

Protection of Areas from Fishing

As stated in the recent report from *Global Fishing Watch*,[6] the IUU fishing around the world "has escalated rapidly as the chance for profit outweighs concerns about the health and sustainability of our oceans." Among others, the annex of European report COM/2016/0134 final—2016/074 COD defines regulated areas for fishing.[7] These areas have potentially associated temporal fishing constraints specifying periods during which fishing is legal and periods during which fishing is prohibited.

Figure 1.1c shows the trajectory of the Komarovo, a trawler registered in Russia. It appears to be fishing five times inside the Dzhugdzhursky State Nature Reserve in September 2013. The Komarovo has fished there at least for 13 days exhibiting different behaviors (e.g., slow motions, fast traveling speed, erratic use of the AIS).

Scenario: Maritime sustainable development use case—protection of areas.

Goal: In order to control fishing in areas at the European level, the operator from the fisheries control agency in charge of monitoring an area to protect it from fishing needs to monitor ships in real time. Entrances, exits, and movements inside the surveyed areas have to be detected. Taking into consideration the identity declared by the ships as well as their type monitoring systems should evaluate their right to be in these areas of interest correlating with the information from the fleet register, fishing licenses, blacklist, and historical data. Detected and

---

[6]www.oceana.org (accessed January 2020).

[7]There are about 6600 marine protected areas covering about 2% of the world's oceans.

visualized fishing patterns together with predicted fishing polygons would help the operator to monitor the fishing grounds.

Actions: Besides the control of fishing itself, there are regions that have to be protected from all activities as they correspond to geographic areas where protected species live (e.g., marine reserves). The operator from the fisheries control agency would also be alerted about ships navigating and stopping in such areas. Detected offenders would be tracked, their trajectory and destination verified. Navy, Coast Guards, or port authorities would control them and verify their freight. Erratic use of the AIS to mask such IUU should result in blacklisting a fishing vessel.

Fishing Pressure on Areas (Density of Fishing)

Regular and legal fishing activities can also have negative impact on the maritime environment and sea resources. There is an increasing concern for the preservation of natural resources against overfishing. Developed by Halpern in 2007, the concept of cumulative impacts mobilizes methods of spatial and quantitative analysis. It considers the potential impact of anthropogenic pressures on the ecosystem components. This has been applied to evaluate spatial and temporal changes in cumulative human impacts on the world's ocean over a period of 5 years [3]. The identification of fishing efforts, based on AIS data, has been also recently considered by [12]. Figure 1.1d shows the estimated fishing activity for scallop dredging vessels per month in Brest bay. Fishing areas and corresponding fishing efforts have been identified by the analysis AIS data and the discovery of typical dredging patterns (speedbased). Algorithms calculating cumulative fishing activities over time and seasons would be valuable for the identification of overfishing areas and consequently for the preservation of marine resources.

Scenario: Maritime sustainable development use case—fishing pressure.

Goal: In order to identify overfishing areas, the operator from the fisheries control agency wants to identify fishing areas, visualize and evaluate changes over time/seasons, and see the cumulative impact.

Actions: Upon detection of intensive fishing areas and regarding concerned species, local, national, and European authorities provide new regulations and update the list of protected areas.

1.1.1.3 Maritime Security

Fishing vessels are often used to conceal illegal activities that may affect maritime security. In particular, illegal immigration or human trafficking and generically illicit activities occurring at sea are of primary concerns of European authorities, as described in the two scenarios below.

Human Trafficking

Human trafficking dramatically increased in the last years, as a large number of migrants risk their lives trying to reach Europe through Mediterranean routes. While the majority of the vessels transporting migrants are small vessels that do not transmit AIS, fishing vessels have been known to be engaged in these activities as well. Accordingly, a fishing vessel detected along a common smuggling route and not engaged in fishing may be identified by the Maritime Security authorities as a "suspicious vessel." For instance, the fact the vessel would lack the fishing gear may suggest it might have been involved in an illicit activity and further investigation may be necessary. In some cases, the presence of a vessel in the area can be licit, but it can still raise an alarm. For instance, the Phoenix, the vessel in Fig. 1.1e, is a former trawler (fishing vessel) reconditioned by an Non-Governmental Organization (NGO) to help migrants. Her presence in an area under surveillance may be an indicator of nearby migrant vessels.

Scenario: Maritime security use case—human trafficking.

Goal: In order to detect and rescue migrants and identify human trafficking activities, the operator from the border control services wants to detect fishing vessels that are heading within migrants routes or loitering in these areas. Additional information on such vessels (fishing gear or previous negative records) should be at hand to help identify them and predict their movements.

Actions: Once a vessel is detected traveling within a human trafficking route, the operator from the border control services would further investigate on the vessel records and decide to launch a mission for escorting it.

Illicit Activities

A number of criminal activities conducted at sea such as piracy, environmental pollution, and drug smuggling are often linked to transnational crime organizations and are difficult to be detected, thwarted, and prosecuted. Predicted and real-time detection of vessels *rendezvous* in areas known for trafficking, or any other "anomalous" behavior as defined by the operator from the border control services, could enhance the ability to detect and intervene in real time and to prosecute illicit activities.

Figure 1.1f displays the path of the rusty refrigerated cargo vessel Dona Liberta, followed from 2011 to 2014 along the coasts of Africa and Europe. During that period, the Dona Liberta was reported to abandon crew members, abuse stowaways, dump oil, and commit other crimes along the way.[8] Port calls were the main means of locating the ship, which often switched off the AIS.

---

[8]http://www.nytimes.com/2015/07/19/world/stowaway-crime-scofflaw-ship.html?_r=0 (accessed January 2020).

Scenario: Maritime security use case—illicit activities.

Goal: In order to prevent and act against illicit activities at sea, the operator from
the border control services wants to detect vessels following a predefined or
customized suspicious behavior, such as heading or loitering in an area known
for trafficking (*rendezvous*), identify the vessels involved, and crosscheck them
against existing records of illicit activities. The operator from the border control
services wants also to predict the next positions of the vessels involved.

Actions: Once a *rendezvous* or another suspicious behavior is detected, the
operator from the border control services may decide to launch a mission for
intercepting the suspicious vessels with helicopters or by sea.

1.1.2   Maritime Situational Indicators

For each scenario, user information needs are expressed through combinations of
MSIs. MSIs match maritime events of interest, such as behavioral patterns and
singularities in vessel traffic data. Twenty-eight (28) MSIs have been formalized
in datAcron (cf. [7]). This list of MSIs is a synthesis of outcomes of workshops
gathering user's requirements on important events and indicators to be detected
reported in the literature (e.g., [1, 18]). These MSIs have been selected according
to their ability to be automatically detected or predicted by processing mainly AIS
data. For instance, any MSI referring to visual sighting has been excluded from this
list. Table 1.1 lists meaningful samples of the associated MSIs to the three use cases
and associated six scenarios. A complete specification is available in [7].

   For instance, the *collision avoidance scenario* addresses the task of protection
of fishing vessels from collision with large vessels (cargos, tankers, ferries). With
the help of dedicated big data analytics techniques, maritime operators should be
able to warn fishing vessels at risk of collision and warn large vessels heading
to fishing areas. As illustrated in Table 1.1, relevant MSIs in this scenario help
the operators identify vessels potentially at risk (e.g., vessels *engaged in fishing*,
loitering or those whose *movement mobility is affected*, with *null speed*, *drifting*, or
that are positioned *on a maritime route*), large vessels doing unexpected maneuvers
(e.g., vessels that suddenly *change course*), or vessels in *proximity to other vessels*,
which could participate in SAR operations.

1.2   Big Mobility Data in the Maritime Domain

The development of algorithms processing maritime data in real time and covering
large areas requires datasets prepared to challenge the components of a system
against big data dimensions, such that *velocity*, *veracity*, *variety*, and *volume* stress

Table 1.1 Relevant Maritime Situational Indicators for the maritime use cases and scenarios (from [16], elaborated from [7, 8])

Use case	Scenario	Description	MSIs, (e.g.)	
Secured fishing	Collision prevention	Protecting fishing vessels from collision with large vessels (cargos, tankers, ferries)	MSI#3	On a maritime route
			MSI#4	Proximity of other vessels
			MSI#6	Null speed
			MSI#12	Change of course
			MSI#21	Movement mobility affected
			MSI#23	Engaged in Fishing
			MSI#26	Loitering
			MSI#27	Dead in water/drifting
	Vessel in distress Man Overboard (SAR)	Provide early assistance to a vessel in distress	MSI#4	Proximity of other vessels
			MSI#6	Null speed
			MSI#16	AIS emission has interrupted
			MSI#21	Movement mobility affected
			MSI#23	Engaged in Fishing
			MSI#25	In SAR operation
			MSI#26	Loitering
			MSI#27	Dead in water/drifting
Sustainable development	Protection of ecological areas	Protect specific areas from illegal fishing activities	MSI#2	Within a given area
			MSI#6	Null speed
			MSI#13	Course not compatible with expected destination
			MSI#16	AIS emission has interrupted
			MSI#18	AIS error detection
			MSI#26	Loitering
	Fishing pressure	Estimate and predict fishing pressure, identify areas at risk	MSI#2	Within a given area
			MSI#4	Proximity of other vessels
			MSI#6	Null speed
			MSI#16	AIS emission has interrupted
			MSI#23	Engaged in fishing
			MSI#26	Loitering
Maritime security	Migration and human trafficking	Detect possible human trafficking involving fishing vessels (or the like)	MSI#13	Course not compatible with expected destination
			MSI#15	No AIS emission/reception
			MSI#16	AIS emission has interrupted
			MSI#17	Change of AIS static information
	Illicit activities	Detect suspicious activities involving fishing vessels	MSI#1	Close to a critical infrastructure
			MSI#4	Proximity of other vessels
			MSI#5	In stationary area

(continued)

Table 1.1 (continued)

Use case	Scenario	Description	MSIs, (e.g.)
			MSI#7 Change of speed
			MSI#12 Change of course
			MSI#13 Course not compatible with expected destination
			MSI#15 No AIS emission/reception
			MSI#17 Change of AIS static information
			MSI#18 AIS error detection
			MSI#26 Loitering
			MSI#28 Rendezvous

the algorithms in realistic conditions as outlined in the fishing monitoring use cases presented in Sect. 1.1.

The current section first discusses the four big data challenges, listing relevant data sources for supporting the monitoring of maritime activities related to fishing (Sect. 1.2.1). The integrated dataset prepared for testing algorithms against these four challenges in this specific operational context is then presented in Sect. 1.2.2, while Sect. 1.2.3 presents enriched datasets encoding specific operational scenarios.

1.2.1 Maritime Big Data Challenges

MSA requires processing in real time a high volume of information of different nature (numerical, natural language statements, objective or subjective assessments, etc.), originating from a variety of sources (like sensors and humans, respectively, hard and soft sources), with a lack of veracity (i.e., being uncertain, imprecise, vague, ambiguous, incomplete, conflicting, incorrect, etc.). The algorithms to be designed in support to MSA should cope with these big data challenges and this ability should be reflected in the quality of the results provided.

1.2.1.1 Variety

Different types of data are available and only if properly combined and integrated they can provide useful information. Different sensor technologies are being developed and the data coming from multiple sources need to be cleaned up from inconsistencies, standardized in format and summarized. The set of sources of data to be processed should cover a wide range of variety to benefit from their complementarity and redundancy:

- *Physical sensors* such as AIS, VMS, coastal or on-board radars as traditionally used for tracking objects, synthetic aperture imagery, cameras, etc.;
- *Automated processors* such as trackers, Automatic Target Recognition algorithms, or classifiers in general;
- *Human sources* including operators or analysts themselves possibly manipulating lower level data (e.g., videos, radar images) to reflect the chain of information processing, from automation to possible subjective assessments, intelligence reports;
- *Databases* as records of past events (e.g., piracy, accidents, illegal fishing activities), records of vessels such as the Lloyds database[9] or blacklisted vessels.

Other sources could be considered as well such as social media, or open-source media.

1.2.1.2 Veracity

Data may lack of veracity which has to be handled by the processing method. For instance, position reports provided by the AIS data are incomplete, intermittent, with errors and the messages can be spoofed (e.g., fields may be intentionally wrongly filled, such as the type of vessel but also the MMSI).

1.2.1.3 Volume

The growing number of sensors (in coastal and satellite networks) makes the coverage of wide areas more effective, but the data processing techniques have to be designed to handle this large volume of data. The need for methods able to scale in time and space the processing of vessel motion data is highly critical for maritime security and safety. The interactive visualization of such a volume of data is also challenging.

1.2.1.4 Velocity

The analysis of streaming data from multiple sensors is essential to detect critical events at sea as soon as they occur. This poses the emphasis on incremental techniques able to process new data as soon as they are added to those already processed, and on methods able to detect critical events by processing data in a continuous way.

---

[9]https://www.lloydslistintelligence.com.

1.2.1.5   Context

Moreover, the context in which maritime data is processed and analyzed is a crucial aspect to be characterized within the use case. It is a relative notion that depends on the user's goal. For instance, the maritime route is contextual information for vessel's destination prediction, while the maritime navigation rules is contextual information for the route extraction problem. Consequently, the user's goal highly influences what contextual information is required for the analysis purposes targeted. We identify the following dimensions of context to be considered:

- *Areas of interest*: It should include all information that could be extracted relatively to the area of interest. For instance, harbor zones characteristics, such as water depths, channels, restricted areas, protected areas, fishing areas, borders, harbors (fishing, recreational, etc.), shipping lanes, ferry lanes, military and Liquid Natural Gas (LNG) anchorage areas, islands, offshore platforms, etc.;
- *Rules*: It should gather together the legislation about navigation or electronic emission such as the AIS transmission or other mandatory reports;
- *Patterns of life*: it should give information about past behaviors, usually followed patterns, routes, etc. It concerns both individual vessels and groups of vessels;
- *Meteorological conditions*: The Meteorological and Oceanographic (METOC) information is mainly about the sea state, the weather, the wind;
- *Traffic density*: This is a current contextual information deduced from the number of vessels in a given area;
- *Time/period*: This could be either the period of the day (night versus day, morning versus afternoon), but also seasonal information;
- *User*: The user's characteristics are considered as part of the context and are crucial components, as information needs are directly derived from the role or mission (see examples of users and missions in Sect. 1.1.1). The user's characteristics include the role and hierarchical position together with a possible communication network, the mission, the decision as the list of possible actions, and the reaction time.

As an illustration of the *velocity* and *volume* challenges, the IMIS Global company provides a stream of ships' positions covering the majority of European waters of about 5,300,000 positioning messages per 24 h (cf. Fig. 1.2).

Dealing with the *variety* and *veracity* of data is as challenging as dealing with their volume, both in terms of processing and visualization. Mastering the processing of *variety* and *veracity* of data is absolutely required where data with known issues of quality are analyzed despite a lack of "ground truth." It is also a means of understanding the quality of the methods that are processing the data, typically: data compression, event detection and prediction, interactive visualization.

With this aim, we performed a board inventory and detailed characterization of available data sources that could contribute to improve MSA of the user fulling the tasks described by the use cases (see Sect. 1.1.1). The resulting reference dataset is specifically designed to challenge research solutions in terms of variety and veracity.

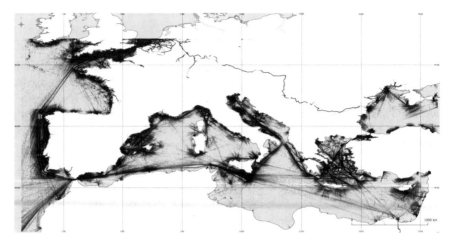

Fig. 1.2 One month of ships' AIS positions in European waters (data from IMIS Global)

Furthermore, it has been enriched (e.g., annotation) or degraded (e.g., contacts have been removed, fields emptied) for the design of operational scenarios as illustrated in Sect. 1.2.3. More details about the methodology of semi-automatic generation of scenarios and associated pseudo-synthetic datasets can be found in [6, 21, 23].

1.2.2 Heterogeneous Integrated Dataset for Maritime Intelligence, Surveillance and Reconnaissance (ISR)

We identified over 40 data sources, classified within 16 categories (cf. [14]). Based on this preliminary study, a representative heterogeneous maritime dataset was built to support the algorithms developments. Table 1.2 summarizes the data prepared and aggregated for this purpose, detailing, for each data piece, type, source, originator, file format (with Comma Separated Value (CSV), JavaScript Object Notation (JSON), Environmental Systems Research Institute, Inc. (ESRI) shapefiles), spatial and temporal extent, size, and approximate average stream rate (with *msg*, messages; *obs*, observations).

As illustrated in Table 1.2, the maritime reference dataset relies on the most widely used maritime reporting system, AIS. The AIS is one of the electronic systems that enable ships to broadcast their position and nominative information via radio communication. In addition to this system, to understand maritime activities and their impact on the environment, spatially and temporally aligned maritime data capturing additional features to ships' kinematic from complementary data sources (environmental, contextual, geographical, etc.) are of great interest. The dataset described below contains ship information collected through the AIS, prepared together with correlated contextual data, spatially and temporally aligned,

Table 1.2 Maritime data (adapted from [15, 20])

Type	Source	Originator	Format	Spatial extent	Temporal extent	Size	Avg stream rate
Surveillance	Automatic Identification System	Naval Academy	CSV	NW of France	Oct 2015–Mar 2016 (6 months)	19.680.743 msg 1.05 GB	76 msg/min
		IMIS Global	CSV	Europe	Jan 2016	80.169.806 msg 8.5 GB	1.830 msg/min
Weather	Sea state	SHOM.fr, Ifremer	CSV	NW of France	Oct 2015–Mar 2016 (6 months)	79.652.684 obs (1463 forecasts) 3.02 GB	1 forecast/3 h
	Weather forecast	UK Met Office (based on NOAA)	CSV	NW of France	Oct 2015–Mar 2016 (6 months)	71.516 obs from 16 stations 5 MB	1 obs/h
Contextual	Geographical	European Environment Agency	ESRI shapefiles	Europe	–	22 features 1.4 GB	–
	Port Registers	World Port Index SeaDataNet	ESRI shapefiles	World	–	5754 ports 70 MB	–
	Ship Registers	European Commission Agence Nationale des Fréquences	CSV	–	–	166.683 distinct ships	–

characterizing the vessels, the area where they navigate and the situation at sea. The dataset contains four categories of data: navigation data, vessel-oriented data, geographic data, and environmental data. It covers a time span of 6 months, from October 1st, 2015 to March 31st, 2016 and provides ship positions over the Celtic Sea, the North Atlantic Ocean, the English Channel, and the Bay of Biscay (France).

For all contextual data, European institutions and projects (e.g., SeaDataNet,[10] Copernicus,[11] EMODnet[12] and Ifremer[13], and the European Commission[14]) have been the preferred data providers and distributors.

Besides, a technical note describing which (licensed) nautical charts and objects are useful for the dataset and the necessary scripts to process them has been shared publicly [13]. Figure 1.4g, c provides examples of such data. Figure 1.4g shows different types of areas that constrain the movement of vessels. Figure 1.4c shows a traffic separation scheme.

1.2.2.1 Navigation-Related Data

The AIS communicates 27 kinds of messages, each one having its own purpose in information transmission (positioning, nominative information, management, etc.). The messages are broadcast in a theoretical range of about 40 nautical miles. These AIS messages are binary messages that comply with the Standard developed by the International Telecommunication Union (ITU)-R.M 1371-5 [5] and the National Marine Electronics Association (NMEA) 4.0 standards.[15] In the reference dataset, two main classes of messages are considered for reporting vessel data: those providing positioning information and those providing nominative information.

Positioning and Nominative Information

Positioning messages (ships' dynamic messages): Several AIS messages provide vessel positions that are acquired automatically by AIS transponders using embedded sensors (typically, Global Positioning System (GPS), gyroscope, loch, compass). To build the dataset, the ITU message types ITU 1, ITU 2, ITU 3, ITU 18, and ITU 19 were selected from which the following fields have been extracted from the vessel data:

---

[10]Pan-European Infrastructure for Ocean & Marine Data Management (SeaDataNet) data search: https://www.seadatanet.org (accessed April 2019).

[11]Copernicus Maritime Environment Monitoring Service: http://marine.copernicus.eu (accessed April 2019).

[12]European Marine Observation and Data Network (EMODnet): http://www.emodnet.eu (accessed April 2019).

[13]Institut français de recherche pour l'exploitation de la mer (Ifremer): https://www.ifremer.fr (accessed April 2019).

[14]Joint Research Centre Data Catalogue: https://data.jrc.ec.europa.eu (accessed April 2019).

[15]https://www.nmea.org/content/nmea_standards/nmea_standards.asp.

- The Maritime Mobile Service Identity (MMSI), which is an international and unique ship identifier;
- The coordinates (longitude and latitude expressed using World Geodetic System (WGS)84 reference system);
- The associated Speed Over Ground (SOG) in knots;
- The true heading in degrees (relative to True North) and the associated Course Over Ground (COG) also in degrees (direction of motion);
- The rate of turn, when available, expressed in degrees per minute;
- The navigational status, which is an integer encoding the current motion status of the ship (e.g., anchored, on the way, sailing, etc.).

These messages constitute an ordered time series. However, AIS messages do not embed the timestamp of the emission. Therefore, each message has been timestamped, with an integer corresponding to a UNIX epoch time upon reception by the receiving workstation. Figure 1.3 illustrates the content of positioning messages reported on a map. The dashed lines correspond to the bounding box of data.

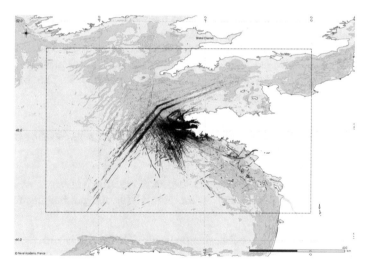

Fig. 1.3 Navigation data (purple background represents fishing areas computed by [19])

Nominative messages (ships' static messages): Several message types provide ship meta-information such as the ship name or voyage-related information. Some of this data is fully static (i.e., set at the initialization of the AIS device onboard) and is not expected to change during ship's life (e.g., ship dimension), while few others can evolve (e.g., the name can change depending on the owner) with varying frequency (e.g., destination should be updated at each voyage). These fields are manually set and thus prone to errors, imprecision, or simply not fed. Nominative information collected for this dataset contains the following vessel data:

- The MMSI and the ship identification number (an integer) provided by the International Maritime Organisation (IMO);
- The international radio call sign (a string of characters);
- The name of the vessel (a string of characters) and associated ship type encoded with an integer;
- The reference point for reported position (biggest ships can reach 400 m in length) and overall dimensions of the ship expressed with four integers describing a rectangle based on the reference point.

In addition, voyage-related information contains:

- The destination (next port of call) of the current trip (a string of characters manually entered) and associated Estimated Time of Arrival (ETA) expressed in a *month-day-hour-minute* format (using the Coordinated Universal Time with time zone);
- The draught of the ship for the current voyage (expressed in meters by a real number between 0.1 and 25.5).

These fields have been extracted from ITU messages of types ITU 5, ITU 19, and ITU 24. Similarly to dynamic messages, each message has been timestamped with UNIX epoch time at the receiving workstation.

Positioning messages (search and rescue dynamic messages): Some SAR aircraft operating at sea are also equipped with AIS transceivers. These particular emitters provide positioning information using message type ITU 9. The information extracted from SAR messages contains:

- The MMSI;
- The coordinates (longitude and latitude expressed using WGS84 reference system);
- The associated SOG in knots;
- The COG expressed in degrees (with respect to the True North);
- The altitude of the search and rescue aircraft (an integer ranging from 0 to 4094 m).

Each message has been also timestamped with UNIX epoch time at the receiving workstation.

Positioning messages (aids to navigation dynamic messages): Dynamic data of the AIS also include locations of Aids to Navigation (AtoN), typically buoys and lighthouses. An aid to navigation can be physical or virtual. Virtual Aids to Navigation (AtoN) do not exist physically and can be useful in time-critical situations and in marking/delineating dynamic locations or areas where navigational conditions change regularly. This equipment communicates information through the message type ITU 21. Information extracted from AtoN messages contains:

- The MMSI;
- The type of the aid to navigation encoded with an integer and associated name (a string of characters);

- The coordinates (longitude and latitude expressed using WGS84 reference system);
- The nature of AtoN (a Boolean value to distinguish between virtual and physical AtoNs). Figure 1.4d shows a selection of aids to navigation.

Each message has been also timestamped with UNIX epoch time at the receiving workstation.

AIS Status, Codes, and Types

Some data fields of AIS messages are encoded, usually with integer codes. To facilitate the understanding and the analysis of AIS data, these enumerations and the associated information have been included in the dataset, in CSV files.

- Status: The navigational status provided by positioning messages is encoded by an integer corresponding to 16 different statuses, whose names are themselves described by predefined strings of characters (e.g., moored, under way);
- Country Codes: Each ship is registered in a country (flag). The country of each ship is encoded in the first three digits of each MMSI number (e.g., "227" is used for France). Country codes and names (string of characters) are included in the dataset.
- Types: The ship type is encoded by an integer in nominative messages (e.g., "30" is the code used for fishing vessels). The correspondence between 38 codes (integer) and the ship types (string of characters) as described in [5] is provided in this dataset. Additionally, a list of 233 refined ship types has been provided for advanced classification of the vessel data (this extended list is based on the details in "MarineTraffic"[16]);
- AtoN: The type of aid to navigation (e.g., floating vs. fixed buoy, light, beacon) is encoded by an integer in the ITU 21 dynamic messages. This CSV data file provides a textual description (nature and type) for these codes.

AIS Receiver

Receptor location: The AIS messages have been collected using a single terrestrial receiver. It is a two-dimensional geometry of type point (i.e., the altitude is not considered) with coordinates (longitude and latitude) expressed using the WGS84 reference system.

Theoretical coverage of the receiver: Each terrestrial AIS receiver has a theoretical coverage that depends on its location and the surrounding topography. The theoretical coverage of the AIS receiver is given as a geometrical polygon with

---

[16]MarineTraffic: https://help.marinetraffic.com/hc/en-us/articles/205579997-What-is-the-significance-of-the-AIS-SHIPTYPE-number- (accessed January 2020).

coordinates (longitude and latitude) expressed using the WGS84 reference system. The receiver's theoretical coverage has been calculated using the definition of sea areas from IMO resolution A801(19). It computes a circle of radius R nautical miles where R is equal to the transmission distance between a ship's Very High Frequency (VHF) antenna at a height of four meters above the sea level and the VHF antenna of the coastal station (at a height of H meters) which lies at the center of the circle. The coverage has been computed with an antenna at 70 m above the sea level (66 m being the altitude of the location, plus four additional meters for the height of the building where the antenna is located).

1.2.2.2   Vessel Data

Other nominative vessel information is available in official registers. For the vessels navigating in the area covered by the dataset, two of the most relevant institutional registers are included in this dataset.

Community Fishing Fleet Register (European Commission)[17]

The European Commission freely provides a list of all fishing vessels that fly a flag of one of the countries of the Union. This fleet register concerns fishing vessels which represent only a fraction of the navigating vessels. Nevertheless, this register can provide a fair reference sample to some algorithms and could be extended later with additional vessel types when available. The data fields that can be matched with AIS data are the international reference call sign, the name, and the vessel length. The fleet register also includes few technical details like length, gear type (61 different types), year of construction, and engine power of the vessels.

Civilian Ships Registered by ANFR (Agence Nationale des Fréquences)[18]

The fleet register provided by the French Frequencies Agency (ANFR) gathers a large number of French-registered vessels. The data fields include several normative information (MMSI, IMO numbers, registration numbers, ship name) that can be matched with AIS fields. The dataset also provides characteristics of the ship (type, length, and tonnage). Communication facilities (e.g., VHF) and information about ship license (i.e., active, not active, and associated dates) are also described.

---

[17]Community Fishing Fleet Register, Fleet Register On the Net: ec.europa.eu/fisheries/fleet/index.cfm (accessed January 2020).

[18]Données radio maritime: data.gouv.fr/fr/datasets/donnees-radio-maritime (accessed July 2018).

1.2.2.3 Geographic Data

Geographic data provide complementary information to vessel movement data about topographic or regulatory context of vessel navigation.

Ports

A lot of ships (e.g., tankers, cargos, passengers, ferries) have origins and destinations corresponding to ports. Having official lists of ports is therefore crucial. It can be used, for instance, to disambiguate the *destination* field of AIS message 5 (filled manually). The dataset encompasses a detailed list of local ports and two worldwide lists (passing traffic around Brittany include world destinations).

- *Ports of Brittany*.[19] This dataset proposed by the Brittany region gathers the location of 222 ports around Brittany with names and coordinates expressed in the WGS84 system.
- *World Port Index*.[20] It is a publication of the National Geospatial-intelligence Agency and contains location and physical characteristics of, and the facilities and services offered by major ports and terminals worldwide. About 3700 ports throughout the world are included in this dataset. Figure 1.4a depicts an excerpt of the World Port Index (WPI).
- *SeaDataNet Ports Gazetteer*.[21] This dataset is provided by the Pan-European Infrastructure for Ocean and Marine Data Management[22] and focuses on halieutic ports. It contains names and coordinates of almost 5000 fishing ports throughout the world.

European Coastline (European Environmental Agency) [7]

The knowledge of the coastline is essential for the understanding of maritime movements. This dataset is a high resolution (1:100,000 scale) coastline of European shores (polylines and polygons, as shapefile), created by the European Environmental Agency (EEA) enabling highly detailed spatial analysis, such as the assessment of the proximity of vessels to coastlines or islands in support to maritime safety.

---

[19]Région Bretagne, Ports de Bretagne: data.gouv.fr/fr/datasets/ports-appartenant-a-la-region-breta gne (accessed July 2018).

[20]National Geospatial-intelligence Agency, World Port Index: msi.nga.mil/NGAPortal/MSI.port al?_nfpb=true&_pageLabel=msi_portal_page_62&pubCode=0015 (accessed July 2018).

[21]SeaDataNet Ports Gazetteer: seadatanet.maris2.nl/v_bodc_vocab_v2/welcome.asp (accessed July 2018).

[22]EMODnet, the European Marine Observation and Data Network: emodnet.eu (accessed July 2018).

It is a data derived from two sources: EU-Hydro[23] and the Global Self-consistent, Hierarchical, High-resolution Geography Database (GSHHG).[24]

Sea Areas (International Hydrographic Organization)

The dataset contains the main seas of the world (101) as polygons representing contiguous bodies of water, regionalized and divided into maritime basins.[25] It can be useful to determine if two vessels are in the same region of the world. Areas covered (polygons concerned) by the navigation data are the Celtic Sea, the North Atlantic Ocean, the English Channel, and the Bay of Biscay.

FAO Major Fishing Areas (Food and Agriculture Organization)[26]

This dataset contains the worldwide fishing regions established by the Food and Agriculture Organization (FAO) of the United Nations (UN). The boundaries were determined in consultation with international fishery agencies considering the distribution of the natural resources, national practices and boundaries, international conventions. It can be used for example to assess the declared provenance of fish against the areas in which a vessel effectively sailed or exhibited a fishing behavior.

Exclusive Economic Areas (Flanders Marine Institute)[27]

All the Exclusive Economic Zones (EEZs) of the world are gathered in this dataset. This can be useful to determine the quality of some at-sea operations as well as the competent court in some activities. Exclusive Economic Zones boundaries are given as polygons and polylines. Areas beyond these boundaries can be classified as high seas. Figure 1.4b depicts an excerpt of the EEZs.

---

[23]EU-Hydro: land.copernicus.eu/pan-european/satellite-derived-products/eu-hydro (accessed July 2018).

[24]National Oceanographic and Atmospheric Administration, Global Self-consistent, Hierarchical, GSHHG [online]. ngdc.noaa.gov/mgg/shorelines (accessed July 2018)

[25]International Hydrographic Organization, IHO Marine Regions: marineplan.es/ES/fichas_kml/iho.html, available at marineregions.org/downloads.php (accessed July 2018).

[26]FAO Major Fishing Areas: fao.org/fishery/area/search/en (accessed July 2018).

[27]Flanders Marine Institute, Exclusive Economic Areas v10: marineregions.org/downloads.php (accessed July 2018).

European Maritime Boundaries (European Environment Agency)[28]

The dataset provided by the European Environmental Agency contains maritime boundaries in Europe that include territorial waters, bi- or multi-lateral boundaries as well as contiguous and Exclusive Economic Zone (EEZ)s.

Natura 2000 Areas (European Environment Agency)[29]

The European database on Natura 2000 collects the reporting of the European protected areas, as an ecological network developed for the preservation of species and habitats (terrestrial and maritime areas). The dataset is managed by the European Environmental Agency and is built upon data submitted by European member states. The dataset contains descriptive data (e.g., the list of all species and habitat types) and spatial data (borders of sites). The version included in the dataset covers 2017 reporting.

European Fishing Areas (European Commission Joint Research Centre) [19]

This dataset was created by the European Commission Joint Research Centre (JRC)[30] and provides the fishing grounds in European waters, which can be used to assess the behavior of fishing vessels. An assessment on the fishing pressure can also be extracted from this database. Fishing grounds have been derived from ship AIS positions collected along 1 year (from September 2014 until September 2015) and emitted by selected categories of fishing vessels exhibiting a fishing behavior. Raw AIS data are originated from Volpe Center of the U.S. Department of Transportation, the U.S. Navy, and MarineTraffic. Figure 1.4e depicts an excerpt of the JRCs fishing effort layer.

Fishing Constraints

This dataset contains two geographic areas where shellfish fishing activity is forbidden in the time window of the dataset.

---

[28]Maritime Boundaries: eea.europa.eu/data-and-maps/data/maritime-boundaries (accessed July 2018).

[29]Natura 2000 data—the European network of protected sites: natura2000.eea.europa.eu. Data available at eea.europa.eu/data-and-maps/data/natura-9 (accessed July 2018).

[30]JRC—AIS derived high resolution fishing effort layer for European trawlers of more than 15 m long 2014–2015: https://data.jrc.ec.europa.eu/dataset/jrc-fad-ais1415 (accessed December 2018).

1.2.2.4   Environmental Data

Weather data and ocean data from forecast models and from observations (e.g., insitu sensor data) which are openly available from several providers can help validate analysis results and explain abnormal behavior. For example, sea and weather conditions can force vessels to change direction or modify their normal route. They can also be used to characterize seasonal trends in traffic routes and to contextualize vessels kinematics such as speed.

Ocean Conditions [2]

The ocean conditions were extracted from an hindcast database built with a WAVEWATCH III model [2] and provided by Institut français de recherche pour l'exploitation de la mer (IFREMER).[31] Initially built for the design of marine energy converters, it can also be used to study ships behavior. Therefore, only the parameters relevant to this context were selected and stored in 6 CSV files (one file per month). The values are provided every 3 h for each point of a regular spatial grid over the dataset bounding box. The grid spacing is 2 arc min, for both latitude and longitude. The parameters include coordinates (longitude and latitude expressed using the WGS84 reference system); bottom depth in meters; sea surface height above sea level in meters (tidal effect); significant height of wind and swell waves; mean wave length in meters; wave mean direction; and a timestamp expressed using epoch time. Figure 1.4f shows an excerpt of the WAVEWATCH III dataset.

Weather Observations[32]

The coastal weather observations were recorded by 16 stations located in the south of England and along the French coasts. The data have been cleaned and formatted in the same way for the 16 selected stations, providing the most relevant fields such as temperature, atmospheric pressure, wind direction/speed, and horizontal visibility, every hour over the 6 months period (Fig. 1.4h). The unformatted or potentially biased fields, especially those based on the human perception, have been removed.

---

[31]Institut français de recherche pour l'exploitation de la mer (IFREMER): ifremer.fr (accessed July 2018).

[32]Weather history and forecast: rp5.ua/archive.php (accessed July 2018).

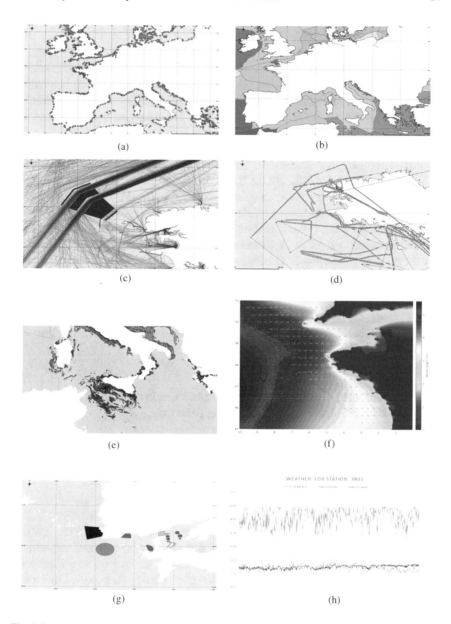

Fig. 1.4 Selection of maritime data. (**a**) Ports for the WPI in blue and from SeaDataNet Port Index (in red). (**b**) Exclusive Economic Zones (EEZs). (**c**) TSS. (**d**) Aids to Navigation (AtoN) (hexagons) and Search and Rescue (planes). (**e**) European Commission Joint Research Centre (JRC) AIS derived high resolution fishing effort layer for European trawlers of more than 15 m long, 2014–2015. (**f**) Wave height (m) and direction (°) forecast example based on Wavewatch III model (NOAA), 2016/01/09—00h00 UTC. (**g**) Movement constraints (light blue), speed restrictions (green), Waiting zones for oil tankers (pink) and cargos (yellow), anchorage areas (purple), crustacean sanctuaries (dark blue). (**h**) Air temperature (black), mean wind speed (green) and relative humidity (blue) for coastal station #3803 over 6 months

1.2.3   Generating Operational Scenario for Experiments

To support the evaluation of the algorithms, the reference dataset has been enriched with contextual historical information. Raw data have been processed to extract maritime patterns and clusters, specifically maritime routes, which facilitate the operators assessment. Additionally, a library of functions has been designed and implemented to systematically degrade and enrich the AIS batch dataset. The modifications applicable to each AIS field have been categorized along *kinematic*, *coverage*, and *spoofing* dimensions. This library of functions provides a rich set of basic constructs to build different modification patterns in a non-unique manner [6]. Having different ways to produce equivalent patterns or to provide a wide diversity in similar patterns is desirable to create realistic synthetic datasets.

A semi-automatic process was designed to generate dedicated scenarios [6, 23] following several steps of data enrichment, data injection, resulting in a story matching the maritime surveillance challenges. As an example, the situation representing a collision avoidance scenario consists of a set of real AIS tracks enriched with specific events such as:

- A collision between a real vessel trajectory and a synthetic vessel trajectory;
- A collision between a real vessel trajectory and the shifted trajectory in time and space of another real vessel trajectory;
- A synthetic near-collision;
- The shift in time of a real tugging case;
- The shift in time and space of specific trajectories in order to simulate a given behavior (e.g., the individuation of fishing patterns);
- The simulation of a *rendezvous* behavior.

In the collision avoidance scenario, the mission of the VTS operator is to prevent and avoid a collision involving fishing vessels. The maritime security system can also enhance the situational awareness of the captains of the involved vessels, anticipating that a vessel will be required to "give way" to a fishing vessel.

Figure 1.5a, b illustrates scenario data generated for experiments. For the first experiment, the chosen area is a part of the Brest roadstead, where traffic entering and exiting both the Brest port and the roadstead cross. The second experiment takes place in the Four channel, a maritime channel located off the West coast of Brittany, between the mainland and a series of islands (Béniguet, Quénémès, Molène, Ushant), leading to the entrance of the Brest roadstead.

1.3   Conclusions

Facing the huge *volume* of *various* information with high *velocity*, which often lacks *veracity*, a system to automatically process both historical and timely information would greatly support the Vessel Traffic System (VTS) operator in monitoring and

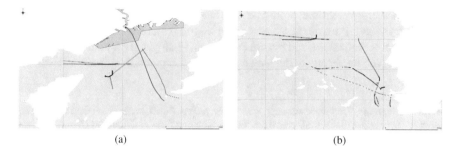

(a)   (b)

Fig. 1.5 Examples of two collision avoidance scenarios built from real and simulated AIS data. (**a**) All contacts points from the first scenario (gray: Brest harbor, each MMSI is assigned a color). (**b**) All contacts points from the second scenario (each MMSI is assigned a color)

performing several types of analysis of situations at sea. This has been the aim of the 3-year *Big Data Analytics for Time-Critical Mobility Forecasting* (datAcron) project[33] that ran from January 2016 to December 2018 and whose main research objectives addressed the development of highly scalable methods for advancing:

1. Spatiotemporal data integration and management solutions;
2. Real-time detection and forecasting accuracy of moving entities' trajectories;
3. Real-time recognition and prediction of important events concerning these entities;
4. General visual analytics infrastructure supporting all steps of the analysis through appropriate interactive visualizations;
5. Producing streaming data synopses at a high-rate of compression.

The operational scenarios and challenges together with the supporting datasets described in this chapter were aimed at supporting the design and development of big data analytics tools for maritime surveillance. Six scenarios corresponding to three use cases of Secure fishing, Sustainable development, and Maritime security were proposed, while the collision avoidance scenario under the Secure fishing use case was retained as the main one. A set of 28 Maritime Situational Indicators was proposed to both capture user information needs and provide targets to event detectors to be designed by the datAcron partners and integrated in the final prototype. The scenario was described with relevant MSIs, for which big data analytics solutions—presented in subsequent chapters of this book—were proposed and implemented. The algorithms thus designed have been exercised on an heterogeneous dataset gathering timely AIS data and other contextual data. The open heterogeneous maritime dataset has been made publicly available to enable other further maritime experimentation in realistic operational settings, challenging specifically the *variety* dimension of big data.

---

[33] datAcron project website: http://www.datacron-project.eu (accessed January 2020).

The datAcron maritime prototype has been setup following a human-centric approach involving maritime experts and tying together scenario definition, data preparation, visualization, and human evaluation. Finally, the evaluation of the maritime prototype has been conducted according to a methodology further described in [22].

Acknowledgments This work was supported by the project *Big Data Analytics for Time-Critical Mobility Forecasting* (datAcron), which has received funding from the European Union's Horizon 2020 research and innovation program under grant agreement No. 687591. The authors would like to thank Christophe Claramunt, Melita Hadzagic, LCdr Eric Reardon, Cdr Mike Ilteris, and Karna Bryan for their participation to the original ideas of the use cases and scenarios and cadets of the French Naval Academy for their participation in the design of scenario datasets.

References

1. Andler, S., Fredin, M., Gustafsson, P.M., van Laere, J., Nilsson, M., Svenson, P.: SMARTracIn - a concept for spoof resistant tracking of vessel and detection of adverse intentions. In: Proceedings Volume 7305, Sensors, and Command, Control, Communications, and Intelligence (C3I) Technologies for Homeland Security and Homeland Defense VIII; 73050G, Orlando, FL (2009)
2. Boudiere, E., Maisondieu, C., Ardhuin, F., Accensi, M., Pineau-Guillou, L., Lepesqueur, J.: A suitable metocean hindcast database for the design of marine energy converters. Int. J. Marine Energy **3–4**, e40–e52 (2013). https://doi.org/10.1016/j.ijome.2013.11.010
3. Halpern, B.S., Frazier, M., Potapenko, J., Casey, K.S., Koenig, K., Longo, C., Lowndes, J.S., Rockwood, R.C., Selig, E.R., Selkoe, K.A., Walbridge, S.: Spatial and temporal changes in cumulative human impacts on the world's ocean. Nat. Commun. **6**, 7615 (2015)
4. Heumann, J.: Tips for writing good use cases. IBM - Transforming software and systems delivery (2008). White paper
5. International Telecommunication Union (ITU): Technical characteristics for an automatic identification system using time division multiple access in the VHF maritime mobile frequency band (recommendation m.1371-5) (2014). https://itu.int/rec/R-REC-M.1371-5-201402-I
6. Iphar, C., Jousselme, A.L., Ray, C.: Pseudo-synthetic datasets in support to maritime surveillance algorithms assessment. In: Proceedings of the VERITA Workshop, 19ieme Journées Francophones Extraction et Gestion des Connaissances (EGC) 2019, Metz, 21–25 January (2019)
7. Jousselme, A.L., Ray, C., Camossi, E., Hadzagic, M., Claramunt, C., Bryan, K., Reardon, E., Ilteris, M.: Maritime use case description, H2020 datAcron d5.1 (2016)
8. Jousselme, A.L., Reardon, E., Hadzagic, M., Camossi, E., Bryan, K., Ilteris, M., Ray, C., Claramunt, C.: A maritime use case in support to collaborative research. In: Vespe, M., Mazzarella, F. (eds.) Proceedings of the Maritime Knowledge Discovery and Anomaly Detection (KDAD) Workshop. European Commission Joint Research Centre, European Commission Publication Office, Ispra, Varese (2016). https://doi.org/10.2788/025881. https://bluehub.jrc.ec.europa.eu/static/KDAD/KDAD_Proceedings.pdf
9. Le Guyader, D., Ray, C., Brosset, D.: Defining fishing grounds variability with automatic identification system (AIS) data. In: 2nd International Workshop on Maritime Flows and Networks (WIMAKS'16) (2016)
10. Lee, J., South, A.B., Jennings, S.: Developing reliable, repeatable, and accessible methods to provide high-resolution estimates of fishing-effort distributions from vessel monitoring system

(VMS) data. ICES J. Marine Sci. **67**(6), 1260–1271 (2010). https://doi.org/10.1093/icesjms/fsq010

11. Mazzarella, F., Vespe, M., Damalas, D., Osio, G.: Discovering vessel activities at sea using AIS data: mapping of fishing footprints. In: 17th International Conference on Information Fusion, FUSION 2014, Salamanca, 7–10 July 2014, pp. 1–7 (2014). http://ieeexplore.ieee.org/xpl/freeabs_all.jsp?arnumber=6916045

12. Natale, F., Gibin, M., Alessandrini, A., Vespe, M., Paulrud, A.: Mapping fishing effort through AIS data. PloS One **10**(6), e0130746 (2015)

13. Ray, C., Grancher, A.: Integration of Nautical Charts in Maritime Dataset (2018). https://doi.org/10.5281/zenodo.1182539

14. Ray, C., Camossi, E., Jousselme, A.L., Hadzagic, M., Claramunt, C., Batty, E.: Maritime data preparation and curation, H2020 datAcron d5.2 (2016)

15. Ray, C., Iphar, C., Dréo, R., Kleynhans, W., Camossi, E., Jousselme, A.L., Zocholl, M., Batty, E., Roche, Q., Metzger, A.: Maritime data preparation and curation (final), H2020 datAcron d5.4 (2018)

16. Ray, C., Camossi, E., Dréo, R., Jousselme, A.L., Iphar, C., Zocholl, M., Hadzagic, M.: Use case design and big data analytics evaluation for fishing monitoring. In: OCEANS 2019, Marseille, 17–20 June (2019)

17. Ray, C., Dréo, R., Camossi, E., Jousselme, A.L., Iphar, C.: Heterogeneous integrated dataset for maritime intelligence, surveillance, and reconnaissance. Data in Brief **25** (2019). https://doi.org/10.1016/j.dib.2019.104141

18. Roy, J., Davenport, M.: Exploitation of maritime domain ontologies for anomaly detection and threat analysis. In: Proceedings of the 2010 International Waterside Security Conference (WSS). IEEE (2010). https://doi.org/10.1109/WSSC.2010.5730278

19. Vespe, M., Gibin, M., Alessandrini, A., Natale, F., Mazzarella, F., Osio, G.C.: Mapping EU fishing activities using ship tracking data. J. Maps **12**(Suppl. 1), 520–525 (2016)

20. Vouros, G.A., Vlachou, A., Santipantakis, G.M., Doulkeridis, C., Pelekis, N., Georgiou, H.V., Theodoridis, Y., Patroumpas, K., Alevizos, E., Artikis, A., Claramunt, C., Ray, C., Scarlatti, D., Fuchs, G., Andrienko, G.L., Andrienko, N.V., Mock, M., Camossi, E., Jousselme, A.L., Garcia, J.M.C.: Big data analytics for time critical mobility forecasting: recent progress and research challenges. In: Böhlen, M.H., Pichler, R., May, N., Rahm, E., Wu, S., Hose, K. (eds.) Proceedings of the 21th International Conference on Extending Database Technology, EDBT 2018, Vienna, 26–29 March 2018, pp. 612–623. OpenProceedings.org (2018). https://doi.org/10.5441/002/edbt.2018.71

21. Zocholl, M., Camossi, E., Jousselme, A.L., Ray, C.: Ontology-based design of experiments on big data solutions. In: Khalili, A., Koutraki, M. (eds.) Proceedings of the Posters and Demos Track of the 14th International Conference on Semantic Systems (SEMANTiCS 2018), Vienna, 10–13 September 2018. CEUR Workshop Proceedings, vol. 2198. CEUR-WS.org (2018). http://ceur-ws.org/Vol-2198/paper_117.pdf

22. Zocholl, M., Iphar, C., Camossi, E., de Rosa, F., Jousselme, A.L., Ray, C., Dréo, R.: User-centric assessment of maritime situation awareness solutions. In: OCEANS 2019, Marseille, 17–20 June (2019)

23. Zocholl, M., Iphar, C., Pitsikalis, M., Jousselme, A.L., Artikis, A., Ray, C.: Evaluation of maritime event detection against missing data. In: Piattini, M., da Cunha, P.R., de Guzmán, I.G.R., Pérez-Castillo, R. (eds.) Quality of Information and Communications Technology - 12th International Conference, QUATIC 2019, Ciudad Real, 11–13 September 2019, Proceedings, Communications in Computer and Information Science, vol. 1010, pp. 275–288. Springer (2019). https://doi.org/10.1007/978-3-030-29238-6_20

Chapter 2
The Perspective on Mobility Data from the Aviation Domain

Jose Manuel Cordero and David Scarlatti

Abstract Air traffic management is facing a change of paradigms looking for enhanced operational performance able to manage increasing traffic demand (number of flights and passengers) while keeping or improving safety, and also remaining environmentally efficient, among other operational performance objectives. In order to do this, new concepts of operations are arising, such as trajectory-based operations, which open many new possibilities in terms of system predictability, paving the way for the application of big data techniques in the Aviation Domain. This chapter presents the state of the art in these matters.

2.1 Introduction

The current air traffic management (ATM) system worldwide has reached its limits in terms of predictability, efficiency, and cost effectiveness. Nowadays, the ATM paradigm is based on an airspace management that leads to demand imbalances that cannot be dynamically adjusted. This entails higher air traffic controllers' (ATCO) workload, which, as a final result, determines the maximum system capacity.

With the aim of overcoming such ATM system drawbacks, different initiatives, dominated by Single European Sky ATM Research SESAR in Europe and NextGen in the USA, have promoted the transformation of the current environment towards a new trajectory-based ATM paradigm. This paradigm shift changes the old fashioned airspace management to the advanced concept of trajectory-based operations (TBO). In the future ATM system, the trajectory becomes the cornerstone upon which all the ATM capabilities will rely on. The trajectory life cycle describes the different stages from the trajectory planning, negotiation, and agreement, to

J. M. Cordero
CRIDA (Reference Center for Research, Development and Innovation in ATM), Madrid, Spain
e-mail: jmcordero@e-crida.enaire.es

D. Scarlatti (✉)
Boeing Research & Development Europe, Madrid, Spain
e-mail: David.Scarlatti@boeing.com

© Springer Nature Switzerland AG 2020
G. A. Vouros et al. (eds.), *Big Data Analytics for Time-Critical Mobility Forecasting*, https://doi.org/10.1007/978-3-030-45164-6_2

the trajectory execution, amendment, and modification. The envisioned advanced decision support tools (DSTs) required for enabling future ATM capabilities will exploit trajectory information to provide optimized services to all ATM stakeholders (airlines, air navigation service providers (ANSPs), air traffic control (ATC), etc.).

The proposed transformation requires high fidelity aircraft trajectory prediction capabilities, supporting the trajectory life cycle at all stages efficiently.

Current Trajectory Predictors (TPs) are based on deterministic formulations of the aircraft motion problem. Although there are sophisticated solutions that reach high levels of accuracy, all approaches are intrinsically simplifications to the actual aircraft behaviour, which delivers appropriate results for a reasonable computational cost. TPs outputs are generated based on a priori knowledge of the planned flight plan, the expected command and control strategies released by the pilot or the flight management system (FMS)—to ensure compliance with ATC restrictions and user preferences (all together known as aircraft intent), a forecast of weather conditions to be faced throughout the trajectory, and the aircraft performance. This model or physics-based approach is deterministic: It returns always the same trajectory prediction for a set of identical inputs.

Although the use of the concept of aircraft intent [1] together with very precise aircraft performance models such as Base of Aircraft Data (BADA) [2] has helped to improve the prediction accuracy, the model-based approach requires a set of input data that typically are not precisely known (i.e., initial aircraft weight, pilot/FMS flight modes, etc.). In addition, accuracy varies depending on the intended prediction horizon (look-ahead time). In summary one can identify current TP as an area of improvement with consequent benefits supporting TBO.

Recent efforts in the field of aircraft trajectory prediction have explored the application of statistical analysis and machine learning techniques to capture non-deterministic influences that arise when an aircraft trajectory prediction is requested by a DST. Linear regression models [3, 4] or neural networks [5, 6] have returned successful outcomes for improving the trajectory prediction accuracy on the vertical plane and for traffic flow forecasting. Generalized linear models [7] have been applied for the trajectory prediction in arrival management scenarios and multiple linear regression [8, 9] for predicting estimated times of arrival (ETA). Although most of these efforts include as input dataset the available surveillance data, there is no consensus on the additional supporting data required for robust and reliable trajectory predictions. Such additional supporting data may include filed or amended flight plans, airspace structure, ATC procedures, airline strategy, weather forecasts, etc.

The outcome of these recent efforts provides promising results in terms of accuracy prediction [10]; however, there is still a lack of global vision on how to apply data-driven approaches to real ATM scenarios, and what the expected improvement will be. The disparity of the datasets used for validating different methods makes difficult the comparison among those studies, and, therefore, prevents from extending the applicability of such techniques to more realistic and complex scenarios.

Another strong limitation found in the current state-of-the-art research is that the proposed data-driven approaches are mostly limited to individual trajectory predictions. The trajectories are predicted one by one based on the information related to them, ignoring the expected traffic at the prediction time lapse, hence disregarding contextual aspects on the individual predictions. Consequently, the network effect resulting from the interactions of multiple trajectories is not considered at all, which may lead to huge prediction inaccuracies. The complex nature of the ATM system impacts the trajectory predictions in many different manners. Capturing this complexity and being able to devise prediction methods that take the relevant information into account will improve the trajectory prediction process: This is a considerable leap from the classical model-based approaches.

2.2 Trajectory Prediction Approaches in the Aviation Domain

A new strategy for trajectory prediction in Aviation is to exploit available trajectory information to predict future trajectories based on the knowledge acquired from historical data. This innovative approach is in contrast to the classic model-based approach in which different models are involved in the computation of aircraft motion.

First of all, it is required to have a common understanding of what a trajectory is. Basically, a trajectory in the Aviation Domain is a chronologically ordered sequence of aircraft states described by a list of state variables. Most relevant state variables are airspeeds (true airspeed (TAS), calibrated airspeed (CAS) or Mach number (M)), 3D position (latitude (ϕ), longitude (λ) and geodetic altitude (h) or pressure altitude (Hp)), the bearing (χ) or heading (ψ), and the instantaneous aircraft mass (m). Additionally, a predicted trajectory can be defined as the future evolution of the aircraft state as a function of the current flight conditions, a forecast of the weather conditions, and a description of how the aircraft is to be operated from this initial state and on.

According to the formulation of the motion problem, there are two possible model-based alternatives:

2.2.1 Kinematic Trajectory Prediction Approach

This solution does not consider the causalities of motion, only takes into account the speeds, altitude, and lateral profiles that may represent the evolution of the aircraft position with time. The accuracy of kinematics Trajectory Predictors (TP) strongly relies on the accuracy of datasets used to model the aircraft's performance and how well they match the actual aircraft's behaviour in all possible flight conditions. The main advantage is that kinematic TPs are usually orders of magnitude faster than other alternatives.

2.2.2   Kinetic Trajectory Prediction Approach

This formulation describes the forces and momentums that cause the aircraft motion. For ATM applications, a simplified 3 degrees of freedom (DoF) approach (point-mass model (PMM)) is typically assumed because it provides enough information to support further decision-making processes. More sophisticated 6 DoF approaches, applied, for instance, in simulators, increase the fidelity to the predicted trajectories by modelling the aircraft attitude, which is of no interest for ATM purposes. To pose a well-formulated kinetic problem, models of the aircraft performance, weather conditions, and aircraft intent (description of command and control directives that univocally turns into in a unique trajectory when applied to aircraft by the pilot or the flight management system (FMS)) are required.

Even though there might be available extremely accurate aircraft performance models, such as BADA models released by EUROCONTROL, in conjunction to accurate weather forecasts, such as those generated by the Global Forecast System (GFS) provided by the National Oceanic and Atmospheric Administration (NOAA), there are intrinsic errors that produce unavoidable deviations between predicted and actual trajectories. Those deviations are the result of representing a stochastic process (prediction of an aircraft trajectory affected by stochastic sources) by a deterministic approach (formulation of a kinematic or kinetic aircraft motion problem).

The concept of data-driven trajectory prediction is a completely different approach than those mentioned above. It does not consider any representation of any realistic aircraft behaviour, only exploits trajectory information recorded from the ground-based surveillance infrastructure or by onboard systems (e.g., flight data recorder (FDR) or quick access recorder (QAR) data) and other contextual data that may impact the final trajectory. This decoupled solution from the mathematical formulation of the aircraft motion should capture variations of the trajectory that cannot be derived directly from the filed flight plans (i.e., intended trajectories), both during the strategic (before departure) and tactical phases (after departure). These discrepancies usually come from air traffic control interventions to ensure optimum traffic management and safe operations (e.g., delays added due the effect of adverse weather). If these interventions respond to a pattern, big data analytics and machine learning algorithms might potentially identify them once the proper system features are considered.

Thus, the preparation of available trajectory data is crucial to train the algorithms in accordance to the expected data-driven TP performance accuracy. Several solutions aim at predicting some aircraft state variables (time at a fix/waypoint) for a representative scenario. In general, different generic prediction methods can be applied in different possible scenarios envisioned in the future trajectory-based operations environment, in which the ATM paradigm will evolve from current tactical airspace-based to a strategic trajectory-based traffic management.

Subsequently we provide a literature review of prominent techniques applied to the problem of predicting an aircraft trajectory leveraging historical recorded flight data.

2.2.3 Data-Driven Trajectory Prediction Approaches

The following list of approaches describes the current state-of-the-art techniques applied to aircraft trajectory prediction driven by data.

Statistical Prediction of Aircraft Trajectory: Regression Methods vs Point-Mass Model [11] This approach proposes a statistical regression model combined with a total energy model (simplified version of the classical point-mass model for aircraft) to predict the altitude of a climb procedure with a 10-min look-ahead time starting from an initial flight level (FL180). The input dataset are radar tracks and meteorological data. The study uses the already flown aircraft positions, the observed calibrated airspeed (CAS) at the current altitude, the temperature deviation with respect to the International Standard Atmosphere (ISA) conditions, and the predicted conditions at different levels of pressure. The main assumption of this approach is that the climb procedure is represented by a CAS/Mach transition for all predicted trajectories. Three techniques were assessed: linear regression, neural networks, and locally weighted polynomial regression, being the latter the one that provides higher accuracy with respect to reference recorded data.

Data Mining for Air Traffic Flow Forecasting: A Hybrid Model of Neural Network and Statistical Analysis [6] This approach employs a combination of feed forward and back propagation neural networks combined with statistical analysis to predict the traffic flow. The basic information required that represents a forecasted traffic sample is the estimated time of arrival (ETA) at designated fixes and airports. Initially, a 5-step data mining process is proposed as preliminary stage to process the radar tracks to generate the input dataset to the neural network. The analysis of historical data suggests that the traffic flow series can be classified in 7 classes from Sunday to Saturday; thus, the applied algorithms uses 7 back propagation neural networks that are trained separately. A relevant outcome of the study is that 1 hidden layer of approximately 5–10 neurons provides best results. The accuracy of the predictions degrades with the look-ahead time.

Using Neural Networks to Predict Aircraft Trajectories [5] This work deals with the problem of predicting an aircraft trajectory in the vertical plane (altitude profile with the time). Two separate approaches have been analysed: the case of strategic prediction considering that the aircraft is not flying yet; and the case of tactical prediction in which flown aircraft states are used to improve the prediction. The study is focused on predicting trajectories for a unique aircraft type. The prediction algorithm is based on a feed forward neural network with a single hidden layer. The neural network is parameterized to learn from the difference between the Requested Flight Level (RFL), which defines the cruise altitude, and the actual altitude. This strategy facilitates capturing of the evolution of the Rate of Climb (ROC) with the altitude. Two neural networks methods (standard and sliding windows) were studied according to the data availability (i.e., tactical or strategically prediction) to predict the aircraft altitude separately. A main conclusion of this work is the higher number of samples describing the trajectories building the training set, the better prediction results.

A Methodology for Automated Trajectory Prediction Analysis [12] According to this approach the prediction process is split in separated stages according to the flight phases. This facilitates the process of identifying the recorded flights (described by actual radar tracks) that show unpredictable modifications of their aircraft intent, removing these outliers from the training dataset. This process is referred to as segmentation. This process is of high interest when preparing a dataset to be fed to machine learning algorithms for trajectory prediction. This methodology relies on the definition of rules for segmenting trajectories and removing outliers from a trajectory dataset.

Trajectory Prediction for Vectored Area Navigation Arrivals [9] This work introduces a new framework for predicting arrival times by leveraging probabilistic information about the trajectory management patterns that would be applied by an air traffic controller (ATCO) to ensure safe operations (i.e., avoiding breaches of separation minima) and manage the traffic efficiently. The likelihood of those trajectory management patterns is computed from the patterns of preceding aircraft. This work considers a dataset of recorded radar tracks, representing trajectories of aircraft of the same wake vortex category. This homogenizes the dataset by removing the variability in arrival times because of the variability of aircraft types. The proposed machine learning algorithm predicts the ETA at the runway considering the time at entry waypoint (fix). The major patterns of vectored trajectories are found by clustering recorded radar tracks for the airspace of interest. The clusters are built upon the computation of the relative Euclidian distance of a trajectory from the other. However, time misalignment among trajectories can result in large distances. To solve this issue, the dynamic time warping (DTW) measure is applied, providing with the optimal alignment of two trajectories. Multiple linear regression models for travel time are designed for each of those identified patterns. Finally, among all identified patterns, the most suitable according to the patterns of trajectory management, flown by the preceding traffic, is chosen.

A 4-D Trajectory Prediction Model Based on Radar Data [7] This work proposes a four-dimensional trajectory prediction model that makes use of historical and real-time radar tracks. Both strategic and tactical prediction processes are designed according to the available datasets. The strategic prediction is used as the baseline against which the tactical predictions are compared to detect deviations and improve prediction accuracy by updating the trajectory prediction. The process is designed in two stages: prediction of total flying time, and prediction of flying positions and altitudes. The former prediction is performed by using a multiple regression method that relates the influences of traffic flow and wind conditions. The latter prediction requires from a process to normalize the flying positions and altitudes of different trajectories (i.e., different recorded radar tracks) to the same time interval. The conclusion from this work is that high prediction accuracy can be achieved, although at the cost of modelling the trajectories individually.

A Machine Learning Approach to Trajectory Prediction [7] A supervised learning regression problem, which implements the so-called generalized linear models (GLM) to trajectory prediction for sequencing and merging of traffic, following fixed arrival routes, is described and evaluated using actual aircraft trajectory and meteorological data. This study selects two aircraft types according to the availability of Automatic Dependent Surveillance-Broadcast (ADS-B) tracks. The first aircraft is a narrow body aircraft in the ICAO wake vortex medium category and the second aircraft is a wide body aircraft in the wake vortex heavy category. Trajectories of flights that were vectored off the arrival route or showed signs of speed control were removed from the dataset. To determine which regressors to include in the GLM, a stepwise regression approach is applied. Stepwise regression provides a systematic approach to add or remove regressors from the GLM based on their statistical significance in explaining the output variable. Due to the scarce availability of input variables obtained from current surveillance systems, only arrival time predictions for aircraft following fixed arrival routes in combination with continuous descent operations (CDO) were made.

An Improved Trajectory Prediction Algorithm Based on Trajectory Data Mining for Air Traffic Management [10] This work uses data mining algorithms to process historical radar tracks and to derive typical trajectories coming from the original tracks by applying clustering algorithms (i.e., Density-Based Spatial Clustering of Application with Noise (DBSCAN)). For predicting a trajectory, the typical trajectory is used to feed a hybrid predictor that instantiates an interacting multiple model Kalman filter. The use of the typical trajectory ensures that the associated flight intent represents better the intended trajectory and, therefore, the errors of long-term prediction diminish.

Aircraft Trajectory Forecasting Using Local Functional Regression in Sobolev Space [8] According to this approach, a time window between 10 and 30 min is considered, in which an aircraft trajectory prediction is to be generated. The proposed algorithm based on local linear functional regression exploits 1 year radar tracks over France as primary source to learn from. The learning process is designed in two separated stages: localization of data using k nearest neighbours; and solving of regression using wavelet decomposition in Sobolev space. The paper describing this approach concludes that this method returns efficient results with high robustness, although the proposed approach does not consider the effect of the weather conditions (especially the wind) in the prediction.

Terminal Area Aircraft Intent Inference Approach Based on Online Trajectory Clustering [13] This work proposes a two-stage process to obtain an inferred estimation of the aircraft intent that represents a flown trajectory. The first stage is devoted to identify the associated intent model, while the second one computes the specific intent based on the knowledge of the referred model. The intent modelling is formulated as an online trajectory clustering problem where the real-time intended routes are represented by dynamically updated cluster centroids extracted from radar tracks without flight plan correlations. Contrary, the intent identification is implemented with a probabilistic scheme integrating multiple flight attributes (e.g., call sign, destination airport, aircraft type, heading angle, and the like). This work suggests that the detection of outlier trajectories based on the clustering process requires a detailed analysis and a review considering the actual ATCO interventions on the considered flights.

New Algorithms for Aircraft Intent Inference and Trajectory Prediction [14] Considering the requirements of aircraft tracking and trajectory prediction accuracy of current and future ATM environments, a hybrid estimation algorithm, called the residual mean interacting, is proposed, with the objective to predict future aircraft states and flight modes using the knowledge of air traffic control (ATC) regulations, flight plans, pilot intent, and environment conditions. The intent inference process is posed as a discrete optimization problem whose cost function uses both spatial and temporal information. The trajectory is computed thanks to an intent-based trajectory prediction algorithm. Using ADS-B messages, the algorithm computes the likelihood of possible flight modes, selecting the most probable one. The trajectory is determined by a sequence of flight modes that represent the solvable motion problems to be integrated to obtain the related trajectory.

Predicting Object Trajectories from High Speed Streaming Data [15] This approach introduces a machine learning model, which exploits geospatial time series surveillance data generated by sea vessels, in order to predict future trajectories based on real-time criteria. Historical patterns of vessels movement are modelled in the form of time series. The proposed model exploits the past behavior of a vessel in order to infer knowledge about its future position. The method is implemented within the MOA toolkit [16] and predicts the position of any vessel within the time range of 5 min. In that context, online vessel's records are processed as they arrive and treated as a single trajectory which directly feeds the forecasting model without taking into account contextual information (i.e., vessel types, geographic area, and other explicit parameters). As this method becomes suitable for real-time applications, it does not contribute to improving the accuracy of predictions and it allows for model replicability and scalability to any prediction model of moving objects' trajectories.

Aircraft Trajectory Prediction Made Easy with Predictive Analytics [17] This approach proposes a novel stochastic approach to aircraft trajectory prediction problem, which exploits aircraft trajectories modelled in space and time by using a set of spatiotemporal data cubes. Airspace is represented in 4D joint data cubes consisting of aircraft's motion parameters (i.e., latitude, longitude, altitude, and time) enriched by weather conditions. It uses the Viterbi algorithm [18] to compute the most likely sequence of states derived by a Hidden Markov Model (HMM), which has been trained over historical surveillance and weather conditions data. The algorithm computes the maximal probability of the optimal state sequence, which is best aligned with the observation sequence of the aircraft trajectory.

2.3   Aviation Datasets

As may have been apparent from state-of-the-art methods, the trajectory prediction process requires different datasets to compute the prediction that represents the aircraft motion. Those datasets are basically grouped in the following categories [19]:

Initial Conditions Data, representing the initial aircraft state for which the trajectory will be predicted; mainly including location, altitude, speed, and time, and if possible, aircraft mass.

Surveillance Information, which might not be required in all prediction use cases, but it is necessary in any data-driven trajectory prediction system, being an essential component of Aviation datasets. It is highly dependent on local implementations, but in general a radar track file consists on tabular data rows with a timestamp key and several rows of geospatial information for each one of these timestamps. The usual update interval is 5 s (radar rotation time).

Alternatively, ADS-B surveillance data is generic and so independent from local systems. This data source refers to the ADS-B messages broadcasted by many airplanes (practically all airliners) using their transponders. These messages are received by ground-based receivers and can be used to reconstruct the trajectory of the flight. There are several types of messages that can be found but the relevant ones are these about aircraft identification and position.

Flight Plan (FP) declaring the intended route, cruise altitude and speed, as well as estimated times at different waypoints. FPs also contain additional information, not directly used for predicting a trajectory such as alternative airports or, potentially, aircraft equipage.

Flight plans contain the information that triggers a lot of operational decision, both in planning and execution phase, and both on the Air Navigation Service Provision (ANSP) side, and in the Airline one. The flight plan is the specified information provided to air traffic services units, relative to an intended flight or portion of a flight of an aircraft.

Weather Information, describing the atmosphere temperature and pressure, and the wind field faced by the aircraft along the trajectory. Multiple sources provide weather data to air traffic systems like satellite, met radar, and the aircraft itself. Some examples are METAR, NOAA models, SIGMET, or TAF:

METAR (Meteorological Terminal Aviation Routine Weather Report) is a format for reporting weather information. METARs typically come from airports or permanent weather observation stations. Reports are generated once an hour or half hour, but if conditions change significantly, a report known as a special (SPECI) may be issued. Some METARs are encoded by automated airport weather stations located at airports, military bases, and other sites. Some locations still use augmented observations, which are recorded by digital sensors, encoded via software, and then reviewed by certified weather observers or forecasters prior to being transmitted. Observations may also be acquired and reported by trained observers or forecasters who manually observe and encode their observations prior to transmission. Raw METAR is the most common format in the world for the transmission of observational weather data. It is highly standardized through the ICAO, which allows it to be understood throughout most of the world. METAR information includes runway visual range, dew point, visibility, and surface winds.

NOAA models are used mainly to obtain the weather conditions at the position an aircraft is at any given time of the flight. Weather models use a Grid with a specific resolution. Forecast models can be run several times a day. Forecast models have a time resolution, or "forecast step", depending on the use case. Data for weather models is typically distributed in "GRIB" format files. GRIB (GRIdded Binary or General Regularly distributed Information in Binary form) format allows to compress a lot the weather data and includes metadata about the content of the file, so it is very convenient for transferring the data. The data can be extracted with many available tools.

SIGMET (Significant Meteorological Information) is a weather advisory that contains meteorological information concerning the safety of all aircraft. This information is usually broadcast on the Automatic Terminal Information Service at ATC facilities, as well as over VOLMET (French origin vol (flight) and météo (weather report)) stations. A new alphabetic designator is given each time a SIGMET is issued for a new weather phenomenon, from N through Y (excluding S and T). SIGMETs are issued as needed, and are valid up to 4 h. SIGMETs for hurricanes and volcanic ash outside the CONUS are valid up to 6 h.

Terminal aerodrome forecast (TAF) is a format for reporting weather forecast information. TAFs are issued every 6 h for major civil airfields: 0000, 0600, 1200, 1800 UTC, and generally apply to a 24- or 30-h period, and an area within approximately five statute miles (or 5NM in Canada) from the center of an airport runway complex. TAFS are issued every 3 h for military airfields and some civil airfields, and cover a period ranging from 3–24 h. TAFs complement and use similar encoding to METAR reports. They are produced by a human forecaster based on the ground. For this reason there are considerably fewer TAF locations than there are airports for which METARs are available. TAFs can be more accurate than Numerical Weather Forecasts, since they take into account local, small scale, geographic effects.

Airspace can be divided in a set of ways, with a different number of segregation/-compartments, called sectors. Each sector is controlled by a single controller, thus the open sectors' configuration depends on airspace demand. A sector configuration is a particular configuration of "open" sectors segregating an airspace. For example, the 9A sector configuration denotes that a particular airspace is divided into 9 sectors, in a particular way. 9B also mean 9 sectors, but divided in a different way. Typically, due to low traffic at nights, the configuration set at those times is a 1A, meaning that a single sector (thus, a single controller) is in place. This leads to the fact that configurations available are fixed, but configuration "in place" varies during day, adapting capacity resources (air traffic controllers, mainly, as more sectors open mean more capacity, but also more controllers) to the expected demand.

It must be noted that, in the case of data-driven trajectory predictions, different inputs need to be considered. For instance, information about aircraft performance is not necessary because the aircraft motion will be predicted by learning from historical recorded tracks, not by solving a mathematical formulation of the aircraft motion problem. In addition, data related to the day of operation, airline, airspace sector configuration, or average delay at departure airport could be of interest to obtain accurate data-driven predictions.

These datasets represent the usual information used to predict a trajectory driven by data as summarized in most of the trajectory prediction approaches described. However, there are gaps that reduce the capability of predicting completely the evolution of the aircraft state vector with the time. For example, there is no available information about the aircraft mass. This information is of high commercial sensitiveness and, therefore, airspace users (i.e., airlines) are often reluctant to share it to protect their business strategies.

Aeronautical data is heavily regulated, especially in Europe according to Eurocontrol Standards. For example, flight plan filing information follows ICAO FPL2012 format, radar information is provided following ASTERIX standard (Asterix Cat62 for fused data), datalink between airlines dispatcher and aircraft follows A702-A format, airspace information is mostly provided in AIXM format. Thus, research results can be applied nationwide in Europe, while the highest quality data is usually at the local side, with national service providers. Is to be mentioned that all datasets to be used as input on any investigation need to be linked amongst them to ensure coherent geographical and temporal alignment, which is not always due to complexity of different formats, volume, and (lack of) veracity of data.

Alignment of the different data sources ensures common geographical and temporal coverage, which is paramount for datasets usage and effective data-driven learning. The data sources need to be combined usually using an ad-hoc reference to ensure that they will refer to the same time and space, as well as to enable links (associations) between them when necessary (for instance, radar tracks with flight plan for a particular flight). The specific linkage criteria will depend on the data sources composing the dataset, as well as the datasets features, ensuring a temporal and spatial common reference. Typically UTC time is the main reference for temporal alignment, using or correcting the different data sources to fit it. Regarding spatial alignment, geographic coordinates are usually the best cross index. Combined indexes using flight callsign, date, time, and aircraft type are usually used. The particular combination method, however, will depend on the specific dataset (and the different data sources it originates from). A significant challenge is in terms of aligning subjective phenomena (such as those described in SIGMET, related to sectors), with quantitative measures of NOAA grid, for instance.

Two drawbacks can be found for these datasets:

- Data-driven algorithms typically work better with great number of data points, but surveillance data is not always available at high resolution. This is for instance the case for QAR data: The number of data points available per flight may be insufficient.
- Surveillance data only includes positions of the aircraft, however there are other variables in a trajectory that may be easier to predict than the coordinates (they may show more clear patterns) and which can be derived from the position with some extra information (e.g., heading, bearing, or ground speed).

To overcome these difficulties, an enhanced dataset generated from the original raw data can be obtained and, then, this can be exploited by the big data analytics and machine learning algorithms. A technique is proposed in which the raw surveillance data can be enhanced, adding much more data points and much more variables; all being compatible with the real flight.

The following paragraphs detail how we can produce enhanced datasets exploiting raw data, so as to include additional information not being originally available.

2.4 Reconstructed Trajectory

A main drawback of data-driven TP based on surveillance datasets is the low granularity and diversity of available data. Even considering ADS-B or QAR, which contain broader information than typical latitude-longitude-altitude-time included in radar tracks, the availability of accurate information about airspeeds, ground speed is almost ineffective, while there is no availability of the aircraft mass, which is the key state variable to compute other related kinetic state variables.

However, making use of the aircraft intent (AI) instance inferred from the raw data, as subsequent paragraphs explain, it is possible to launch an aircraft mass inference and a trajectory reconstruction process [20, 21] that populates the state vector with times (increased granularity) and state variables (state vector enrichment) not included in the original surveillance-based trajectory representation.

2.4.1 Aircraft Intent

The aircraft intent (AI) can be defined as a set of instructions to be executed by the aircraft in order to realize its intended trajectory. These instructions represent the basic commands issued by the pilot of the FMS to steer the operation of the aircraft. The pilot can issue instructions by, for example, directly controlling the stick and the throttle, commanding the autopilot and the auto-throttle or programming the FMS. Instructions can be instantaneous, if they are considered to be issued at a specific instant in time, or continuing, if they are issued throughout a finite time interval. For example, consider an instruction requiring the flaps to be deflected a certain angle. In this case, it can be assumed that the time taken by the pilot to move the flap deployment lever is very short, so that the instruction can be considered instantaneous. Consider now a pilot taking control of the stick and commanding it during a certain interval of time. In this case, the resulting instruction would be continuing.

The Aircraft Intent Description Language (AIDL) is a formal language designed to describe AI instances in a rigorous but flexible manner. The AIDL contains an alphabet and a grammar. The alphabet defines the set of instructions used to close each of the DoF of the mathematical problem of the aircraft motion. The grammar contains both lexical and syntactical rules. The former govern the combination of instructions into words of the language, which are called operations, and the latter govern the concatenation of words into valid sentences, i.e. sequences of operations [22].

The AIDL captures the mathematics underlying trajectory computation into a rigorous, flexible, and simple logical structure that allows both human and computers to correctly describe meaningful operating strategies without the need to understand the underlying mathematics. In addition, the flexibility of the language allows defining aircraft intent with different levels of detail (e.g., aircraft intent formats employed by different TPs) using a common framework [1, 23].

The relationship between AI instance and (predicted) trajectory is unique; thus, once an AI instance is well formulated, a unique trajectory can be computed once the aircraft performance model (APM) corresponding to the actual aircraft is available and (resp. forecasted) weather conditions are known. Based on this property, it is possible to derive the AI instance that represents an actual trajectory from the chronologically ordered sequence of surveillance reports that identifies it.

Figure 2.1 exemplifies a descent trajectory from cruise altitude (FL320) up to capturing a geodetic altitude of 4500 ft. During this flight segment, the speed is also reduced from Mach 0.88 to 180kn calibrated airspeed (CAS). The lateral profile is described by a fly-by procedure around a waypoint of coordinates N37o 9' 45.72" W3o 24' 38.01". The associated AI instance is determined once the 6 threads (3 motions + 3 configurations) are well defined:

- Configuration Profiles. The flight is executed at clean configuration, meaning that high lift devices (HL), landing gear (LG), and speed breaks (SB) are held retracted. This is specified by the instruction Hold HL (HHL), Hold LG (HLG), and Hold SB (HSB).
- Motion profiles (described for each one of the 3 degrees of freedom, which allow representation of the trajectory).

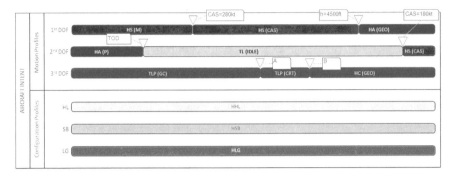

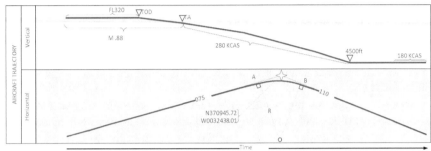

Fig. 2.1 Relationship between aircraft intent and trajectory

- 1st DoF. The cruise Mach is held up until the CAS reaches 280kn by applying a Hold Speed (HS) instruction, and then this CAS value is held up to 4500 ft altitude. From this instant, the altitude is maintained constant (Hold Altitude (HA) instruction).
- 2nd DoF. Cruise altitude is constant up to the Top of Descent (TOD) when the descent starts by setting the engine regime (Throttle Law (TL) instruction) to idle. This setting ends when CAS reaches 180kn. Then, the speed is maintained constant.
- 3rd DoF. The lateral path is described by the geodesic defined from the initial location and the established waypoint—indicated with an asterisk—(Track Lateral Path (TLP) instruction), a circular arc of radius R that determines the fly-by procedure up to capturing the exiting geodesic defined by a constant heading (Hold Course (HC) instruction).

Applying inference algorithms and techniques [18], and based on the assumption that the aircraft motion can be represented as a point-mass model of 3 DoF, it is possible to compute the AI instance that best describes an actual trajectory. Using therefore the raw surveillance data, and matching them with the weather forecasts that represent the atmosphere conditions of the day of operation and with the aircraft type that actually executed the planned trajectory, we can enhance the available surveillance dataset by adding this valuable information that cannot be immediately derived from the raw data. This additional set of information will enable additional hybrid data-driven capabilities, in which big data analytics and machine learning algorithms can be used to predict the most suitable AI instance, and then, compute it by using a model-based TP to obtain a 4D description of the trajectory. Figure 2.2 shows a schematic representation of the whole process.

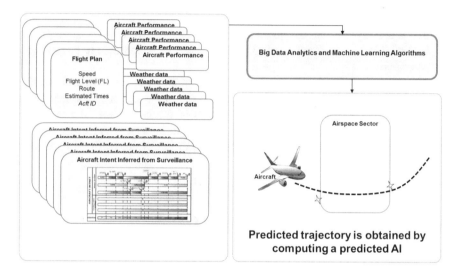

Fig. 2.2 Data-driven trajectory prediction based on aircraft intent (AI) instances

It is necessary to note that the AI representation of this kind of data is compliant to the well-established notion of semantically annotated or enriched trajectories, in the mobility data management and mining literature. Instead of a sequence of space-time information (as in a raw trajectory), in an enriched trajectory the motion is represented as a sequence of semantically meaningful episodes (typically in human mobility these are stops, e.g. "at home", "at office", "for shopping", and moves, e.g. "walking", "driving", etc., which results in detecting homogeneous fractions of movement. Extracting and managing semantics from (raw) trajectory data is a promising channel that leads to significant storage savings. Maintaining semantic information turns out to be quite useful in terms of context-aware movement analysis. In fact, semantic aware abstractions of motion enable applications to better understand and exploit mobility: for instance, concerning human mobility, analysis methods may identify those locations where some activity (work, leisure, relax, etc.) takes place, infer how long does it take to get from one place of interest to another using a specific transportation means, conclude about the frequency of an individual's outdoor activities, calculate indices related to environmentally friendly or sustainable mobility, and so on. Similarly, in our context, aircraft' routes may be transformed to sequences of critical points (see Chap. 4 for details) where certain events take place (e.g., "take-off", "climb out", "descent", "landing", or any of the AIDL instructions mentioned above).

The main advantages of the aircraft intent (AI)-based approach are:

- This formulation based on the notion of enriched (or semantic) trajectory is suitable to be used with highly sophisticated analytics AI/ML algorithms that can potentially capture in better ways hidden patterns;
- The complete description of the 4D trajectory is obtained from a mathematical model that provides the evolution of all possible states with time, contrary to the case of using only raw data in which every state variable needs to be predicted separately.
- The aircraft intent decouples the influence of the aircraft type and weather conditions, providing purely information about how the aircraft is operated along a time interval. This could help the process of finding command and control patterns that are common to all aircraft flying within the same airspace volume, although they fly dissimilar trajectories due to the effect of those decoupled factors.

2.4.2   The Trajectory Reconstruction Process

As already pointed out, making use of the aircraft intent instance inferred from the raw data, it is possible to launch a trajectory reconstruction process [20, 21] that populates the state vector with times (increased granularity) and state variables (state vector enrichment) not included in the original surveillance-based trajectory representation.

Figure 2.3 depicts the enriched list of aircraft state variables obtained from the trajectory reconstruction and enrichment process such as the Mach, CAS, TAS, VG (ground speed), FC (fuel consumption), wind components (Wx, Wy), or OAT (outside air temperature), not usually available in the input datasets used by the algorithms proposed in the literature.

The reconstruction process requires an aircraft performance model and also a model of the actual weather conditions faced by the aircraft along a real trajectory. Thanks to such a process, the heading (true with respect to the geographic North), speed (e.g., Mach number) and altitude (geopotential pressure altitude) profiles that univocally define each trajectory can be obtained for any of the recorded tracks. These heading, speed, and altitude profiles will be used as input to the big data analytics algorithms that will generate a prediction of the evolution of these three state variables with the same granularity as that selected for reconstructing the original training dataset. The remaining variables will be computed by building an AI instance upon those three predicted variables. According to the AIDL rules, it is possible to describe a trajectory by setting three non-dependent motion constraints. Thus, the evolution of those three state variables along the trajectory determines univocally the trajectory to be predicted, and, therefore, AIDL-based TP can be

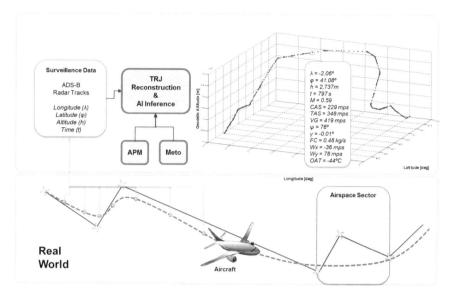

Fig. 2.3 Trajectory reconstruction and enrichment process

used to solve the aircraft motion problem and generate the related prediction. This approach can be seen as a hybrid solution that requires—given the AI instance built—the computation of the complete state vector that defines a 4D trajectory.

The main advantage of this method is twofold:

- The usage of extended and enriched datasets leads to better trained algorithms, and should turn into better trajectory predictions;
- The hybrid approach reduces significantly the training effort because only three independent state variables are to be predicted out of the complete aircraft state vector.

2.5 Aviation Operational Scenarios: Big Data Challenges and Requirements

The current air traffic management (ATM) is nowadays changing its point of view from a time-based operations concept to a trajectory-based operations (TBO) one, which means a better exchange, maintenance, and use of the aircraft trajectories for a collaborative decision-making environment, involving all the stakeholders in the process. In addition to that, real-time tracking and forecasting of trajectories, and early recognition of events related to aircraft are essential for operations. Essentially, the trajectory becomes the cornerstone upon which all the ATM capabilities will rely on. The trajectory life cycle describes the different stages from the trajectory planning, negotiation, and agreement, to the trajectory execution, amendment, and modification. This life cycle requires collaborative planning processes, before operations. The envisioned advanced decision support tools required for enabling future ATM capabilities will exploit trajectory information to provide optimized services to all ATM stakeholders.

To address these challenges the knowledge of more accurate and more predictable trajectories is needed. Thus, the more accurate and rich information on trajectories and related events we have, and as we increase our abilities to predict trajectories and forecast events regarding moving entities' behaviour, the more we will advance situational awareness, and consequently the decision-making processes.

Once the decision-making process has been improved, there are direct consequences in safety, efficiency, and economy in the ATM domain. For instance, by having a better understanding of the air navigation data (historical data of flight plans, sector configurations, and weather), the number of published regulations could be more accurately forecasted to improve the adherence to scheduled trajectories, with less delays and operational costs.

Due to the complexity of the ATM system, the current techniques for predicting trajectories are limited to a short-term horizon, while the event detection and forecasting abilities are limited. This is also due to the lack of methodologies to exploit the big amount of data from heterogeneous data sources with lack of veracity for (actual, historical, and planned) trajectories and other contextual aspects (e.g., airspace sector configurations, regulations and policies, weather patterns, for instance).

Efficiency in the air traffic management system requires minimizing costs for both the airspace users (mainly airlines) and the operators (ANSPs). In general, one key enabler for reducing costs is the predictability of the system. In particular, from the point of view of the ANSP, maintaining the balance between the demand (number of users trying to use limited resources like airports, airspace sectors...) and the capacity (number of users which can safely use the mentioned resources) is one of the main challenges. For the airline, flying according to the plan, avoiding delays or extra fuel consumption represents the ideal way to achieve daily operations, which however cannot be met.

The role of the trajectory in this efficiency enhancement endeavour is obvious: it defines which resources of the air traffic management system will be used by each flight (airports, airways, sectors...), and it defines the achievable schedules, as well as the implied costs.

Big data technology presents opportunities to increase predictability capacities which are based mainly on complex theoretical models of the different components of the air traffic management system. Exploitation of very large historical and streaming data sources for positioning, contextual aspects, and weather is now possible, thanks to state of the art in data management.

Surveillance is an ever-increasing data source since new technologies are deployed (like ADS-B) which allow to collect data more widely (space-based ADSB-B promises global coverage) and more frequently. Weather data, identically, each time is offered with more resolution both geographical and temporal. Contextual data like flight plans, waypoints, or airways is increasing, linked to the traffic growth, year after year. While each dataset is big, correlating and jointly exploiting all of them together is what makes big data technology necessary.

The aircraft trajectory must be understood not only as the 4D collection of points but also, including events relevant for the traffic management and the airline operations. So, predicting the aircraft trajectory implies predicting these events too, and vice versa. The amount of information involved in this trajectory prediction process requires advanced visual analytics aids in order to understand the patterns of the predicted trajectories and events, inspect the exact reasons for deviating from plans towards either making adjustments to the actual system, or tune trajectory and event detection and prediction methods for more accurate results.

Accurate predictions of trajectories will further advance adherence to flight plans (intended trajectories) reducing many factors of uncertainty, allowing stakeholders to do better planning of the operations, reducing risk of disruptions.

In this context, the Demand and Capacity Balancing (DCB) operational problem has been addressed, as it is a cornerstone of ATM operations: how to be able to accommodate the existing traffic demand with the available airspace capacity. The DCB problem considers two important types of objects in the ATM system: aircraft trajectories and airspace sectors. Sectors, as already explained, are air volumes segregating the airspace, each defined as a group of airblocks. These are specified by a geometry (the perimeter of their projection on earth) and their lowest and highest altitudes. Airspace sectors configuration (one is active at any time) changes frequently during the day, given different operational conditions and needs. This happens transparently for flights.

The capacity of sectors is of utmost importance: this quantity determines the maximum number of flights flying within a sector during any time period of specific duration (e.g., in any 20 min period). The demand for each sector is the quantity that specifies the number of flights that co-occur during any time period within a sector. The duration of these periods is equal to the duration of periods used for defining capacity. Demand must not exceed sector capacity for any time interval.

There are different types of measures to monitor the demand evolution, with the most common ones being Entry Count and Occupancy Count. In this work Entry Count it is considered, as this is the one normally used by network managers at real world operations.

The Entry Count (EC) for a given sector is defined as the number of flights entering in the sector during a time period, referred to as an Entry Counting Period. This Entry Counting Period is defined given a "picture" of the entry traffic, taken at every time "step" value along a period of fixed duration: The Step value defines the time difference between two consecutive Entry Counting Periods. The Duration value defines the time difference between the start and end times of an Entry Counting Period. For example, for a 20-min step value and a 60-min duration value, entry counts correspond to pictures taken every 20 min, over a total duration of 60 min.

2.5.1   Regulations Detection and Prediction

The objective of this operational case is to demonstrate how regulations detection and prediction capability is useful for reproducing Flow Management behaviour. This behaviour is mainly represented by the applied regulations that the system must learn to reproduce and to anticipate in specific problematic situations, as it would happen in a realistic scenario.

Regulation is a measure that a flow manager takes to solve a specific situation, in a punctual moment in a certain sector and it is applied over those flights that have not yet took off. Thus, regulations are consequence of specific situations as those in which there is an excess of demand vs capacity in sectors, or those caused by different weather conditions, among others. In this case we are interested on regulations that impose effective delays to flights still on ground, given that these flights are planned to cross a volume of airspace where demand will exceed capacity.

The main consequence of a regulation is the reschedule of an ETOT (Estimated take off time) by a CTOT (Calculated take off time), that is a new time to take off after the scheduled one, causing a delay. ETOTs that are replaced by the CTOTs concern only those flights that were going to fly in the affected sector (i.e., a sector with an imbalance between demand and capacity), during a punctual moment.

Hence, there are three objectives:

1. Investigate the available historical data in order to identify patterns in the emergence of the regulations. The patterns thus identified should suggest possible approaches to regulation prediction.
2. Develop a method or methods for regulation prediction based on the patterns identified.
3. Verify the method(s) by comparing predictions based on available historical data (without regulation data) with the real regulations.

It is not known at the beginning what kinds of patterns can exist. It is therefore necessary to analyse data from various perspectives using interactive visual displays as well as various filters and data transformations. The possible types of patterns are:

Temporal Patterns, such as regularities with respect to the daily and weekly time cycles.

Spatial Patterns, determining how regulations emerging in a certain area affect flights associated with certain origin and/or destination.

Spatiotemporal Patterns, identifying different temporal patterns of regulations in different areas.

Dependencies Among Regulations, identifying kinds of regulations with certain properties that lead (after some time) to other regulations.

Once regulations and their cause (e.g.: weather, ATC capacity, accident/incident, etc.) are known, flight plans have to be checked on how these regulations affect them.

2.5.2 Demand and Capacity Imbalance Detection and Prediction

The objective of this operational case is to demonstrate the detection/prediction of demand and capacity imbalances by means of indicators monitoring.

Those indicators are based on real demand (Hourly Entry Count) and declared capacity (maximum number of flights allowed to enter in a sector during 1 h) of the current configuration of airspace. These indicators are calculated using the initial flight plans (deregulated traffic) instead of the real (finally flown) flight plans. The main reason for using the initial flight plans relays on the fact that if a flight has been regulated with delay, then the detected excess of demand may have been resolved.

Although in theory an imbalance could be produced by an excess of capacity compared with the demand, it should be an unusual situation that is out of scope.

The final objective is to reconstruct the system's behaviour in handling and resolving demand capacity imbalances. This will allow us, in particular, to investigate propagation of the consequences of the regulations, as delaying some flights in a given entry time period may lead to increasing the demand in a next entry time period, in the same or another sector. It may also be useful to investigate the consequences of regulations on various entry time period lengths: E.g., what would happen if the currently adopted time period length of 1 h is replaced by a 30-min period. Furthermore, it may be also reasonable to compare the use of fixed time periods with the use of a sliding time period. In the latter approach, the demand is calculated not from the beginning of an hour but from the time when each flight enters a sector.

2.5.3 Trajectory Prediction: Preflight

This operational case of study objective is to demonstrate how predictive analytics capability can help in trajectory forecasting. For a given flight plan, a forecasted trajectory will be obtained and compared with the real one finally flown (as recorded in the historical dataset).

The prototype will be used to select the flight plans desired for the evaluation. These need to be "searchable" by callsign, aircraft model, airline, origin and destination airports, estimated time of departure (ETD), estimated time of arrival (ETA), equipage, cruise level, cruise speed. Thus, one may select a number of flight plans (typically all) and request a predicted trajectory for each of them. As we need to cover large fleets for large geographical regions, scalability issues emerge. Therefore, the trajectory prediction abilities should be able to scale effectively.

2.5.4   Trajectory Prediction: Real Time

The objective of this operational case is to analyse how predictive analytics capability can help in trajectory forecasting in real time. For a given flight plan and the current surveillance data arriving to the platform, a forecasted trajectory will be obtained and updated continuously. This real-time need will be paramount in the new TBO setting, in particular in a highly automated scenario where decisions will be taken with the support of machine learning systems which will need to rely on accurate, and very updated, trajectory forecasts.

2.6   Conclusions

The vision of the future ATM system evolving towards higher levels of automation, as a key driver to enhanced ATM performance, is expressed in successive releases of the European ATM Master Plan. This emerges both, as a mid-term need (with EUROCONTROL as Network Manager forecasting increases in traffic of +50% in 2035 compared to 2017, meaning 16 million flights across Europe) and as a long-term need (2035+).

The effects of collapsed sectors can be observed, for instance, in the yearly Performance Review Report (PRR), addressed by EUROCONTROL Performance Review Commission, which allocates a high share of the overall Air Traffic Flow Management (ATFM) delays to this reason (over 90% in some airspaces). It was significantly bad in 2018 when AFTM delays across Europe more than doubled, due to the increase in traffic among other factors, a trend expected to keep. In general, all performance analysis and studies lead to the idea that the ATM system is very close to, or already at, a saturation level.

Effective automation that will enable an increase in capacity is considered one of the pillars of future ATM, but this means facing some difficulties and challenges. This has been evident in recent times with some potentially optimistic implementation of automation features, which allegedly may have impacted the situational awareness and reaction capabilities of the operators.

Complementarily, new opportunities have arisen for the enhancement of the ATM approach to automation, in particular with the widespread introduction of artificial intelligence/machine learning (AI/ML) techniques in society in general. These techniques bring to the ATM research domain new opportunities, in particular as key enabler to reach the necessary higher levels of automation.

On the other hand, predictability is considered as the main driver to enhance operational performance key performance areas (KPAs), such as capacity, efficiency, and even safety. Trajectory prediction, in particular within the TBO concept of operations, is the paramount enabler for this new stage of ATM operations. This chapter addresses the state of the art, as well as the main operational scenarios where these capabilities bring significant benefits.

References

1. Lopez Leones, J., Vilaplana, M., Gallo, E., Navarro, F., Querejeta, C.: The Aircraft Intent Description Language: a key enabler for air-ground synchronization in Trajectory-Based Operations. In: IEEE/AIAA 26th Digital Avionics Systems Conference (2007)
2. BADA, Base of Aircraft Data. https://simulations.eurocontrol.int/solutions/bada-aircraft-performance-model/
3. Hamed, M.G., et al.: Statistical prediction of aircraft trajectory: regression methods vs point-mass model. In: 10th USA/Europe Air Traffic Management Research and Development Seminar (ATM 2013), 10–13 June 2013
4. Kun, W., Wei, P.: A 4-D trajectory prediction model based on radar data. In: 27th Chinese Control Conference, 16 July 2008
5. Le Fablec, Y., Alliot, J.M.: Using neural networks to predict aircraft trajectories. In: IC-AI (1999)
6. Cheng, T., Cui, D., Cheng, P.: Data mining for air traffic flow forecasting: a hybrid model of neural network and statistical analysis. In: Proceedings of the 2003 IEEE International Conference on Intelligent Transportation Systems, vol. 1, pp. 211–215 (2003)
7. de Leege, A.M.P., Van Paassen, M.M., Mulder, M.: A machine learning approach to trajectory prediction. In: AIAA Guidance, Navigation, and Control (GNC) Conference 19–22 August, Boston, MA (2013)
8. Tastambekov, K., et al.: Aircraft trajectory forecasting using local functional regression in Sobolev space. Transp. Res. C: Emerg. Technol. 39, 1–22 (2014)
9. Hong, S., Lee, K.: Trajectory prediction for vectored area navigation arrivals. J. Aerosp. Inf. Syst. 12, 490–502 (2015)
10. Yue, S., Cheng, P., Mu, C.: An improved trajectory prediction algorithm based on trajectory data mining for air traffic management. In: International Conference of Information and Automation (ICIA), 6 June 2012
11. Hamed, M.G., et al.: Statistical prediction of aircraft trajectory: regression methods vs point-mass model. In: 10th USA/Europe Air Traffic Management Research and Development Seminar (ATM 2013) (2013)
12. Gong, C., McNally, D.: A methodology for automated trajectory prediction analysis. In: AIAA Guidance, Navigation, and Control Conference and Exhibit (2004)
13. Yang, Y., Zhang, J., Cai, K.: Terminal area aircraft intent inference approach based on online trajectory clustering. Sci. World J. 2015, 671360 (2015)
14. Yepes, J.L., Hwang, I., Rotea, M.: New algorithms for aircraft intent inference and trajectory prediction. J. Guid. Control Dynam. 30(2), 370–382 (2007)
15. Zorbas, N., Zissis, D., Tserpes, K., Anagnostopoulos, D.: Predicting object trajectories from high-speed streaming data. In: Proceedings of IEEE Trust-com/BigDataSE/ISPA, pp. 229–234 (2015)
16. Bifet, A., Holmes, G., Kirkby, R., Pfahringer, B.: MOA: massive online analysis. J. Mach. Learn. Res. 11, 1601–1604 (2010)
17. Ayhan, S., Samet, H.: Aircraft trajectory prediction made easy with predictive analytics. In: Proceedings of ACM SIGKDD, pp. 21–30 (2016)
18. La Civita, M.: Using aircraft trajectory data to infer aircraft intent. U.S. Patent No. 8,977,484, 10 Mar 2015
19. Mondoloni, S., Swierstra, S.: Commonality in disparate trajectory predictors for air traffic management applications. In: IEEE/AIAA 24th Digital Avionics Systems Conference (2005)
20. Luis, P.D., La Civita, M.: Method and system for estimating aircraft course. U.S. Patent Application No. 14/331,088, 2015
21. D'Alto, L., Vilaplana, M.A., Lopez, L.J., La Civita, M.: A computer based method and system for estimating impact of new operational conditions in a baseline air traffic scenario. European Patent No. EP15173095.9, 22 June 2015

22. Lopez Leones, L.J.: Definition of an aircraft intent description language for air traffic management applications. PhD thesis, University of Glasgow (2008)
23. Vilaplana, M.A., et al.: Towards a formal language for the common description of aircraft intent. In: IEEE/AIAA 24th Digital Avionics Systems Conference (2005)

Part II
Visual Analytics and Trajectory Detection and Summarization: Exploring Data and Constructing Trajectories

The second part of this book focuses on big data quality assessment, processing and exploration, focusing on the detection and compression of trajectories according to the requirements and objectives presented in the first part of the book. This, second part of the book, presents novel technologies, appropriate to serve with compressed and high quality data the mobility analytics components that are presented in subsequent chapters. In doing so, preparatory workflows regarding data sources' quality assessment via visual analytics methods are presented in the first chapter. These are considered to be essential to understand inherent features of the data, affecting the ways data should be processed and managed. In addition to this, the second chapter presents methods for online construction of data synopses, towards addressing big data challenges presented by surveillance, mostly, data sources in the Maritime and ATM domains.

Chapter 3
Visual Analytics in the Aviation and Maritime Domains

Gennady Andrienko, Natalia Andrienko, Georg Fuchs, Stefan Rüping, Jose Manuel Cordero, David Scarlatti, George A. Vouros, Ricardo Herranz, and Rodrigo Marcos

Abstract Visual analytics is a research discipline that is based on acknowledging the power and the necessity of the human vision, understanding, and reasoning in data analysis and problem solving. It develops a methodology of analysis that facilitates human activities by means of interactive visual representations of information. By examples from the domains of aviation and maritime transportation, we demonstrate the essence of the visual analytics methods and their utility for investigating properties of available data and analysing data for understanding real-world phenomena and deriving valuable knowledge. We describe four case studies in which distinct kinds of knowledge have been derived from trajectories of vessels and airplanes and related spatial and temporal data by human analytical reasoning empowered by interactive visual interfaces combined with computational operations.

G. Andrienko (✉) · N. Andrienko
Fraunhofer Institute IAIS, Sankt Augustin, Germany

City University of London, London, UK
e-mail: gennady.andrienko@iais.fraunhofer.de; natalia.andrienko@iais.fraunhofer.de

G. Fuchs · S. Rüping
Fraunhofer Institute IAIS, Sankt Augustin, Germany
e-mail: georg.fuchs@iais.fraunhofer.de; stefan.rueping@iais.fraunhofer.de

J. M. Cordero
CRIDA (Reference Center for Research, Development and Innovation in ATM), Madrid, Spain
e-mail: jmcordero@e-crida.enaire.es

D. Scarlatti
Boeing Research & Development Europe, Madrid, Spain
e-mail: david.scarlatti@boeing.com

G. A. Vouros
University of Piraeus, Piraeus, Greece
e-mail: georgev@unipi.gr

R. Herranz · R. Marcos
Nommon Solutions and Technologies, Madrid, Spain
e-mail: ricardo.herranz@nommon.es; rodrigo.marcos@nommon.es

© Springer Nature Switzerland AG 2020
G. A. Vouros et al. (eds.), *Big Data Analytics for Time-Critical Mobility Forecasting*, https://doi.org/10.1007/978-3-030-45164-6_3

3.1 Introduction

Visual Analytics (VA) has been defined as 'the science of analytical reasoning facilitated by interactive visual interfaces' [25, p. 4]. Visual analytics is a research discipline that is based on acknowledging the power and the necessity of the human vision, understanding, and reasoning in data analysis and problem solving. An essential idea of visual analytics is to combine the power of human reasoning with the power of computational processing. It thus aims at developing methods, analytical workflows, and software systems that can support the unique capabilities of humans by providing appropriate visual displays of data and involving as much as possible the capabilities of computers to store, process, analyse, and visualise data.

As facilitators of human understanding and reasoning, VA techniques and tools can greatly support analysts in all stages of a typical analytical process. They can be used for the following purposes:

- gain awareness of properties and problems of available data and understand how the data need to be corrected, transformed, enriched, and/or complemented to become suitable for the intended analysis;
- comprehend the phenomena reflected in the data, grasp essential features, relationships, patterns, trends, and understand how to represent these in models;
- create valid and useful models of the phenomena by involving human critical thinking in model design, preparation, configuration, evaluation, comparison, and iterative improvement.

A substantial body of research in VA has been focusing on data and problems related to mobility and transportation [6]. This chapter includes several examples of applying visual analytics approaches to data and tasks in the domains of air and maritime transportation. The aim is to demonstrate how interactive visual displays in combination with relatively simple computational techniques support the involvement of human understanding and reasoning in the analytical process.

3.2 Related Work

A particularly active sub-field of research in visual analytics deals with the analysis of movement data [3, 6], with approaches including trajectory-centred techniques [4, 5, 8], representation and analysis of overall mobility patterns [13, 27], discovery of interactions between moving objects [15], and support of domain-specific decision-making processes [8, 9, 18] in complex transportation systems.

A comprehensive survey of the visual analytics research dedicated to mobility and transportation has been published recently [6]. Particularly, there have been research works focusing on the aviation or maritime transportation domains.

VA approaches have been proposed for various specific problems in air traffic analysis. Methods for detection of holding loops, missed approaches, and other aviation-specific events and patterns were implemented in a system integrating a moving object database with a visual analytics environment [22]. Albrecht et al. [1] calculate air traffic density and, considering aircraft separation constraints, assess the conflict probability and potentially underutilized air space. The traffic density and conflict probability are aggregated over different time scales to extract fluctuations and periodic air traffic patterns. Hurter et al. [14] propose a procedure for wind parameter extraction from the statistics of the speeds of planes that pass the same area at similar flight levels in different directions. Buchmüller et al. [10] describe techniques for studying the dynamics of landings at Zurich airport with the goal to detect cases of violating the rules imposed for decreasing the noise in populated areas. The detected violations can be examined in relation to weather conditions and air traffic intensity. Sophisticated domain-specific analyses can be done by applying clustering to interactively selected relevant parts of aircraft trajectories [8]. Andrienko et al. presented an approach to detection of deviations of the routes of actual flights from the planned routes and exploration of the distributions of the deviations over space, time, set of flights, trajectory structures, and spatiotemporal contexts [9].

Related to the maritime domain, a state-of-the-art survey [2] uses a set of vessel trajectories as a running example to show how different visual analytics techniques can support understanding of various aspects of movement. Andrienko et al. [7] use vessel movement data to demonstrate the work of an interactive query tool called TimeMask that selects subsets of time intervals in which specified conditions are fulfilled. This technique is especially suited for analysing movements depending on temporally varying contexts. Scheepens et al. [24] have designed special glyphs for visualizing maritime data. Tominski et al. [26] apply a 3D view to show similar trajectories as bands stacked on top of a map background. The bands consist of coloured segments representing variation of dynamic attributes along vessel routes. Lundblad et al. [19] employ visual and interactive techniques for analysing vessel trajectories together with weather data. Variants of dynamic density maps combined with specialized computations and techniques for interaction [17, 23, 28] were proposed to support exploration of the density and other characteristics of maritime traffic. Kernel density estimation can be used to compute a volume of the traffic density in space and time [12], which can be represented visually in a space-time cube [16] with two dimensions representing the geographical space and one dimension the time.

Apart from these researches aiming at understanding of the phenomena, events, and processes pertinent to the domains of aviation and maritime traffic, there has been research focusing on exploration of properties of movement data and the use of VA techniques for detection of various quality problems that may occur in such data [5]. In the next section, we shall demonstrate several examples of visual detection of some quality problems in aviation and maritime data.

3.3   Visual Exploration of Data Quality

Possible quality problems in movement data include errors in spatial positions of objects, gaps in spatiotemporal coverage, low temporal and/or spatial resolution, use of the same identifiers for multiple objects, and others [5]. It is essential to reveal such problems before starting to use the data in the planned analysis. The use of inappropriate data or reliance on unchecked assumptions concerning data properties can lead to invalid analysis results and wrong conclusions.

Let us present several examples of quality problems that may occur in data concerning vessel or aircraft movement. Figure 3.1 demonstrates obvious errors in recorded spatial positions: here, many points from trajectories of vessels are located on land. Such wrong records need to be removed from the data, e.g., by filtering.

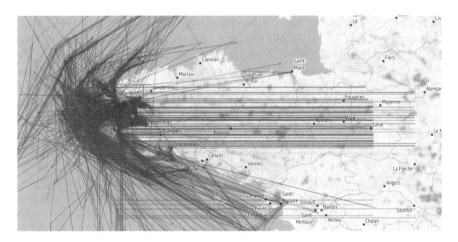

Fig. 3.1 Some trajectories of vessels include positions located on land far from the sea

Positioning errors in trajectories are not always so obvious. A good indication of a recorded position being out of the actual path of a vessel is an unrealistically high value of the computed speed in the previous position. The computed speed is the ratio between the distance to the next position and the length of the time interval between the positions. If wrong position records detected in this way occur occasionally in the data, they are not difficult to filter out. However, trajectories containing many positions supposedly reached at unrealistic speeds require special investigation. Thus, it may happen that the same identifier is assigned to two or more simultaneously moving objects. Connecting consecutive points of such trajectories results in zigzagged or more complex shapes, as demonstrated in Fig. 3.2. Such problems may occur due to errors in manually entered data fields, such as flight call signs.

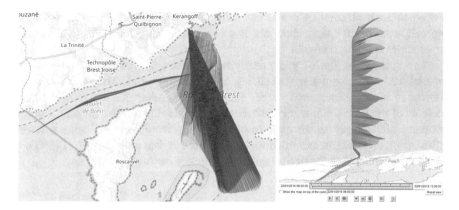

Fig. 3.2 Visual investigation of a trajectory with a high number of erroneous positions. On the left, the trajectory is shown on a map, and on the right in a space-time cube, where the base represents the geographic space and the vertical dimension represents the time

Zigzagged shapes may also result from incorrect integration of data from multiple sources, such as different radars. Thus, the shape in Fig. 3.3 resembles a mixture of two flight trajectories, but it is unlikely that there were two simultaneous flights following parallel routes and keeping a constant distance between them, as could be deduced from the shape. It is more likely that the same flight is represented twice in the dataset, and at least one set of records contains systematically shifted positions with respect to the real flight trajectory.

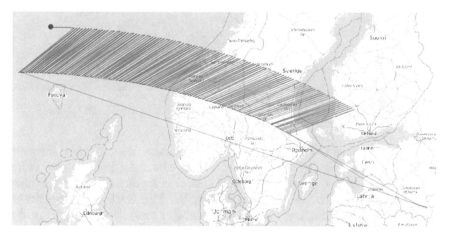

Fig. 3.3 An unrealistic shape of a trajectory indicates either systematically occurring positioning errors (displaced positions) or a mixture of movements of two airplanes

When such positioning errors are identified, it is necessary either to devise special, case-specific algorithms for data correction or to discard the problematic trajectories from the analysis.

The example in Fig. 3.4 prominently demonstrates the problem of gaps in the spatial coverage of a dataset consisting of flight trajectories. The trajectories are drawn in a semi-transparent mode. Respectively, darker colours reflect higher density of the flights. Some regions where we expect flights to be frequent, appear as completely empty on the map. In other regions, the density of flights is lower than in neighbouring areas. Obviously, large pieces of data describing the flights are missing, which makes the dataset unsuitable for any meaningful analysis.

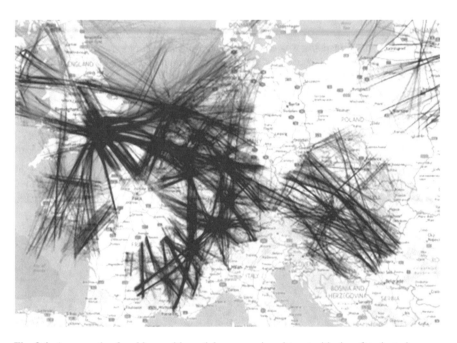

Fig. 3.4 An example of problems with spatial coverage in a dataset with aircraft trajectories

Another example in Fig. 3.5 comes from the maritime domain. Here, there are spatiotemporal gaps in some trajectories, that is, absence of position records for long time intervals of vessel movement. These gaps appear as long straight lines when the trajectories are drawn on a map (Fig. 3.5, top). Such segments must be excluded from the trajectories when it is necessary to analyse the paths of the vessels or to aggregate the trajectories into overall traffic flows; otherwise, the results will be wrong and misleading. A suitable way to exclude spatiotemporal gaps is to divide the trajectories with the gaps into several smaller trajectories, so that the point preceding a gap is treated as the end of the previous trajectory and the following point is treated as the beginning of the next trajectory. A gap is defined by choosing appropriate thresholds for the spatial and temporal distances between consecutive trajectory points. Suitable thresholds can be chosen based on domain-specific knowledge, such as usual speeds of vessel movement and normally expected frequencies of position reporting, and taking into account the statistics of the distances in the data.

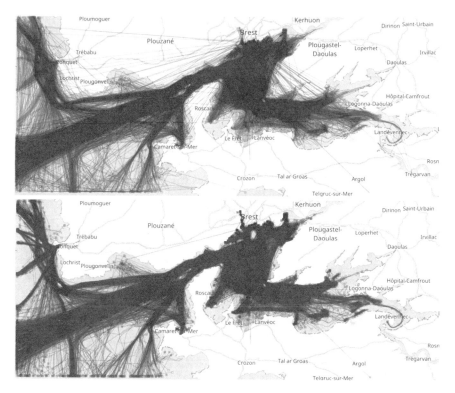

Fig. 3.5 Top: Long straight line segments in vessel trajectories correspond to *spatiotemporal gaps*, i.e., long time intervals in which position records for the vessels are missing. Bottom: The result of dividing the trajectories by the spatiotemporal gaps in which the spatial distance exceeded 2 km and the time interval length exceeded 30 min

Many trajectory analysis methods assume that temporal resolution of position records is constant. Very often datasets do not comply with this requirement. Figure 3.6 presents an example of a dataset where the sampling rates of 15, 30, and 60 s occur the most frequently, but other values occur as well. Hence, before applying an analysis method that assumes equal-length time steps between positions or attribute values, it is necessary to re-sample the data to make the time steps equal; otherwise, the method results may be invalid.

Figure 3.7 demonstrates that errors may occur not only in positions or identifiers but also in attribute values associated with the positions. In this example, the values of the attribute reporting the navigation status of vessels are unreliable. Hence, if the analysis requires the navigation status to be taken into account, it is

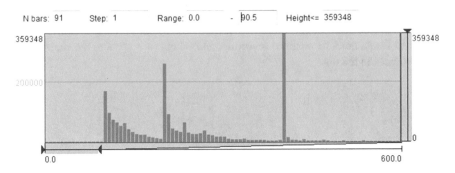

Fig. 3.6 Variability of sampling rates in trajectories

necessary to determine the actual status based on movement characteristics rather than attribute values. For example, when analysis requires extraction of stops, they can be identified by finding parts of trajectories where the positions are nearly the same during a chosen minimal stop time.

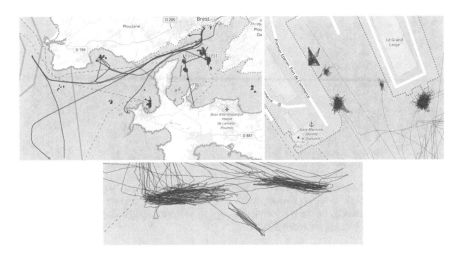

Fig. 3.7 Fragments of vessel trajectories with wrongly reported navigation status. The reported status in the upper left and bottom images is 'at anchor', whereas the vessels were actually moving. On the upper right, the reported status is 'under way using engine', while the vessels remained at the same places and should have reported 'at anchor'

The examples shown in this section do not cover all possible kinds of errors and problems that may occur in movement data, in particular, in trajectories of aircraft and vessels. Our intention was to demonstrate the utility of visual displays for detecting existing problems, understanding their likely reasons and possible impacts on the analysis, and finding suitable remedies.

3.4 Examples of Visual Analytics Processes

This section briefly presents several case studies intended to demonstrate the use of visual analytics approaches for gaining understanding of different phenomena in the air and maritime traffic. As mentioned in the introduction, visual analytics combines interactive visual displays with computational techniques for data selection, transformation, and automated derivation of various analytical artefacts that can supply relevant information for human reasoning.

A case study performed in the maritime traffic domain mostly aims at demonstrating the use of data transformations for supporting visual analysis and reasoning. The transformations include extraction of relevant parts of trajectories, detection and extraction of particular events, and spatiotemporal aggregation of events and movements.

As an example of using a computational analysis method, two case studies include clustering of flight trajectories based on geometric similarity of the routes. The general approach is to use a density-based clustering algorithm with a special distance function that matches corresponding points and segments of trajectories according to their spatial proximity. The specifics of the case studies we undertook was that not all parts of trajectories might be relevant to the analysis goals. Thus, in studying route choices, the initial and final parts of trajectories were irrelevant because these parts depend on the wind direction and not subject to choice by airlines. In studying the separation scheme of the approach routes to multiple airports of London, we needed to disregard the holding loops as inessential parts of the routes. To be able to apply clustering only to task-relevant parts of trajectories, we adapted the distance function so that it could account for results of interactive filtering of trajectory segments [8]. The main idea is that the distance function receives two trajectories to compare together with two binary masks specifying which points of the trajectories to take into account and which to ignore.

One case study in the aviation domain demonstrates exploratory analysis made with three different kinds of data, planned flight trajectories, geometries of airspace configurations, and temporal succession of the configurations. The aim of the analysis was to understand the relationships between characteristics of the air traffic and the choices of the airspace configurations for controlling the traffic. The analytical process involved seeking evidence to support or refute hypotheses generated by the analyst based on patterns observed.

All examples presented in this section do not merely demonstrate application of visual analytics techniques to data. They also highlight the importance of human perception, interpretation, understanding, and analytical reasoning and show how visual analytics techniques provide inputs to these cognitive processes.

3.4.1   Detection and Analysis of Anchoring Events in Maritime Traffic

In this example, visual analytics approaches are used for exploration and analysis of trajectories of vessels that moved between the bay of Brest and the outer sea [21]. The specific analysis task is to study when, where, and for how long the cargo vessels were anchoring and understand whether the events of anchoring may indicate waiting for an opportunity to enter or exit the bay (through a narrow strait) or the port of Brest. The dataset consists of about 18M positions of 5,055 vessels during 6 months from the 1st of October, 2015 till March 31, 2016. The exploration of the data properties revealed many problems, some of which have been demonstrated in the previous section.

After cleaning the data, we selected the task-relevant data subset consisting of trajectories of 346 cargo vessels that passed the strait connecting the bay of Brest to the outer sea at least once. From these trajectories, we took only the points located inside the bay of Brest, in the strait, and in the area extending to about 20 km west of the strait. To exclude the long straight line segments corresponding to periods of position absence (Fig. 3.5, top), we divided the trajectories into sub-trajectories by the spatiotemporal gaps with distance thresholds 2 km in space and 30 min in time (Fig. 3.5, bottom). Next, we further divided the trajectories by stops (segments with low speed) within the Brest port area and then selected from the resulting trajectories only those that passed through the strait and had duration of at least 15 min. An outcome of these selections and transformations is a set 1718 trajectories suitable for the further analysis. Of these trajectories, 945 came into the bay from the outer area, 914 moved from the bay out, and 141 trajectories include incoming and outgoing parts.

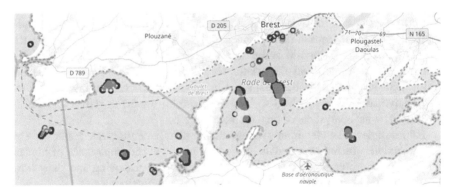

Fig. 3.8 Delineation of anchoring zones: violet points show positions from all trajectories marked as anchoring in the data, tan polygons outline anchoring zones containing dense enough concentrations of the anchoring points outside the port and major traffic lanes

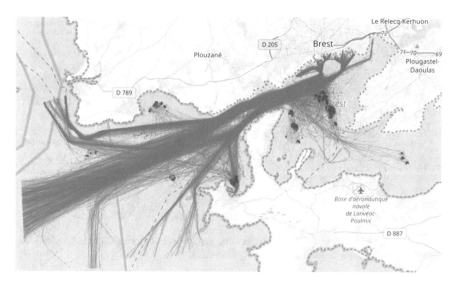

Fig. 3.9 The trajectories selected for analysis with the anchoring events (stops) marked in red

The analysis goal requires us to identify the *anchoring events*. As we cannot rely completely on the navigation status in position records (see Fig. 3.7), we apply the following heuristics. First, we identify areas where many trajectories had records marked as anchoring. We consider these areas as only anchoring zones (Fig. 3.8), thus ignoring occasional records marked as anchoring but located in unusual places. Next, we assume that any sufficiently (at least 5 min) long stop in an anchoring zone corresponds to anchoring. So, we get a set of 212 anchoring events (we shall further call them shortly 'stops') that happened in 126 trajectories. Fig. 3.9 shows these trajectories in light blue and the positions of the stops in red.

Since we want to understand how the stops are related to passing the strait between the bay and the outer sea, we find the part corresponding to strait passing in each trajectory. For this purpose, we interactively outline the area of the strait as a whole and, separately, two areas stretching across the strait at the inner and outer ends of it. The segments of the trajectories located inside the whole strait area are treated as *strait passing events*. For these events, we determine the times of vessel appearances in the areas at the inner and outer ends of the strait. Based on the chronological order of the appearances, we determine the direction of the strait passing events: inward or outward with respect to the bay. Then we categorise the stops based on the directions of the preceding and following events of strait passing.

The pie charts on the map in Fig. 3.10 represent the counts of the different categories of the stops that occurred in the anchoring areas. The most numerous category 'inward;none' (105 stops) includes the stops of the vessels that entered the bay, anchored inside the bay, and, afterwards, entered the port. The category 'outward;inward' (36 stops) contains the stops of the vessels that exited the bay,

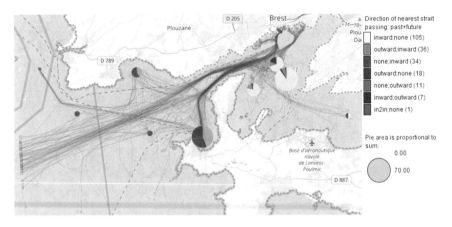

Fig. 3.10 The pie charts represent the counts of the stops in the anchoring zones categorized with regard to the directions of the strait passing by the vessels

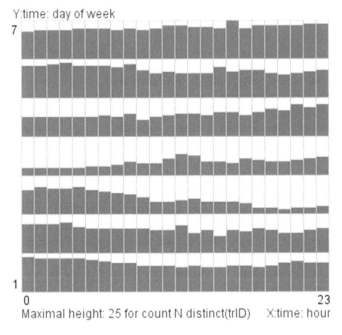

Fig. 3.11 A 2D time histogram represents the counts of the anchoring events by the hours of the day (horizontal axis) and days of the week (vertical axis) by the heights of the corresponding bars

anchored in the outer area, then returned to the bay and came in the port. 34 stops took place before entering the bay ('none;inward'), 18 happened after exiting the bay ('outward;none'), and 11 before exiting the bay ('none;outward'). In 7 cases, vessels entered the bay from the outside, anchored, and then returned back without visiting the port ('inward;outward'), and there was one stop that happened after entering the strait at the inner side and returning back ('in2in;none').

We see that the majority of the stop events (yellow pie segments) happened after entering the bay and, moreover, a large part of the stops that took place in the outer area happened after exiting the bay and before re-entering it (orange pie segments). It appears probable that the vessels stopped because they had to wait for being served in the port. Most of them were waiting inside the bay but some had or preferred to wait outside. Hence, the majority of the anchoring events can be related to waiting for port services rather than to a difficult traffic situation in the strait.

Additional evidence can be gained from the 2D time histogram in Fig. 3.11. It shows us that the number of anchoring vessels reaches the highest levels on the weekend (two top rows) and on Monday (the bottom row). It tends to decrease starting from the morning of Wednesday (the third row from the bottom of the histogram) till the morning of Thursday (the fourth row), and then it starts increasing again. The accumulation of the anchoring vessels by the weekend and gradual decrease of their number during the weekdays supports our hypothesis that the stops may be related to the port operation.

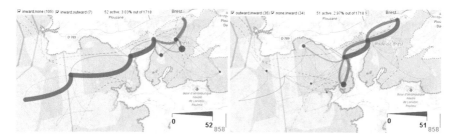

Fig. 3.12 Subsets of the trajectories under study are represented in an aggregated form on flow maps. Left: The trajectories having stops after entering the bay of Brest. Right: The trajectories having stops before entering the bay

To refine our conclusions, we also look at the routes of the vessels that made stops after entering the bay of Brest and before that. In Fig. 3.12, the trajectories of the vessels having stops after (left) and before (right) entering the bay have been aggregated into flows between interactively defined areas. The flows are represented by curved lines with the widths proportional to the move counts and the curvature increasing in the direction of the flow. The image on the left shows us that most of the vessels that stopped after entering the bay came from the outer sea. After stopping, they eventually moved into the port of Brest. Evidently, the reason for the stops was waiting for the port services. The image on the right shows that a large fraction of the vessels that anchored in the outer area before entering the bay previously came from the port. After stopping, they returned in the bay and moved back into the port. A likely explanation could be that these vessels were unloaded in the port and had to move to the outside area for waiting until the next cargo to be transported is ready for loading.

3.4.2 Exploring Separation of Airport Approach Routes

This case study was conducted using 5045 trajectories of actual flights that arrived at 5 different airports of London during 4 days from December 1 to December 4, 2016. The goals were, first, to reconstruct the major approach routes, second, to determine which of them may be used simultaneously and, third, to study how the routes that can be used simultaneously are separated in the three-dimensional airspace, i.e., horizontally and vertically.

A suitable approach to identifying the major approach routes is clustering of the trajectories by route similarity. A problem we had to deal with was the presence of holding loops in many trajectories (Fig. 3.13). It was necessary to identify the loops in the trajectories and filter them out so that they could not affect the clustering. We have found a combination of query conditions involving derived attributes of trajectory segments, such as sum of turns during 5 min, which allowed us to separate the loops from the main paths and filter them out [8]. The clustering was then applied to the remaining parts of the trajectories.

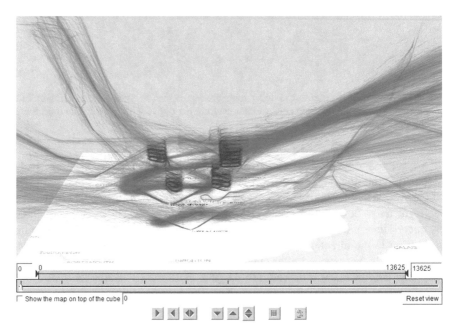

Fig. 3.13 Holding loops in the trajectories of the flights arriving to London are marked in red

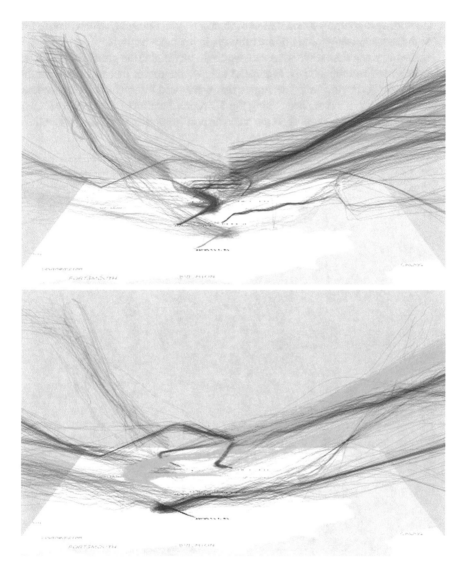

Fig. 3.14 The routes that were used on the first day till 18:25 (top) and on the following days after the wind change (bottom)

By means of clustering, we have identified 34 distinct routes, 16 of which were used only on the first day out of four. A major change in the use of the routes happened at about 10 AM on the second day, when the east-west component of the wind direction changed from the western to the eastern. This refers to all airports except Stansted, where the approach routes changed on the first day at about 18:25 in response to a change of the north-south component of the wind. This was due to the northeast-southwest orientation of the runway in Stansted, which is different from the east-west orientation of the runways in the other airports.

Knowing when each route was used, we could investigate the groups of the routes that were used simultaneously. Figure 3.14 shows the routes that were used on the first day till 18:25 (top) and the routes that were used after 10:00 on the second day, i.e., after the wind change. Using the 3D representation of the trajectories, we observe that the routes coming to the same airport from different sides join in their final parts.

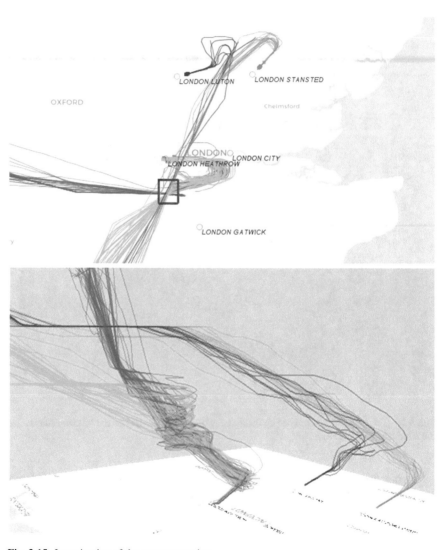

Fig. 3.15 Investigation of the route separation

Some routes going to different airports intersect or overlap on the 2D map. To investigate whether they are separated vertically, we repeatedly applied a spatial filter for selecting various groups of intersecting and overlapping trajectories. An example is shown in Fig. 3.15. The filter (Fig. 3.15, top) selects two partly overlapping routes ending at Luton and Stansted (pink and orange, respectively) that apparently intersect two routes ending at Heathrow. In a 3D view (Fig. 3.15, bottom), we see that the former two routes overlap also in the vertical dimension but there is no intersection with the routes to Heathrow due to differences in the flight levels. Our interactive investigation shows that it is a general pattern: where segments of different routes overlap in the horizontal dimension, their altitude ranges overlap as well, and routes intersecting in 2D are separated vertically. Hence, relevance-aware clustering of trajectories and interactive exploration with the use of temporal and spatial filters and a combination of a geographic map and a 3D view helped us to understand how air traffic services organise and manage a huge number of flights following diverse routes within a small densely packed air space.

3.4.3 Revealing Route Choice Criteria of Flight Operators

In this study, we wish to reveal the criteria used by airlines in choosing particular flight routes from many possible routes connecting a given origin-destination pair. This translates to a significant improvement in terms of predictability at pre-tactical phase (in particular for routes near local airspace boundaries, for which subtle route changes might imply the appearance or disappearance of hotspots), among other potential applications. As a representative example, we consider the flights from Paris to Istanbul. This example provides rich information for the study: there are many flights conducted by multiple airlines, which take diverse routes crossing the air spaces of different European countries whose navigation charges greatly vary. Some airlines may prefer such flight routes that minimize the navigation costs by avoiding expensive airspaces or travelling shorter distances across such airspaces. One of the questions in the study was to check if indeed some airlines are likely to have such preferences.

We apply our analysis to trajectories constructed from flight plans, because the route choices are made at the stage of planning. We use the plans of 1717 flights performed during 5 months from January to May, 2016. Additionally, we use a dataset specifying the boundaries of the navigation charging zones in Europe and the unit rate in each. The map background in Fig. 3.16 represents the navigation rates by proportional darkness of shading. The labels show the exact values, in eurocents per mile. On top of this background, coloured lines represent the result of clustering of the trajectories by route similarity excluding the initial and final parts. On the bottom left, the area around Paris is enlarged; the initial parts of the trajectories are shown in dashed lines. The lines are coloured according to their cluster membership. Through clustering, we have revealed 9 major routes. The most frequent was route

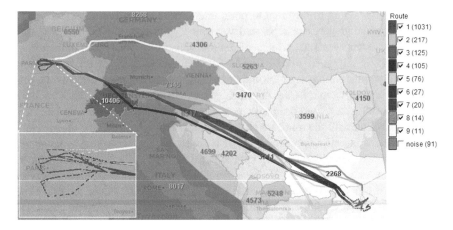

Fig. 3.16 Trajectories according to flight plans have been clustered by route similarity to reveal the major flight routes from Paris to Istanbul. The initial and final parts of the trajectories, which are represented by dashed lines, were disregarded in the clustering

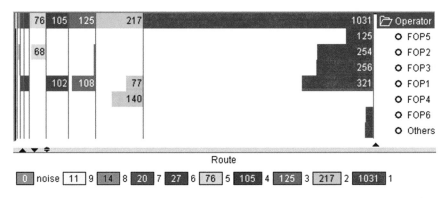

Fig. 3.17 Route choices by 6 major flight operators labelled FOP1 to FOP6. The length of each coloured bar represents the frequency of using the corresponding route by the flight operator specified in the respective row of the graph. The topmost row corresponds to all operators in total

1 shown in red; it was used 1031 times, i.e., in 60% of the flights. Route 2 (green) was used 217 times (12.6% flights), and the others were much less frequent.

It can be observed that the green route goes through cheaper airspaces than the other routes. This is the "cheapest" route among all, with the total navigation cost ranging from 434.9 to 492.8 euro, with the median 459.4 euro. The most popular route 1 costs from 472.2 to 547.3 euros, with the median 515.6 euros. Route 2 is the longest among all, except route 9 (yellow) that was taken only 11 times; however, the difference from route 1 is not dramatic, only about 12 km.

The graph in Fig. 3.17 shows how many times each of the 6 major flight operators (airlines) conducting flights from Paris to Istanbul chose each of the routes. The operators are labelled FOP1 to FOP6. It can be seen that FOP4 used only the cheapest route 2. This route was also occasionally used by FOP1, who conducted the largest number of flights (41.9% of all) but not by any other airline. Possibly, this route has disadvantages that overweigh the navigation cost saving. Apart from the path length difference, which is not very large, it may be lower flight levels or frequent deviations from the flight plans. Indeed, the flight levels on route 2 are lower than on route 1 by about 6 levels on the average and the difference between the third quartiles is 20. We have also calculated the deviations of the actual flights from the planned routes (i.e., the distances between the corresponding points in the planned and actual trajectories) and found that they are higher on route 2 than on route 1 by about 0.8 km on the average while the third quartiles differ by 3.2 km. Route 2 may also have other disadvantages that are not detectable from the available flight data.

Hence, we see that the navigation costs is not the main route choice criterion for most airlines, but it has high importance for some airlines.

Further details on analysis and modelling route choice preferences can be found in paper [20].

3.4.4 Understanding Airspace Configuration Choices

A sector configuration is a particular division of an airspace region into sectors, such that each sector is managed by a specific number of air traffic controllers (typically two, Executive and Planning Controllers). The number of active sectors depends, on the one hand, on the expected traffic features (such as number of flights within a time interval and their associated complexity/workload given the traffic complexity) and, on the other hand, on the available number of controllers for that given shift (which depends on the strategical demand forecast, which diverges from actual flights for a set of reason).

On the other hand, often there are multiple ways to divide a region into a given number of sectors. The choice of a particular division depends on the flight routes within the region. Sector configurations schedule is continuously refined as getting closer to operation, when the available flight plan information is progressively refined. The flight plan information available the day before operation, while is sure to change in tactical phase, already allows to prepare a schedule of sector configurations for the next shifts of the air traffic controllers.

Ideally, configurations should be chosen so that the demand for the use of the airspace in each sector does not exceed the sector capacity while making efficient and balanced use of resources (controllers). In reality, demand-capacity imbalances happen quite often for a set of reasons (deviations of actual flights from flight plans, weather conditions, etc.), causing flight regulations and delays. In search for predictive models that might support enhanced pre-tactical planning (able to forecast deviations), researchers would like to understand how configuration choices are made by airspace managers. They would also like to find a way to predict which configuration will be used at each time moment during the day of operation, considering uncertainty caused by operational factors in search for a more accurate sector configuration schedule in the day before operation (or earlier), allowing better management of demand-capacity imbalances. However, it is unclear what features should be used for building a predictive model. We utilised visual analytics approaches to gain understanding of the configuration system, patterns of change, and probable reasons for preferring one configuration over another. We performed interactive visual exploration of configurations used in several regions.

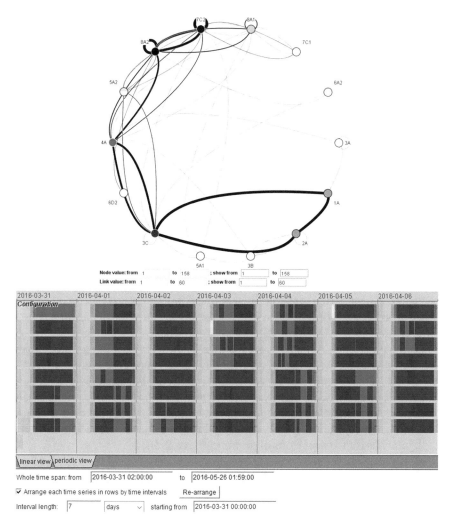

Fig. 3.18 Top: A state transition graph shows changes of airspace configurations in one region during a month. Bottom: The configurations are represented by differently coloured bar segments in a periodic time view. The rows correspond to time intervals of 1 week length

As an example, the upper image in Fig. 3.18 shows the configurations that were used in one of the regions in Spain (namely, LECMCTAS) during 1 month. The configurations are denoted by labels starting with a digit showing the number of sectors in which the region is divided. Almost for each number of sectors, there are two or more variants, some of which are used quite rarely. The lower image shows the use of the different configurations over time. The configurations are represented by coloured segments of horizontal bars. The light colours correspond to small numbers of sectors and dark blue to dark purple colours to 7 and 8 sectors, respectively. The positions of the segments correspond to the times when the configurations were used. The rows correspond to time intervals of 1 week length. The temporal bar graph shows that the changes of the configurations happen quite periodically. The configurations with small numbers of sectors are used in nights, when the air traffic is low. The configurations with 7 and 8 sectors are usually used from 07:30 till 22:30.

While the choices between configurations differing in the number of sectors can be explained by differences in the traffic volume, the reasons for choosing between configuration variants with the same number of sectors are not obvious. To understand how configurations differ from each other, we used a 3D view as shown in Fig. 3.19. The example in Fig. 3.19 shows two configurations in which the region is divided into 8 sectors, CNF8A1 on the left and CNF8A2 on the right. The sectors are represented by distinct colours. The configurations are almost identical, except the vertical division of the sub-region on the west. In CNF8A1, the sub-area is divided into two sectors at the flight level 325, and in CNF8A2 at the flight level 345. These two configurations are often used interchangeably during a day.

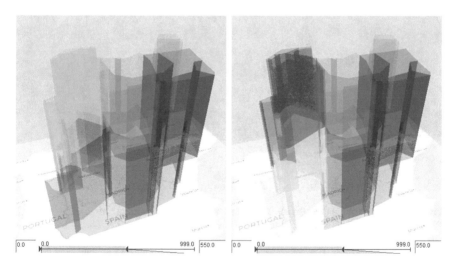

Fig. 3.19 Two configurations with the same number of sectors differ only in the vertical division of the sub-region on the west

The density graph in Fig. 3.20, in which the horizontal dimension represents time and the vertical dimension flight level, shows the traffic intensity in the western sub-region in 1 day when the configuration CNF8A1 was used in time interval from 12:30 till 14:00 and CNF8A2 in the remaining time from 07:30 till 22:30. These times are marked in the graph by vertical lines. The horizontal lines mark the flight levels 325 and 345. The flight intensity is represented by shading from light yellow (low) to dark red (high). The upper image shows the temporal density of all trajectory positions within the western sub-region and the lower image shows the density of the positions where the flight level changed with respect to the previous positions.

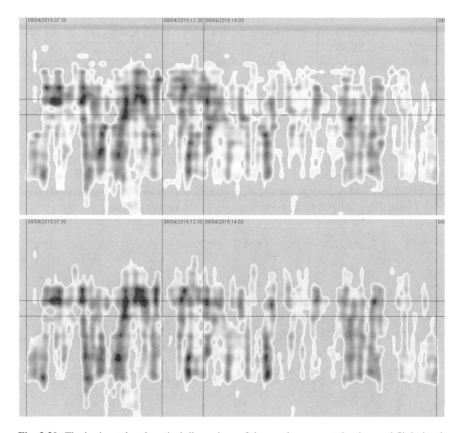

Fig. 3.20 The horizontal and vertical dimensions of the graph represent the time and flight level, respectively. The vertical lines mark the times 07:30, 12:30, 14:00, and 22:30. The horizontal lines mark the flight levels 325 and 345. The shading shows the variation of the traffic intensity in the western sub-region; top: all trajectory segments; bottom: segments where the flight level changed with respect to the previous position

A reasonable hypothesis for explaining the choice between different subdivisions would be that the traffic managers strive to balance the workload among the operators controlling different sectors, according to the behaviour of the specific traffic. Indeed, we see that the traffic intensity at the flight levels above 345 decreased after 12:30, and the division level was lowered from 345 to 325. However, after 14:00, when the division level returned to 345, there was no corresponding increase of traffic at the higher levels; so, our hypothesis would not be supported by this exclusive factor. Another possible decision rationale would be to choose such a division level that fewer flights have to cross this level while they are within the area. However, this hypothesis is not supported by the lower image in Fig. 3.20, where we see many intersections of both level 325 and level 345 at the time of using either of the two configurations. Hence, the vertical distribution of the flights does not explain the reasons for preferring one configuration over the other, and further investigation is needed. Domain experts suggest that the sector configuration change was motivated by controller workload, not always precisely represented by traffic counts or intensity. For this model, controller workload was not an input, so this factor could only be taken into account indirectly through traffic.

3.5 Discussion and Conclusion

This chapter fulfils several purposes. First, it introduces the concept of visual analytics as a methodology of data analysis where the key role belongs to the human reasoning. The methodology involves the use of interactive visual representations of information for facilitating the cognitive activities of human analysts. Second, the chapter provides multiple examples that can help readers grasp the essence of the VA methodology and see its utility in investigating properties of data (Sect. 3.3) and in analysing data for understanding real-world phenomena (Sect. 3.4). Third, it describes analysis scenarios in the domains of maritime and air traffic that resulted in gaining valuable knowledge concerning the design and planning of the business activities in these domains. This kind of knowledge can be potentially used for building predictive models and/or for improvement of the businesses.

We would like to emphasise that it is humans, not machines, who can generate new knowledge. Although the term 'knowledge discovery' is commonly applied to computational techniques for data analysis, their outcomes can not be called 'knowledge' in the sense usually meant by humans. They require human apprehension and reasoning for being transformed to knowledge. Therefore, it is absolutely necessary that human reasoning is involved in analyses aimed at gaining new knowledge and finding possible or new, better ways to solve problems. Visual analytics techniques, which support human reasoning, have therefore high importance and high potential.

This potential has been illustrated by four different cases corresponding to diverse operating environments and different data sources. The results have been discussed and validated with domain experts to ensure applicability to operational needs. Particularly, in the domain of air traffic management (ATM), the analysis scenarios demonstrated the value of the VA methods to identify decision criteria as key aspects of the ATM system, able to feed predictive or analytic models applicable in planning phase. The scenarios especially highlighted the power of these techniques to derive knowledge from spatiotemporal patterns. The VA techniques also proved their utility for assessment of data quality. The domain experts admitted that in some cases, as well as in data quality assessment, similar results can be achieved by means of non-visual techniques, but at a significantly higher cost of data preparation and analysis. Visual analytics techniques have proven as time-efficient for these purposes.

In the aviation domain, several Single European Sky ATM Research (SESAR) projects concluded that visual analytics is an important instrument for data analysis and modelling. The white paper [11] supports the use of visual analytics for performance modelling. The improvement in data quality and reliability at planning stages that SESAR new concepts will deliver (i.e., by means of shared business trajectory (SBT), reference business trajectory (RBT), and Trajectory-Based Operations) will only enhance the benefits demonstrated by reducing data uncertainty. However, current day data are already usable by this kind of techniques, delivering applicable results.

Acknowledgments This research was supported by Fraunhofer Cluster of Excellence on "Cognitive Internet Technologies" and by EU in SESAR project TAPAS (Towards an Automated and exPlainable ATM System.

References

1. Albrecht, G., Lee, H.T., Pang, A.: Visual analysis of air traffic data using aircraft density and conflict probability. https://doi.org/10.2514/6.2012-2540
2. Andrienko, N., Andrienko, G.: Visual analytics of movement: an overview of methods, tools and procedures. Inf. Vis. **12**(1), 3–24 (2013). https://doi.org/10.1177/1473871612457601
3. Andrienko, G., Andrienko, N., Jankowski, P., Keim, D., Kraak, M., MacEachren, A., Wrobel, S.: Geovisual analytics for spatial decision support: setting the research agenda. Int. J. Geogr. Inf. Sci. **21**(8), 839–857 (2007). https://doi.org/10.1080/13658810701349011
4. Andrienko, G., Andrienko, N., Bak, P., Keim, D., Wrobel, S.: Visual Analytics of Movement. Springer (2013). https://doi.org/10.1007/978-3-642-37583-5
5. Andrienko, G., Andrienko, N., Fuchs, G.: Understanding movement data quality. J. Locat. Based Serv. **10**(1), 31–46 (2016). https://doi.org/10.1080/17489725.2016.1169322
6. Andrienko, G., Andrienko, N., Chen, W., Maciejewski, R., Zhao, Y.: Visual analytics of mobility and transportation: state of the art and further research directions. IEEE Trans. Intell. Transp. Syst. **18**(8), 2232–2249 (2017). https://doi.org/10.1109/TITS.2017.2683539
7. Andrienko, N., Andrienko, G., Camossi, E., Claramunt, C., Cordero Garcia, J.M., Fuchs, G., Hadzagic, M., Jousselme, A.L., Ray, C., Scarlatti, D., Vouros, G.: Visual exploration of movement and event data with interactive time masks. Vis. Inf. **1**(1), 25–39 (2017). https://doi.org/10.1016/j.visinf.2017.01.004
8. Andrienko, G., Andrienko, N., Fuchs, G., Garcia, J.M.C.: Clustering trajectories by relevant parts for air traffic analysis. IEEE Trans. Vis. Comput. Graph. **24**(1), 34–44 (2018). https://doi.org/10.1109/TVCG.2017.2744322

9. Andrienko, N., Andrienko, G., Cordero Garcia, J.M., Scarlatti, D.: Analysis of flight variability: a systematic approach. IEEE Trans. Vis. Comput. Graph. **25**(1), 54–64 (2019). https://doi.org/10.1109/TVCG.2018.2864811

10. Buchmüller, J., Janetzko, H., Andrienko, G., Andrienko, N., Fuchs, G., Keim, D.A.: Visual analytics for exploring local impact of air traffic. Comput. Graph. Forum **34**(3), 181–190 (2015). https://doi.org/10.1111/cgf.12630

11. Cordero Garcia, J., Herranz, R., Marcos, R., Prats, X., Ranieri, A., Sanchez-Escalonilla, P.: Vision of the future performance research in SESAR. White paper. SESAR 2020 (2018)

12. Demšar, U., Virrantaus, K.: Space–time density of trajectories: exploring spatio-temporal patterns in movement data. Int. J. Geogr. Inf. Sci. **24**(10), 1527–1542 (2010). https://doi.org/10.1080/13658816.2010.511223

13. Huang, X., Zhao, Y., Ma, C., Yang, J., Ye, X., Zhang, C.: TrajGraph: a graph-based visual analytics approach to studying urban network centralities using taxi trajectory data. IEEE Trans. Vis. Comput. Graph. **22**(1), 160–169 (2016). https://doi.org/10.1109/TVCG.2015.2467771

14. Hurter, C., Alligier, R., Gianazza, D., Puechmorel, S., Andrienko, G., Andrienko, N.: Wind parameters extraction from aircraft trajectories. Comput. Environ. Urban Syst. **47**, 28–43 (2014). https://doi.org/10.1016/j.compenvurbsys.2014.01.005. Progress in Movement Analysis - Experiences with Real Data

15. Konzack, M., McKetterick, T., Ophelders, T., Buchin, M., Giuggioli, L., Long, J., Nelson, T., Westenberg, M.A., Buchin, K.: Visual analytics of delays and interaction in movement data. Int. J. Geogr. Inf. Sci. **31**(2), 320–345 (2017). https://doi.org/10.1080/13658816.2016.1199806

16. Kraak, M.J.: The space-time cube revisited from a geovisualization perspective. In: Proceedings of the 21st International Cartographic Conference, pp. 1988–1996 (2003)

17. Lampe, O.D., Hauser, H.: Interactive visualization of streaming data with kernel density estimation. In: IEEE Pacific Visualization Symposium, PacificVis 2011, Hong Kong, 1–4 March 2011, pp. 171–178 (2011). https://doi.org/10.1109/PacificVis.2011.5742387

18. Lu, M., Lai, C., Ye, T., Liang, J., Yuan, X.: Visual analysis of route choice behaviour based on GPS trajectories. In: 2015 IEEE Conference on Visual Analytics Science and Technology (VAST), pp. 203–204 (2015). https://doi.org/10.1109/VAST.2015.7347679

19. Lundblad, P., Eurenius, O., Heldring, T.: Interactive visualization of weather and ship data. In: Proceedings of the 13th International Conference on Information Visualization IV2009, pp. 379–386. IEEE Computer Society, Washington (2009)

20. Marcos, R., Cantu Ros, O., Herranz, R.: Combining visual analytics and machine learning for route choice prediction. Application to pre-tactical traffic forecast. In: Proceedings of the 7th SESAR Innovation Days, Belgrade (2017)

21. Ray, C., Dreo, R., Camossi, E., Jousselme, A.L.: Heterogeneous integrated dataset for maritime intelligence, surveillance, and reconnaissance (2018). https://doi.org/10.5281/zenodo.1167595

22. Sakr, M., Andrienko, G., Behr, T., Andrienko, N., Güting, R.H., Hurter, C.: Exploring spatiotemporal patterns by integrating visual analytics with a moving objects database system. In: Proceedings of the 19th ACM SIGSPATIAL International Conference on Advances in Geographic Information Systems, GIS '11, pp. 505–508. ACM, New York (2011). https://doi.org/10.1145/2093973.2094060

23. Scheepens, R., Willems, N., van de Wetering, H., Andrienko, G.L., Andrienko, N.V., van Wijk, J.J.: Composite density maps for multivariate trajectories. IEEE Trans. Vis. Comput. Graph. **17**(12), 2518–2527 (2011). https://doi.org/10.1109/TVCG.2011.181

24. Scheepens, R., van de Wetering, H., van Wijk, J.J.: Non-overlapping aggregated multivariate glyphs for moving objects. In: IEEE Pacific Visualization Symposium, PacificVis 2014, Yokohama, 4–7 March 2014, pp. 17–24 (2014). https://doi.org/10.1109/PacificVis.2014.13

25. Thomas, J., Cook, K.: Illuminating the path: the research and development agenda for visual analytics. IEEE, Los Alamitos (2005)

26. Tominski, C., Schumann, H., Andrienko, G., Andrienko, N.: Stacking-based visualization of trajectory attribute data. IEEE Trans. Vis. Comput. Graph. **18**(12), 2565–2574 (2012). https://doi.org/10.1109/TVCG.2012.265

27. von Landesberger, T., Brodkorb, F., Roskosch, P., Andrienko, N., Andrienko, G., Kerren, A.: MobilityGraphs: visual analysis of mass mobility dynamics via spatio-temporal graphs and clustering. IEEE Trans. Vis. Comput. Graph. **22**(1), 11–20 (2016). https://doi.org/10.1109/TVCG.2015.2468111
28. Willems, N., van de Wetering, H., van Wijk, J.J.: Visualization of vessel movements. Comput. Graph. Forum **28**(3), 959–966 (2009). https://doi.org/10.1111/j.1467-8659.2009.01440.x

Chapter 4
Trajectory Detection and Summarization over Surveillance Data Streams

Kostas Patroumpas, Eva Chondrodima, Nikos Pelekis, and Yannis Theodoridis

Abstract In this chapter, we present Synopses Generator, a stream-based processing framework that can provide *online* summarized representations of trajectories specifically for sailing vessels and flying aircraft. Assuming that surveillance data monitoring their locations over a large geographical area is available in a streaming fashion, this novel methodology drops any predictable positions (along trajectory segments of "normal" motion characteristics) with minimal loss in accuracy. Effectively, it can keep only those positions conveying salient mobility events (annotated as stop, change in speed, heading, or altitude, etc.), identified when the mobility pattern of a given vessel or aircraft changes significantly. Moreover, this framework specifies parametrized conditions for detecting such mobility features, as well as suitable heuristics that can eliminate inherent noise and can provide succinct *trajectory synopses* in one pass over the incoming streaming positions. A prototype implementation on top of Apache Flink and Kafka has been set up in modern cluster infrastructures to enable parallelization of the trajectory summarization process against such big mobility data. A comprehensive experimental evaluation has been conducted against various surveillance data in the maritime and aviation domain, and offers concrete evidence of its timeliness, scalability, and compression efficiency, with tolerable concessions to the quality of resulting trajectory approximations. The resulting compressed trajectories can be particularly useful in efficient online or offline post-processing (e.g., mobility analytics, statistics, pattern mining, etc.) while also facilitating their comparison irrespectively of differing update frequencies.

K. Patroumpas
Athena Research Center, Marousi, Greece
e-mail: kpatro@athenarc.gr

E. Chondrodima · N. Pelekis (✉) · Y. Theodoridis
University of Piraeus, Piraeus, Greece
e-mail: evachon@unipi.gr; npelekis@unipi.gr; ytheod@unipi.gr

© Springer Nature Switzerland AG 2020
G. A. Vouros et al. (eds.), *Big Data Analytics for Time-Critical Mobility Forecasting*, https://doi.org/10.1007/978-3-030-45164-6_4

4.1 Introduction

Detecting trajectories from a large number of moving entities like vessels (maritime domain) or aircraft (aviation domain) is a challenging task, especially if it has to be carried out in *online* fashion against multiple, heterogeneous, voluminous, fluctuating, and noisy data streams. In such scenarios, trajectories must be constructed from online surveillance data: terrestrial or satellite Automatic Identification System (AIS) messages[1] from vessels, and Automatic Dependent Surveillance–Broadcast (ADS-B) messages[2] from aircraft or tracklogs from Air Traffic Control (ATC) radars.

To address this challenge, in this chapter we present *Synopses Generator*, a stream-based framework employing single-pass techniques for succinct, lightweight representation of trajectories without harming the quality of the resulting approximation. Instead of retaining every incoming position for each object, this framework drops any predictable positions along trajectory segments of "normal" motion characteristics. Except for adverse weather conditions, traffic regulations, local manoeuvres close to ports and airports, congestion situations, accidents, etc., most vessels and aircraft normally follow almost straight, predictable routes at open sea and in the air, respectively. It turns out that a large amount of raw positional updates may be suppressed with minimal losses in accuracy, as they hardly contribute additional knowledge about their actual motion patterns. Instead of resorting to a costly trajectory simplification algorithm, we opt to reconstruct their traces approximately from judiciously chosen *critical points* along their trajectories.

Effectively, this Synopses Generator keeps only those positions conveying salient *mobility events* (stop, slow motion, change in heading, change of speed, change of altitude, etc.). As a first step, *trajectory construction* is being applied, i.e., offering distinct sequences of timestamped positions per moving object. This process also involves discarding any inherent *noise* detected in the streaming positions due to, e.g., delayed arrival of input messages, duplicate positions, crosswind or sea drift, discrepancies in Global Positioning System (GPS) measurements, etc. In a second step, any predictable positions along "normal" segments are dropped with minimal loss in accuracy. As exemplified in Fig. 4.1, effectively this framework keeps only those positions detected as mobility events when the pattern of movement changes significantly. The derived stream of the so-called *trajectory synopses* must keep in pace with the incoming raw streaming data so as to get incrementally annotated with semantically important mobility features once they get detected. Thanks to the computation of such synopses, the derived motion paths remain lightweight for efficient real-time processing without sacrificing accuracy, and can be actually compared to each other irrespectively of the actual reporting frequency that may

---

[1]http://www.imo.org/OurWork/Safety/Navigation/Pages/AIS.aspx.
[2]https://www.faa.gov/nextgen/programs/adsb/.

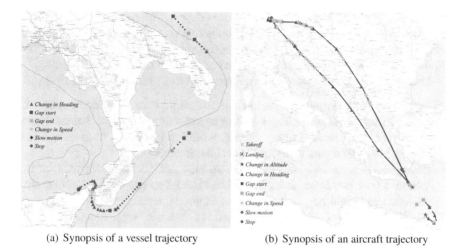

(a) Synopsis of a vessel trajectory   (b) Synopsis of an aircraft trajectory

Fig. 4.1 Example synopses of critical points for (**a**) the maritime and (**b**) aviation use cases

differ amongst objects. No context (weather, areas of interest, characteristics of moving objects, etc.) is taken into account when constructing trajectory synopses.

It should be noted that our Synopses Generator works in a *stream-in-stream-out* fashion, i.e., consuming position updates arriving at varying frequencies from numerous objects and producing a derived stream of distinct subsequences of (expectedly noise-free) timestamped positions as their evolving trajectory. Most importantly, this framework can be used to obtain summarized, connected motion paths concerning each object's whereabouts. In addition, spatiotemporal features (speed, travel time, rate of climb, etc.) of each trajectory can be computed with minimal error. Hence, the resulting trajectory synopses are incrementally emitted online as a derived stream, carrying not just the original timestamp and coordinates of selected locations (i.e., those actually detected as critical points), but also their annotation and several calculated spatiotemporal measurements (speed, travel time, travelled distance, rate of climb/descent, etc.). This information may be further exploited online or offline by other modules, e.g., for recognition and forecasting of complex events, visual and mobility analytics, statistical analysis, clustering, etc. Besides, the detected trajectory features can be archived into a data repository (e.g., an RDF graph or a database), where they can be contextually enriched with other static or streaming information.

Another major objective is to address the issue of *cross-stream* processing of surveillance data by reconciling and aligning parts of synopses from different sources into a unified streaming summary. In particular, cross-stream processing of ADS-B data can be seen as a means of "filling-in" gaps in trajectory representations with information obtained from other sources despite differences in object identification schemes, reporting frequencies, conflicting coordinates, etc. This real-time integration of data from different sources effectively allows the correlation of data from multiple sources in order to provide a coherent trajectory representation.

Overall, the proposed Synopses Generator addresses the following challenges:

- *Timeliness.* Trajectory detection and summarization must be carried out in real-time. Critical points concerning evolving trajectories of vessels and aircraft must be issued at operational latency (ideally within milliseconds, or at most a few seconds) in order to enable immediate action, if necessary.
- *Compression.* Since trajectory synopses per moving object are extracted from the incoming positions by retaining salient movement features only, this online data reduction can yield huge space savings. Empirical results indicate that at least 70–80% of the raw data may be discarded as redundant, while compression ratio can be up to 99% when frequency of position updates is high.
- *Quality.* Such compressed representations are also reliable enough in reconstructing trajectories with small deviations (i.e., tolerable approximation error) from original traces, also coping with imperfections (such as network delays, noise, etc.) inherent in real-world surveillance streams.
- *Scalability.* This framework can be deployed in both centralized and distributed infrastructures and can manage scalable volumes of frequently updated, streaming positions from large fleets of vessels or aircraft moving over a large area.

The remainder of this chapter proceeds as follows. Section 4.2 surveys related work with a particular focus on trajectory management. In Sect. 4.3, we discuss the major characteristics, as well as issues concerning streaming data sources used for maritime and aviation surveillance. Section 4.4 presents basic notions regarding trajectory representation and an overview of the data processing flow as applied by the Synopses Generator. Section 4.5 consolidates our methodology towards effective and efficient trajectory summarization in online fashion, formulates the types of detected critical points, and describes the derived synopses. Section 4.6 discusses the implemented software prototype and core technical aspects of its geostreaming functionality. In Sect. 4.7, we report indicative results from a comprehensive validation of the prototype against real surveillance streams, both in terms of performance and quality of the derived synopses. Section 4.8 presents a cross-stream processing methodology, together with a brief proof-of-concept evaluation. Finally, Sect. 4.9 summarizes the methodology and underscores the substantial benefits from this approach in maritime and aviation surveillance.

4.2   Related Work

Stream Processing frameworks like Apache Flink,[3] Apache Spark streaming,[4] or Apache Storm[5] offer powerful capabilities to ingest, process, and aggregate enormous amounts of streaming information from diverse big data sources. Note that these are general-purpose platforms, which allow customizations and extensions for specific applications. A recent survey [7] discusses their pros and cons, along with an experimental study of their current processing capabilities with a particular focus on analytics. Besides, several spatial extensions have emerged for big data platforms, like SpatialHadoop [5], GeoSpark [29], or Simba [28]. These modules offer support for representation and indexing of spatial entities, as well as basic topological operators for querying, but they focus on point locations and simple spatial features, ignoring entirely any spatiotemporal notions concerning mobility. Hence, trajectories of moving objects cannot be managed by any of these platforms. Only recently, UlTraMan [4] introduces support for storing and querying historical trajectories over Spark. In our prototype implementation of the Synopses Generator framework, we use Apache Flink as a stream processing engine and we add custom functionality for trajectory detection and summarization on top of it. To the best of our knowledge, no other streaming framework has been specifically tailored for surveillance over noisy, intermittent, geostreaming messages from large fleets of vessels and aircraft.

Regarding *trajectory simplification*, state-of-the-art algorithms like [12, 16] operate in batch (*offline*) fashion since they require beforehand knowledge of all points (i.e., complete trajectories). In contrast, due to the high arrival rate of streaming locations in maritime and aviation surveillance, trajectory synopses detecting significant mobility features must be updated in real time (*online*). Ideally, retained samples in these synopses should keep each compressed trajectory as much closer to its original course, chiefly by minimizing approximation error as in trajectory fitting methods [2, 14]. For instance, the sliding window approach in [14] keeps simplifying points along a line until error exceeds a given threshold. The *STTrace* algorithm in [23] uses the concept of safe areas to generate a simplified trajectory by keeping samples that deviate from predefined speed and direction error bounds. Under a similar error-based principle, *SQUISH* [15] drops samples by employing a priority queue in order to achieve a target compression ratio λ. Its adaptive variant called *SQUISH-E* [16] aims not only to maximize λ but also to minimize estimated error below a user-specified bound μ. Time-decaying approximation of streaming trajectories can be carried out in an ageing-aware fashion [24], by gradually evicting older samples and offer greater precision along the most recent segments. Dead-reckoning policies like [27] and mobility tracking protocols in [8] may be employed

---

[3]https://flink.apache.org/.

[4]http://spark.apache.org/streaming/.

[5]http://storm.apache.org/.

on board of the moving objects to relay positional updates only upon significant deviation from the course already known to a centralized server. Recently, a bounded quadrant technique was introduced in *FBQS* [11] by applying an open window over the recent (not yet compressed) trajectory portion and enabling estimation of various compression error bounds. The one-pass, error-bounded *OPERB* algorithm in [10] is based on a novel local distance checking method and involves several optimizations to achieve higher compression. To the best of our knowledge, none of the aforementioned techniques has ever been applied on streaming trajectories of vessels or aircraft, and certainly do not support annotation of the retained samples.

In [20, 21] we had introduced a maritime surveillance framework specifically for tracking vessel trajectories and also recognizing complex events (e.g., suspicious vessel activity). This technique applied a sliding window over the incoming positions and distinguished instantaneous events (e.g., a sudden change in heading) from long-lasting ones deduced after examining a sequence of instantaneous events over a recent time period (e.g., a smooth turn). Detected events were reported periodically (i.e., critical point locations evicted from the window when it slides) with all recent "delta" changes. Extensive tests showed that this summarization method could yield a compression ratio better than 95% over the raw data, and also sustain scalable volumes of streaming vessel positions. Synopses Generator differs substantially and not just because it also deals with aircraft trajectories, i.e., locations having an extra z-ordinate; this is not trivial, as it involves handling of additional events (e.g., change in altitude, takeoff, landing). Most importantly, mobility events can now be emitted at operational latency, i.e., within milliseconds (or a few seconds at most) since the arrival of raw messages instead of relying on the slide step of windows (which can even be hours). We now prescribe more noise elimination filters and support multiple annotations per location in order to capture richer semantics of all mobility-related information at a given location. Finally, the prototype implementation of the Synopses Generator is now specifically designed so as to enable scalability and can be executed in multiple concurrent threads or in distributed cluster infrastructures.

The Synopses Generator framework is distinct from either deterministic or probabilistic techniques towards trajectory prediction. For instance, a stochastic Hidden Markov Model is employed in [1] to make such predictions by taking into account weather observations, aircraft specifications, as well as historical trajectories indexed in a fixed three-dimensional grid in space. Our approach also differs from Complex Event Processing (CEP) methods, like CEP-traj [25] applied over vessel traces, as we only consider spatiotemporal information in detecting mobility events. Overall, our framework is geared towards a data-driven summarization and semantic annotation of streaming trajectories, as a means of reducing the amount of information to be further processed (e.g., for mining or prediction) and associating extra knowledge to the retained locations (with annotations like stop, slow motion, turn, etc.).

4.3 Streaming Data Sources in Maritime and Aviation Surveillance

Maritime and aviation surveillance must cope with data stream imperfections, e.g., the *noise* inherent in vessel or aircraft positions due to sea drift, delayed arrival of messages, discrepancies in GPS signals, etc. Next, we outline some issues that may hinder mobility tracking from surveillance data streams, while the proposed heuristics regarding noise reduction will be discussed in Sect. 4.5.1.1.

4.3.1 Maritime Data Sources

Nowadays, online tracking of vessels across the seas has become commonplace largely thanks to the *Automatic Identification System* (AIS). A vessel equipped with AIS transponders periodically broadcasts messages that include information about its movement. In our Synopses Generator, we only consider *positional* messages (AIS types: 1, 2, 3, 18, 19, 27) concerning timestamped locations of vessels from:

- *terrestrial AIS* (T-AIS) continuously collected by onshore receiving stations; and
- *satellite AIS messages* (S-AIS) arriving in bursts when satellites transfer buffered data into a ground station.

These types of AIS messages include location-based and spatiotemporal information about vessels: longitude, latitude, Speed over Ground (SoG), Course over Ground (CoG), Rate of Turn (RoT), navigational status, and true heading.

Reporting Frequency of AIS messages can be quite low (in messages/hour), especially when vessels are sailing at open seas. The range of AIS signals is typically about 70 km due to the curvature of the Earth; therefore, satellites may be used for monitoring instead of onshore base stations. But due to the small number of available satellites, the update frequency between two consecutive measurements can range from several minutes up to a few hours. Unfortunately, such a very sparse sampling of AIS locations (sometimes, just a handful of positions per hour from a single vessel) can have serious repercussions both on vessel monitoring as well as on trajectory maintenance. Reporting frequency also depends on the vessel type, e.g., boats often switch off their AIS transponders while fishing, hence contact may be lost over several hours or even longer.

Timestamping Timestamp values are suitably expressed in UNIX epochs (i.e., *seconds*) and can be used as time reference for each incoming AIS message. However, the timestamp in each message does not reflect the measurement on board, but is assigned when this message is received by the station(s). Hence, it might be affected by the accuracy of their clock, occasionally leading to disorder in the resulting stream items. More frequent than not, streaming AIS messages do not arrive chronologically. This imperfection is an inherent problem caused by collecting AIS data from various sources (terrestrial or satellite) or by combining streams that are out of sync. More specifically, a message of timestamp τ could arrive with a *lag* $L = \tau_{max} - \tau$ epochs with respect to the latest timestamp τ_{max} seen in the stream thus far. Admitting messages with a delay would distort proper ordering of stream items by time [13].

Vessel Identification Each vessel can be uniquely identified by its 9-digit Marine Mobile Service Identifier (MMSI). Duplicate or invalid MMSIs should be cleaned by the data provider, and any problematic cases or inconsistencies (e.g., blacklisted vessels) should be marked if they must be excluded from any further processing.

Geopositioning Geospatial reference for vessel positions is in geographical coordinates (*longitude*, *latitude*) referenced in the World Geodetic System 1984 (WGS84). Occasionally, vessels report locations in the mainland, sometimes at a long distance from the coast and not along rivers, lakes, canals, etc., so these positions should be eliminated altogether in a preprocessing step.

Deduplication of AIS messages is another concern. It can happen that a position is reported twice from a single vessel, i.e., identical coordinates at the same timestamp. Perhaps this occurs because the same message is collected from multiple base stations. However, duplicate messages may cause trouble in their processing, especially if values in some attributes differ between these messages. As will be discussed in Sect. 4.5.1.1 regarding inherent noise in AIS messages, it may happen that the same message is received by two base stations with deferred clocks, so it may get differing timestamps. Such problems necessitate additional filters against the incoming AIS stream in order to discard duplicates.

4.3.2 Aviation Data Sources

The Synopses Generator framework may accept raw surveillance data concerning timestamped positions of aircraft from various sources:

- *ADS-B messages* broadcast by airplanes and collected by providers. Such messages may be available via APIs or as historical files (in JSON or CSV format).
- *Secondary radar tracks* in log files collected from ATC providers, typically available in historical CSV files that can be replayed for simulations.

As discussed in Chap. 3 of this book, these data sources track aircraft at different fashion and granularity, have diverse attribute schemata, while also differ in their spatial coverage. More specifically:

Reporting Frequency varies widely amongst aviation data providers. Tracklogs from ATC secondary radars have a regular frequency of updates every 5 s, while ADS-B messages are relayed less frequently (on average, less than one message per minute from each aircraft). There is no fixed rate of reporting even for a single flight; instead, this rate fluctuates a lot, with messages sometimes generated with a 5 s difference in time and afterwards having to wait even a minute until the next one. This can potentially hamper precise measurements of spatiotemporal features (speed, rate of climb, etc.) that are indispensable in trajectory detection.

Timestamping Timestamp values in ADS-B data are expressed at the granularity of *seconds* or *milliseconds*, thus leading to different precision in all derived spatiotemporal calculations (speed, rate of climb, etc.). Yet, as previously observed with AIS messages, ADS-B messages may also not arrive chronologically ordered. However, in the aviation use case, such delays in transmission are more critical, since airplanes move at a very high speed; so their actual position could be off by several kilometres even for a lag of a few seconds in their position reports.

Aircraft Identification There is no standard identification scheme for incoming positional messages. Several identifiers may be available in ADS-B messages or radar tracklogs, like airline and flight codes, hexadecimal aircraft identifiers, etc.

Geopositioning Geospatial reference for aircraft is provided via geographical WGS84 coordinates (longitude, latitude). Nevertheless, spatial coverage of the reported positions varies amongst data providers, as some offer good coverage over the entire planet with a few "holes" over the oceans, while other sources have a more cluttered coverage, so it may not be possible to track a long-haul flight in its entirety.

Insufficient Positions Near Airports In general, there is lack of sufficient samples near airports across data sources, which can seriously deter detection of certain critical points in trajectory synopses. In general, it seems that very few positions per flight are reported in airport areas, thus affecting identification of several stages of a flight such as standing, pushback, taxiing, take-off, and landing.

Altitude-Related Information Most data providers report altitude of aircraft in hundreds of feet (e.g., 1700, 1800, 1900, etc.). Such a coarse resolution may cause problems in detecting changes in altitude, whereas it is also susceptible to noise.

4.4 System Overview

In this section, we first discuss some basic concepts regarding representation of streaming trajectories of moving objects and their synopses. Then, we outline the components of Synopses Generator and its processing flow.

4.4.1 Trajectory Representation

We consider that trajectories must be constructed from streaming surveillance data in order to provide representation in real time of the original (raw) positions from vessels or aircraft. Our objective with Synopses Generator is to maintain distinct summarized sequences of timestamped positions per moving object (vessel or aircraft), after excluding any inherent noise detected in streaming positions due to, e.g., delays, duplicate messages, etc. Streaming timestamped positions are three-dimensional (3-d) for vessels, but 4-d for aircraft, so they, respectively, consist of tuples:

- $\langle x_i, y_i, \tau_i \rangle$ for vessels, and
- $\langle x_i, y_i, z_i, \tau_i \rangle$ for aircraft,

where $x_i, y_i \in \mathbb{R}$ are coordinates in the respective axis of the spatial reference system (typically, in WGS84), z_i represents the altitude, and τ_i is the timestamp value. Without loss of generality, we assume all coordinates in the same georeference system; altitude values z are expressed in feet, whereas (x, y) represent GPS longitude/latitude pairs. Overall:

Definition 4.1 *Trajectory* T_o is abstracted as a (possibly unbounded) sequence of tuples $\langle o, p_i, \tau_i \rangle$ for a given moving object o (either vessel or aircraft). Each successive position $p_i \in \mathbb{R}^d$ in the Euclidean space has d-dimensional coordinates recorded at timestamp $\tau_i \in \mathbb{T}$.

Timestamp values τ are considered as discrete, totally ordered time instants from a time domain $\mathbb{T}$, e.g., UNIX epochs at the granularity of milliseconds. Note that timestamps may have different interpretations [17]. In maritime AIS messages, timestamps denote the *transaction time*, i.e., when the message was received by the base station. The case is similar with ADS-B messages and locations from secondary radars in aviation. Ideally, each location should report the *valid time* of the actual measurement in the real-world, but this is not the case in any dataset available in either use case. Thus, in our methodology, we accept the original timestamp values as included in each incoming message, and our processing is based on them in order to identify the sequence of positions per object.

For maintaining trajectory synopses in real time, *"critical points"* along each trajectory must be identified strictly based on mobility features, effectively discarding redundant locations along a "normal" course. A detected event may require one, two, or multiple critical points to be issued (e.g., the start and end point of a slow motion event), but always in a streaming fashion. Formally:

Definition 4.2 A *synopsis* S_o over trajectory T_o of an object o consists of a possibly unbounded, time-ordered sequence of *critical points*, each represented as a tuple $\langle o, p_i, A_i \rangle$, where $\langle o, p_i \rangle \in T_o$ and A_i is a set of *annotations* for mobility events.

For a given object o, it holds that $S_o \subset T_o$; ideally, such summarization should result into $|S_o| \ll |T_o|$ in order to provide a concise synopsis, but with minimal

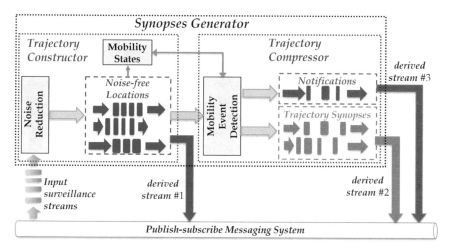

Fig. 4.2 Online processing flow of the Synopses Generator framework

deviation from the original trajectory. Set A_i denotes *multiple annotations* for a single timestamped location, acting like a bitmap where setting a bit signifies that a particular mobility event (e.g., a change in heading) was detected at that position.

4.4.2   Framework Architecture

Figure 4.2 illustrates the processing flow in the Synopses Generator. Streaming messages carrying 3-*d* timestamped positions for vessels (4-*d* for aircraft) may be admitted via external connectors to online surveillance sources or replayed from historical files using custom data feeders from public-subscribe systems (e.g., Apache Kafka[6] in our prototype). The Synopses Generator consists of two major components:

Trajectory Constructor   Constructing trajectories from surveillance data relies on timely admission of the *original positions* being received as input. In effect, a **Noise Reduction** module provides distinct sequences of timestamped positions per object, after excluding any inherent *noise* detected in the streaming positions due to, e.g., delayed or duplicate messages, GPS discrepancies, etc. The resulting *noise-free locations* are issued in a streaming fashion (derived stream #1 in Fig. 4.2). The *m* latest positions per object are maintained as its persistent *state* by the **Mobility States** module. This enables calculation of its mean velocity vector and useful spatiotemporal estimates (rate of turn, rate of climb, etc.) over its recent movement.

---

[6]https://kafka.apache.org/.

Trajectory Compressor Incremental maintenance of *trajectory synopses* from surveillance data offers timely detection of characteristic positions from each moving object. Based on the noise-free locations per trajectory obtained from the previous component, the ***Mobility Event Detection*** module can track major changes along each object's movement *in isolation* from the rest. By tracking vessels or aircraft in real time, this process instantly identifies *"critical points"* along each trajectory, essentially maintaining their synopses in a streaming fashion (indicated as derived stream #2 in Fig. 4.2). To this end, spatiotemporal measurements maintained in each object's state are useful in order to avoid issuing false positives.

Besides, the Mobility Event Detection module also signals out possible disruptions in communication. Typically in real-world surveillance streams, a vessel or aircraft may stop relaying positions for quite a while (e.g., flights over oceans). Of course, such a case can only be detected with some delay with respect to the last known location of the object. Importantly for mission-critical or emergency situations, once such a *communication gap* is spotted, this last reported location is emitted into a derived stream of *notifications* (marked as #3 in Fig. 4.2). We make no attempt to "fill-in" such gaps either via interpolation, extrapolation, or historical patterns, but with a *cross-stream* method that checks against other available data sources.

The entire operation is controlled by a user-specified *parametrization* concerning both noise reduction and mobility event detection on trajectories. Choosing a proper parametrization strongly depends on stream data characteristics (e.g., arrival rate of positions) and the type of moving objects (e.g., fishing vessels, passenger ships) and should be fine-tuned in order to trade compression efficiency with quality of results.

4.5 Online Processing of Streaming Trajectories

In the sequel, we suppose that a tuple $\langle o, p_{cur} \rangle$ has just streamed in, reporting p_{cur} as the latest known location of a given object o. Similarly, let p_{prev} be the previously known position of object o relayed at time $p_{prev}.\tau < p_{cur}.\tau$. For detecting significant motion changes, we employ an *instantaneous velocity vector* $\overrightarrow{v}_{cur}$ per object over its two latest known positions p_{prev}, p_{cur}. We also maintain the *mean velocity* $\overrightarrow{v_m}$ per object to get its short-term course based on m recent positions in its mobility state. In trajectory computations, we typically employ linear interpolation [2], assuming that any two consecutive positions (p_{prev}, p_{cur}) in a trajectory lie in a small area, which can be locally approximated with a Euclidean plane using Haversine distances.

In this section, we first discuss trajectory construction including noise reduction from surveillance data streams. Next, we present our approach for effective and efficient trajectory summarization in online fashion. We formulate the types of detected critical points, as well as the rules and conditions guiding their annotation.

4.5.1 Trajectory Construction

The *Trajectory Constructor* component in the Synopses Generator provides time-ordered point locations for each monitored object o (vessel or aircraft) in a streaming fashion; such a sequence of points can then be used to construct its trajectory T_o. No fictitious tuples like punctuations [26] are embedded to this derived stream, but positions considered as noise are removed. Based on each incoming positional message, trajectory construction involves the following steps:

- Extracts attributes $\langle o, p_i, \tau_i \rangle$ from the raw (AIS or ADS-B) message;
- Identifies o as the object (vessel or aircraft) concerned based on its unique identifier as included in the original message, and fetches its currently maintained trajectory T_o composed of a subset of previously reported "noise-free" locations;
- Computes basic spatiotemporal features of its movement (speed, heading, acceleration, rate of climb, etc.) as derived from each successive pair of incoming positions, which are indispensable for filtering out noise in the data;
- If the resulting position is considered valid (i.e., not qualifying as noise), then this timestamped position should be appended to trajectory T_o.

This process works in a *stream-in-stream-out* fashion. Indeed, it consumes position updates arriving at varying frequencies from numerous objects and maintains their evolving trajectories, while also issuing distinct sequences of noise-free timestamped positions as a derived stream (marked as #1 in Fig. 4.2).

4.5.1.1 Noise Reduction

Despite their high value in maritime and aircraft surveillance, positional data streams are not error-free as discussed in Sect. 4.3. To remedy such imperfections as much as possible, while also avoiding costly offline cleansing, we apply *online, single-pass* empirical filters upon admission of each incoming streaming position. These heuristics examine the instantaneous velocity vector $\vec{v}_{cur}$ of each object as computed by its two most recent observations. Since trajectory compression is our principal objective, we can afford to lose garbled, out-of-sequence positions and not consider correcting their timestamps. In such a streaming context, an object should

Table 4.1 Parameter settings for noise reduction from maritime and aviation surveillance streams

Parameter	Threshold description	Aviation	Maritime
v_{max}	Max speed of movement for any aircraft (in knots)	1000	50
δ_{max}	Max rate of change in speed (in knots/h)	50,000	50,000
$\Delta\phi$	Max difference between successive headings (in degrees)	120°	120°
ρ_{max}	Max rate of turn between successive positions (in °/sec)	10	3
γ_{max}	Max rate of climb between successive positions (in feet/sec)	200	–
h_0	Max altitude to consider an aircraft as landed (in feet)	100	–

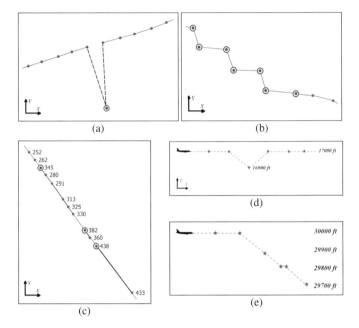

Fig. 4.3 Noise-related situations along trajectories of vessels and aircraft. (**a**) Off-course position. (**b**) Zig-zag distortion. (**c**) Out-of-sequence positions. (**d**) Altitude perturbation. (**e**) Altitude discrepancy

soon relay another fresh, noise-free location, compensating for the removal of any preceding erroneous position(s). Subject to a careful parametrization (Table 4.1), a noisy situation may be identified if at least one of the following conditions apply:

- *Off-course positions* incur an abrupt change both in speed and heading of $\overrightarrow{v}_{cur}$. Figure 4.3a illustrates such an outlier with an abnormal temporary deviation from the known (anticipated) course as abstracted by mean velocity $\overrightarrow{v_m}$ of the object over its previous m positions. In that case, it is most probable that $v_{cur} > v_{max}$, e.g., exceeding a speed limit of $v_{max} = 1000$ knots is not expected by an aircraft.
- Vessels and aircraft normally take a turn smoothly, marked by a series of transmitted locations. So, if the latest update indicates a very abrupt turn (e.g., more than $\Delta\phi \geq 120°$) with respect to its known course, then this message should better be ignored. In case of adverse situations (e.g., a storm) a trajectory may look like a *"zig-zag"* with a series of such abrupt turns as shown in Fig. 4.3b. Dropping those points as noise is not typically correct; yet, in terms of data reduction this is quite desirable, as the object evidently tries to keep to its planned course.
- Sometimes, an object appears to accelerate far too much (over δ_{max} knots/hour), rather unusual for vessels or aircraft. This is typical for *out-of-sequence* messages with twisted timestamps. As shown in Fig. 4.3c, all locations are along the known

course, but due to the late arrival of the three red spots to the base station, the object looks like suddenly retracting backwards at an unrealistic speed.

- Identical locations from the same moving object should qualify as noise. When objects are stationary (e.g., aircraft stationed at a terminal gate), slight agitations in their GPS measurements are acceptable. Instead, coincidental coordinates in succession and only differing slightly on their timestamp values are almost certainly duplicate messages that arrived to multiple base stations.
- A similar problem occurs with conflicts in timestamping, when the same timestamp is assigned to distinct messages from a given object, each reporting different coordinates. To resolve such ties, we arbitrarily choose one of them.
- Upon a sudden surge in the *rate of turn*, e.g., above a given threshold $\rho_{max} \geq 10^o/sec$, the current location should be considered as noise so as to avoid improbable, very sharp turns not expected by vessels or airplanes.
- To filter out *discrepancies in altitude*, we take advantage of the mobility state maintained per aircraft over its latest m positions. Computing a moving average in the rate of climb generally seems to eliminate false indications regarding altitude change, as the one in Fig. 4.3d. These are mainly caused by the lack of precision in altitude measurements usually reported at discrete flight levels (multiples of 100 ft). This rule is also needed to filter out small deviations in altitude not actually qualifying as mobility events, as the one in Fig. 4.3e where the aircraft momentarily appears to be cruising at 29,800 ft while clearly descends.
- Ascents or descents at a rate more than γ_{max} (e.g., 200 ft/s) between two successive locations are erroneous. We discard such temporary steep climbs or drops in altitude caused by false timestamping or imprecise altitude values.
- Of course, an aircraft cannot be stopped in the air, so it is impossible to have a speed $v_{cur} \cong 0$ at an altitude above a threshold h_0 (e.g., 100 ft); up to h_0 an aircraft may be safely considered as landed to the ground.

We experimentally verified that noise may concern a significant portion of received positions, qualifying to one of the aforementioned cases, so data cleaning is a necessary step before any further processing. Most importantly, accepting noisy positions would drastically distort the resulting synopses (as in Fig. 4.3a), and also hamper proper detection of mobility events, as discussed next.

4.5.1.2 Mobility State Maintenance

Position updates $\langle o, p_i, \tau_i \rangle$ pertinent to a given object o are chronologically buffered in memory, and used internally by the Synopses Generator in all computations. So, in order to extract annotations of critical points (to be discussed next in Sect. 4.5.2) the *most recent portion* of its evolving trajectory T_o^ω is available as a distinct sequence of m "clean", time-ordered point locations. We call this trajectory portion the current *mobility state* of object o, which is continuously maintained as a *count-based* sliding window [19] ranging over its m most recent, noise-free locations. Parameter m is a small integer, e.g., $m = 10$, which may depend on the reporting frequency of objects, so it can vary for vessels and aircraft. Anyway, from information available in this state, the mean velocity vector $\vec{v}_{mean}$ of that object can be calculated, as well as several derived spatiotemporal measurements (such as distance, travel time, overall change in heading, rate of climb, etc.). To avoid considering obsolete locations in velocity vector calculations, we also set a maximum time span to ω_{max} for the contents of this window back from current timestamp τ_{cur}.

4.5.2 Trajectory Summarization

The second major component in Synopses Generator concerns a *Trajectory Compressor*, which maintains coherent trajectory synopses online, i.e., summarized, approximate trajectory segments as a *derived, append-only data stream*. Each such synopsis contains detected *critical points* based on mobility features characterizing the observed course of a moving object (e.g., stop, turn, speed change, etc.).

Next, we specify intuitive heuristics that can be used to capture mobility events along vessel and aircraft trajectories. As listed in Table 4.2, some types of events, e.g., stops or changes in heading, are captured on the (x, y) plane. Other events involve computations in the z-dimension, like changes in altitude, while others (communication gaps) rely strictly on temporal features (t-dimension). Depending on the type of the mobility event, either one, two, or multiple critical points may be issued to fully describe it. For example, stops are characterized by their start and end points (to indicate the duration of this event). Instead, when a significant change in heading is observed, then a single critical point is emitted to mark this location as a turning point in the trajectory; and in case of smooth turns, a series of such critical points may be emitted in order to better capture this event.

Table 4.2 Types of mobility events detected along various dimensions in both use cases

	Mobility event	Parameter	Critical point(s)	Use case
(x, y)	Stop	v_{min} (knots)	Start point and end point	Aviation; maritime
	Slow motion	v_σ (knots)	Start point and end point	Aviation; maritime
	Change in heading	$\Delta\theta$ (°)	One or multiple points	Aviation; maritime
	Change of speed	α (%)	Start point and end point	Aviation; maritime
z	Change in altitude	$\Delta\gamma$ (ft/s)	One or multiple points	Aviation
	Takeoff	h_0	Start point	Aviation
	Landing	h_0	Start point	Aviation
t	Communication gap	ΔT (s)	Start point (*notifications*) and End point (*synopses*)	Aviation; maritime

4.5.2.1 Mobility Events on (x, y) Dimensions

Stop indicates a stationary object (e.g., at ports or airports) over a period of time, so actually several nearby positions may be received due to GPS discrepancies. To capture its duration, two critical points should be emitted as shown in Fig. 4.4a:

- *Start of stop* annotates a fresh location once $v_{cur} \leq v_{min}$,

where v_{min} is an appropriately chosen speed threshold for denoting immobility (e.g., 1 knot). Any subsequent locations also validating this condition are not emitted as critical with respect to stop, implying that this phenomenon is still ongoing.

- *End of stop* is identified once $v_{cur} > v_{min}$ and this object was previously stopped (i.e., a *Start of stop* had been issued).

Thus, the stop event and its duration may be inferred by associating this pair of critical points for a given vessel or aircraft.

Slow Motion indicates an object moving at a very low speed (below a suitably chosen threshold v_σ, e.g., 5 knots) over some time. For aircraft, this essentially concerns a subtrajectory on the ground, e.g., while taxiing on runways. Vessels may move at slow speed even at open sea, e.g., boats approaching fishing areas. Provided that sufficient position reports exist, the first and the last positions in this subsequence should be emitted as critical points (Fig. 4.4b) with the following annotations:

- *Start of slow motion* in case that $(v_{cur} \leq v_\sigma) \wedge (v_{prev} > v_\sigma)$.

Subsequent positions also having a speed below v_σ are not marked as critical with respect to slow motion, as long as the vessel or aircraft still continues to move according to the same pattern.

- *End of slow motion* is issued *either* when the object stops (i.e., concurrently with a *Start of stop*) *or* once $(v_{cur} > v_\sigma) \wedge (v_{prev} \leq v_\sigma)$, so it no longer moves slowly.

Change in Heading may be spotted when actual heading has just changed by more than a given *threshold angle* $\Delta\theta$; e.g., if there is a difference of $\Delta\theta > 10°$ w.r.t. the previous heading. In particular, we calculate its current bearing (azimuth) as the angle in degrees (clockwise) between North and the direction of its current velocity vector $\vec{v}_{cur}$. To avoid emitting false indications caused by occasional GPS errors, we propose to detect such changes in heading employing the mean velocity $\vec{v}_m$ from the mobility state of that object. Since angle $\phi = arccos\left(\frac{\vec{v}_{cur} \cdot \vec{v}_m}{\|\vec{v}_{cur}\| \|\vec{v}_m\|}\right)$ indicates the deviation between mean velocity $\vec{v}_m$ and instantaneous bearing $\vec{v}_{cur}$ (from previous p_{prev} to current location p_{cur}), an event regarding

- *Change in Heading* is detected once it holds that $\phi \geq \Delta\theta$.

Although such a change gets actually detected at the current location p_{cur}, the *previous* one p_{prev} must be annotated as critical point; this offers a more precise

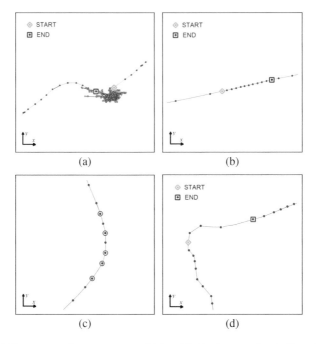

Fig. 4.4 Mobility events (of one, two, or multiple critical points) on (x, y) dimensions detected along trajectories of vessels or aircraft. (**a**) Stop. (**b**) Slow motion. (**c**) Changes in heading. (**d**) Change of speed

trajectory approximation, as at p_{cur} the object already moves along a changed course.

In addition, we also need to check whether the *cumulative* change of instantaneous headings across a few previous positions exceeds threshold $\Delta\theta$ (as we applied in [21] for vessels only). Indeed, due to their large size and aviation regulations, aircraft and vessels usually make smooth turns or manoeuvres, e.g., awaiting permission for landing (a holding situation) or when docking to the port. As illustrated in Fig. 4.4c, a series of such critical points need be issued to achieve a more reliable piecewise approximation of the actual motion path. Of course, the lesser the angle threshold $\Delta\theta$, the more the critical points denoting such slight changes in heading. The last point in the series should be the one at which no significant change in heading is observed anymore, i.e., from now on the object is about to follow a straight course.

Note that a change in heading for aircraft may be also combined with ascent or descent along the z-dimension (as discussed next), e.g., during manoeuvres before landing or at holding patterns.

Change of Speed occurs once current speed v_{cur} deviates by more than $\alpha\%$ from mean speed v_m. Given a parameter α, which quantifies a cutoff threshold for ignoring occasional perturbations in speed estimations due to GPS error, wind drift, etc., a pair of critical points (Fig. 4.4d) should be annotated:

- *Start of speed change* is marked once it holds $|\frac{v_{cur}-v_m}{v_m}| > \alpha$.

Subsequent positions where this condition still holds (i.e., the object keeps speeding up or slowing down) need not emitted as critical with respect to speed change, as long as the object is considered to continue moving by the same pattern.

- *End of speed change* occurs once $|\frac{v_{cur}-v_m}{v_m}| \leq \alpha$, *and also* a critical point was earlier annotated as *Start of speed change*.

This end point indicates that the object's speed practically stabilizes close to the mean v_m based on its mobility state. Speed change may be further distinguished into *acceleration* or *deceleration*, effectively by taking the sign (+/−) of the fraction (and not just the absolute value) in the aforementioned rules.

Ideally, this pair of critical points should be matched (i.e., a starting point followed by one marking the end of this event) so as to characterize a subtrajectory where the object changes its speed significantly. However, often other events suddenly occur (e.g., communication gap) and disrupt the sequence, so the latter point may be missing from the resulting synopsis.

4.5.2.2 Mobility Events on z-Dimension

This set of mobility events concern aircraft trajectories only, since they are based on their reported altitude measurements. More specifically:

Change in Altitude may be detected once there is a significant change at the altitude, i.e., points of transition to a different cruise flight level, as depicted in Fig. 4.5a. Essentially, this aims to capture positions with significant change at the rate of climb (or descent) of the aircraft. Due to occasional spurious jumps in reported altitudes as shown in Fig. 4.3d, we avoid checking directly against altitude values.

Instead, we employ the *rate of climb* (or *rate of descent*) γ, which is the vertical speed of an aircraft (in feet/sec) between its previous and current location when ascending (respectively, descending):

$$\gamma = \frac{p_{cur}.z - p_{prev}.z}{p_{cur}.\tau - p_{prev}.\tau}.$$

Ideally, when $\gamma > 0$ the aircraft is ascending, and with $\gamma < 0$ it is descending; once $\gamma = 0$, then the aircraft flies at a cruise level. However, small fluctuations in altitude (especially when measured in hundreds of feet) may erroneously infer ascent or descent (Fig. 4.3d). Our goal is to identify positions where changes in this rate start or cease to take place. On first sight, by setting a *threshold* $\Delta\gamma$ for

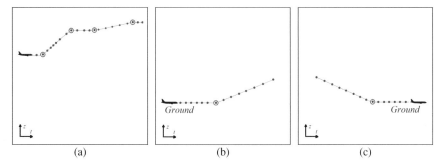

Fig. 4.5 Mobility events on the z-dimension detected along aircraft trajectories. (**a**) Changes in altitude. (**b**) Takeoff. (**c**) Landing

significant changes in this rate (e.g., more than 20 ft/s), a critical point should be issued once it is found that $|\gamma_{cur}| > \Delta\gamma$, where γ_{cur} expresses the rate of climb (or descent) of the aircraft at timestamp $p_{cur}.\tau$. As with changes in heading, note that actually the *previous* location p_{prev} must be annotated as critical, because at current location p_{cur} the aircraft is already on its ascent (or descent). In addition, we should avoid characterizing as critical all locations along such a climb (or descent), as long as there is no significant change in the previously calculated rate γ_{prev}. Moreover, to avoid emitting potentially invalid critical points caused by numeric imprecision in timestamp and altitude values, we may conservatively detect such changes by also checking with the *mean rate of climb* (or *descent*) $|\gamma_{mean}|$ computed over the m most recently reported locations of the aircraft available in its state. Therefore, we annotate the previous location p_{prev} (with the same rationale as for changes in heading) as

- *Change in Altitude* if $(|\gamma_{cur}| > \Delta\gamma) \wedge (|\gamma_{prev}| \leq \Delta\gamma) \wedge (|\gamma_{cur} - \gamma_{mean}| > \Delta\gamma)$.

A series of critical points may be annotated to mark a smooth ascent (or descent), as this can yield a more reliable trajectory approximation along the z-dimension.

Takeoff concerns the last location reporting that the aircraft was still on the ground. ADS-B messages carry a Boolean flag indicating whether ground squat switch[7] is active, i.e., whether the aircraft is *grounded*, although this flag is usually null in real datasets. But, since tracklogs from secondary radars do not include such a flag at all, checking with absolute altitude values is the only option for possibly detecting such an event. To overcome discrepancies in measurements, altitude values should be compared against a threshold h_0 indicating the highest altitude at which an aircraft may be considered as landed to the ground (e.g., $h_0 = 100$ ft). So,

- *Takeoff* is issued if $(!p_{cur}.grounded \wedge p_{prev}.grounded) \vee ((p_{cur}.z > h_0) \wedge (p_{prev}.z \leq h_0))$.

---

[7] A sensor that identifies if the weight of the aircraft is resting on its gear.

Note that the previous location p_{prev} gets annotated as takeoff so as to provide a better trajectory approximation, as illustrated in Fig. 4.5b.

Landing   concerns the location at which an aircraft touches the ground (Fig. 4.5c). With a rationale similar to takeoffs, we check against threshold h_0, and also inspect ADS-B-specific flag *grounded*, so current location p_{cur} is annotated as

- *Landing* in case that $(p_{cur}.grounded \land !p_{prev}.grounded) \lor ((p_{cur}.z \leq h_0) \land (p_{prev}.z > h_0))$.

4.5.2.3   Mobility Events on t-Dimension

Communication Gaps   occur often in long-haul flights or vessel itineraries across oceans when contact with the base stations is lost. As illustrated in Fig. 4.6, when no message is received from the moving object for at least a *time interval* ΔT, e.g., over the past 10 min, its actual course is unknown. Such communication gaps might be also caused by accidents or other suspicious situations (e.g., hijacks). Obviously, this event is concluded when contact is restored, sometimes after a period much longer than ΔT. To maintain *trajectory synopses* in append-only fashion at operational latency, current location p_{cur} is promptly annotated as

- *End of communication gap* if $(p_{cur}.\tau - p_{prev}.\tau > \Delta T)$.

This rule only depends on the elapsed time between current p_{cur} and previously known location p_{prev} of a given aircraft, hence is based on time measurements only.

Still, a *notification* may be issued concerning the location at which contact was lost, i.e., the previous position p_{prev} is annotated as

- *Start of communication gap* once $(\tau_{now} - p_{prev}.\tau > \Delta T)$.

In contrast with other types of critical points, we stress that these latter notifications *cannot* be issued online, but with at least ΔT time units delay from system time τ_{now}. To avoid disrupting temporal ordering of critical points in trajectory synopses, notifications are derived as a *separate stream* (as shown in Fig. 4.2).

4.5.2.4   Discussion

It is worth explaining the difference in semantics amongst three types of critical points (*stop, slow motion, change in speed*) that are all based on speed measurements. Clearly, *stop* aims to identify when an object remains stationary, i.e., its speed is practically zero. During this period of immobility, checking for the other two types of events is switched off in the Synopses Generator. Next, *slow motion* is particularly designed to capture movements along a path at consistently very low speed (below a threshold v_θ, e.g., 5 knots). Such an event may follow or precede a stop (e.g., a vessel leaving or entering a port), as well as slow movements along a

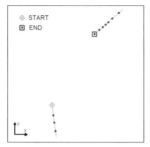

Fig. 4.6 Mobility events on the *t*-dimension marking start/end of a communication gap with a moving vessel or aircraft

path (e.g., a boat in a fishing area), as illustrated in Fig. 4.4b. Finally, a *change in speed* is intended to find cases where an object suddenly accelerates or decelerates while on the move, before its speed stabilizes at another (higher or lower) value after a while. So, this latter type of event captures such "spikes" of speedup or slowdown in order to distinguish subtrajectories at different speed levels, i.e., before the start and after the end critical point of such a change in speed, respectively. It cannot be ruled out that the same location may be marked as both an endpoint of change in speed and of slow motion, e.g., a vessel was slowing down and now it moves at a very low speed. So, events may be consecutive in time, but occasionally there may be overlaps (excluding stops). We intentionally avoid a "crisp" separation of such events in order to extract as much information about mobility features in the resulting trajectory synopses.

4.6   Prototype Implementation

We implemented a software prototype[8] for our application framework in Scala using the DataStream API of *Apache Flink* with *Apache Kafka* as broker for streaming messages. Flink is an open-source stream engine able to execute complex dataflow applications in a data-parallel, pipelined manner. Flink adopts a *tuple-at-a-time* model for its stream operators, offering results at low latency and achieving higher throughput [3, 6] compared to the *micro-batch* approach in Spark Streaming. Flink also supports *exactly-once* semantics for fault tolerance, guaranteeing that each tuple will be processed to ensure correct results. This is particularly important for our application, as losing tuples may completely alter spatiotemporal estimates (e.g., distance, speed) and distort identification of events.

---

[8]Source code available at https://github.com/DataStories-UniPi/Trajectory-Synopses-Generator.

4.6.1 Custom Support for Mobility Features

Flink currently lacks native support for *spatial entities* (e.g., points, polygons), let alone representation of *spatiotemporal* data and handling of mobility features like speed, heading, acceleration, etc. Thus, we had to introduce custom data structures in Scala for maintaining trajectories and support all mobility operations required in our processing flow (Sect. 4.5). We also specified an extensible *attribute schema* in Avro[9] with all spatiotemporal properties involved in aircraft trajectories. Further, we defined specific *topics* in Kafka, for handling messages from input surveillance sources and the derived streams (Fig. 4.2).

Flink offers powerful *partitioning* features for its operators (*Map, Reduce, Filter, Window,* etc.). An operator can have a "state" organized as list (typically, a distributed key-value map) scoped locally to the operator's task. Flink automatically distributes streaming items by a key (in our case, a unique identifier per vessel or aircraft) into parallel instances *of each operator*, each maintaining its own local state. In our case, this feature caused more trouble than benefit, exactly because it is established at operator level. Indeed, our processing strategy needs to maintain a mobility state *per object*, i.e., its most recent trajectory segments. Once each fresh position is admitted, it must be processed by a pipeline of operators (Fig. 4.7) that actually encapsulate our custom methods. Therefore, we absolutely need to keep a *separate persistent state per vessel or aircraft* not per operator, but across all operators in the entire pipeline. As Flink did not provide any flexibility to maintain a globally accessible state, we resorted to implementing a custom hashing scheme in Scala. This is based on vessel or aircraft identifiers (*key*) and keeps a small number m of recent locations per object (its *mobility state*), as well as certain flags that immediately inform on its current status (e.g., is stopped, in slow motion, has changed speed). Of course, this incurs some overhead (once a fresh location arrives, the state must be updated) and housekeeping (discard obsolete items in a count-based window fashion [19]), but otherwise trajectory-aware processing would not have been possible.

Most importantly for real-world applications, Flink provides support for *event time* semantics, i.e., using timestamps generated by the data sources. In our case, timestamp values in relayed positions are assigned upon reception to a base station or when captured by radars, hence a vessel or aircraft may momentarily appear go back and forth along its course due to *delayed messages*. In real-world aviation and maritime surveillance streams, temporal disorder caused by such delayed messages is not the exception, but rather the norm. Note that disorder [9] may occur due to time skew between input streams, induced by operators (e.g., joins), etc. However, establishing order amongst streaming data items is necessary for their effective and consistent processing. To deal with this inherent disorder, we specified a *lag* value,

---

[9]https://avro.apache.org/docs/current/.

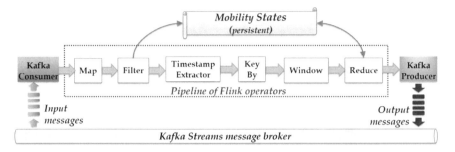

Fig. 4.7 Pipeline of Flink and Kafka operators in the prototype implementation

so the pipeline will accept messages delayed by at most *lag* (say, 5) seconds behind the latest known timestamp and accordingly reorder messages chronologically.

4.6.2 Processing Through a Pipeline of Operators

Figure 4.7 illustrates the pipeline of Kafka and Flink operators utilized in our prototype implementation. More specifically:

Kafka Consumer To simulate streaming input into our application, we have built data feeders in Scala that can parse historical records of vessel or aircraft traces and push positional messages to a specified Kafka topic. Records are replayed either according to their *original* arrival rate or consumed by a user-specified *fixed* stream rate (e.g., 10,000 messages/s). A custom consumer subscribes to a particular Kafka topic and starts receiving timestamped stream messages. It is very important that those messages have timestamps, although temporal order is not a strict prerequisite in the input data, as our prototype can extract timestamps and rearrange items in chronological order even in the case of lagged messages, as explained above.

Map Each incoming message is mapped to the Avro attribute schema. From now on, this message will travel through the operator pipeline as a tuple; annotations may be assigned next, after further checking by subsequent operators.

Filter This discards noisy locations, which are not further considered. In case that a newly arrived location concerns an object currently not yet known to the system, its *mobility state* is initialized; otherwise, this location is appended to the existing state evicting the oldest location therein.

Timestamp Extractor We employ *time watermarks* in Flink that allow processing of delayed stream items according to an application-specified *lag* parameter.

Keyed Windowing Flink offers several types of *windows*, which split a stream into temporary finite "pieces" used in computations. We employed *count-based sliding windows* with a size set to two tuples *keyed by* object identifier, in order to calculate instantaneous spatiotemporal estimates (distance, heading, altitude change, etc.) between two consecutive locations. Thus, spatiotemporal computation is streamlined with the arrival of each fresh message and the resulting estimates can be assigned to the tuple as it is being processed by all subsequent operators in the pipeline.

Reduce The core of our application logic is encapsulated in a series of custom functions under the generic Reduce operator in Flink. Essentially, this Reduce operator takes a pair of consecutive locations (p_{prev}, p_{cur}) and triggers evaluation of the detection rules in Sect. 4.5.2. If a location qualifies as a critical point for an event, the respective bit in its annotation is set; eventually, multiple bits may be set.

Kafka Producers Output is pushed into distinct Kafka topics by three producers, each dealing with a derived stream (*noise-free raw locations, trajectory synopses, notifications* as explained in Sect. 4.4.2). Functionality in each producer is similar, but they differ in the filtering applied to the streaming output of the Reduce operator.

4.7 Empirical Validation

Synopses Generator has been tested extensively against real and synthetic datasets with diverse characteristics in the maritime and aviation use cases. These experiments confirmed its advanced performance in terms of timeliness, scalability, and compression efficiency, as well as the quality of approximation in the resulting synopses. Next, we report some indicative results from this comprehensive validation.

Maritime Dataset (*AIS*) This dataset[10] contains AIS messages from 5055 vessels sailing in the Atlantic Ocean around the port of Brest, Brittany, France, and spans a period from 1 October 2015 to 31 March 2016. After deduplication of the original AIS messages, this dataset yielded 18,495,677 point locations (kinematic AIS messages only), which were used as input at their original arrival rate.

Aviation Dataset (*ADS-B*) This data[11] comes from ADS-B messages collected via the FlightAware API during one full day (10 March 2017). After parsing the original messages in JSON format and keeping position reports only, this dataset yielded 3,701,966 raw point locations from 24,416 airplanes flying over Europe. These positions were consumed in the simulations at their original arrival rate; on average, there is a message every 67.83 s from any given aircraft.

Online detection and summarization of trajectories is very sensitive to proper parametrization. Parameter values listed in Table 4.1 regarding *noise reduction* and

---

[10]Available at https://zenodo.org/record/1167595#.WxVrJy97HOR.

[11]Available at http://chorochronos.datastories.org/?q=content/flightaware-ads-b-march-2017.

Table 4.3 Parameter settings in the Synopses Generator for annotating critical points

Symbol	Description	AIS	ADS-B
v_{min}	Minimum speed (in knots) for asserting movement	0.5	1
v_σ	Maximum speed (in knots) for asserting slow motion	5	5
α	Minimum rate for asserting speed change between locations	25%	10%
ΔT	Minimum time interval (in minutes) for asserting communication gaps	30	2
$\Delta\theta$	Threshold for asserting change in heading (in degrees)	5°	5°
$\Delta\gamma$	Threshold for asserting changes on rate of climb (feet/sec)	N/A	20
m	Positions in each object's mobility state	5	5

Table 4.4 Timeliness of Synopses Generator against real datasets

Dataset	Throughput (messages/s)	Latency (s)
AIS (*maritime*)	16,874.5	0.238
ADS-B (*aviation*)	8470.5	0.321

Table 4.3 for *annotating critical points* represent typical settings deduced after consistent data exploration and consultation with domain experts in both use cases.

4.7.1 Performance Results

First, we examine *timeliness* of the Synopses Generator simulating a streaming operation by accepting data at the original arrival rate. Concerning *throughput*, i.e., the amount of input messages that may be processed per second concerning the entire operator pipeline (Fig. 4.7), Table 4.4 confirms that the framework can cope with thousands of streaming positions per second. Aviation data also involves computations on the z-dimension, so performance is not that high as in the maritime use case. Regarding the *average latency* of incoming messages, i.e., how long each one remains in the operator pipeline, results show that the Synopses Generator achieves *operational* latency, as it can provide results in near real time (less than a second), keeping up with the arrival rate of the incoming surveillance stream. Note that these performance results were obtained in tests against a *centralized* instance (i.e., single thread) of the framework without any data partitioning. Scalability tests (detailed in [22] for the aviation use case) with parallelized execution on multiple threads or in cluster infrastructure indicate even more advanced performance gains.

As every data reduction process, effectiveness of trajectory summarization is a trade-off between compression efficiency and approximation accuracy. Hence, regarding maintenance of *trajectory synopses*, we measured *compression ratio* λ as

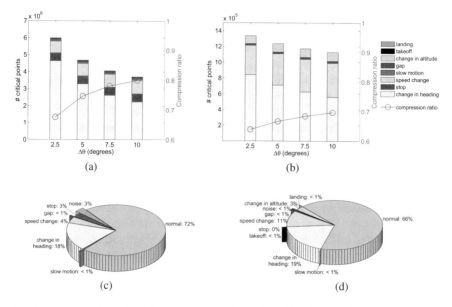

Fig. 4.8 Compression ratio with a varying angle threshold and breakdown of annotated positions. (**a**) Compression ratio (*AIS*). (**b**) Compression ratio (*ADS-B*). (**c**) Annotated positions (*AIS*). (**d**) Annotated positions (*ADS-B*)

the percentage (%) of positions dropped from the approximate trajectory synopses over the raw ones originally obtained, or equivalently:

$$\lambda = 1 - \frac{\#critical\ points}{\#raw\ positions}.$$

The higher this ratio, the more compressed and lightweight the resulting synopses. A compression ratio λ closer to 1 signifies stronger data reduction, as the vast majority of original locations are dropped and few critical points suffice to approximately represent the entire trajectories.

Next, we show how compression ratio varies with respect to angle threshold $\Delta\theta$ used for detecting changes in heading; the pattern is similar with other parameters. With a lower $\Delta\theta$, even slight deviations in direction can be spotted, and thus extra critical points get issued. Bar charts in Fig. 4.8a, b illustrate the breakdown of critical points in each class retained from either dataset. Clearly, every further increase in threshold $\Delta\theta$ suppresses more and more turning points and does not affect the share of any other class of critical points, but incurs extra reduction in the total amount of emitted critical points. Hence, relaxing this parameter value leads to a stronger compression. Compression ratio always remains above 70% in the maritime use case (*AIS*), and with a more relaxed $\Delta\theta$ it reaches as much as 80%, meaning that only 20% of the original positions are retained in the synopses. Eliminating noise also plays an important role in data reduction, as erroneous points need not be retained. Similar conclusions regarding compression efficiency can be drawn from tests against the aviation dataset (*ADS-B*). Compression ratio never exceeds 70%, because the available ADS-B messages have a low reporting frequency, so an incoming position may indicate an important change in mobility and thus has more chances to qualify as critical. Note that most of the annotated critical points in Fig. 4.8b concern changes in heading, speed, and altitude.

Figure 4.8c, d depict an overall account of annotations assigned to all raw locations obtained from each dataset. Clearly, the Synopses Generator eliminates more than two thirds of the incoming positions as redundant, while retaining the rest in the trajectory synopses. The suppressed points are either "normal" locations or noise (which anyway must be discarded). Notice that there seems to be some noise in the maritime dataset (*AIS*), but it is only marginally present in aviation ADS-B messages. However, in tests over other datasets, we have identified that sometimes even 20% of the raw positions may be noisy. Here, it is also notable the almost complete lack of critical points concerning movement of aircraft on or close to the ground (i.e., stop, slow motion, takeoff, landing), as discussed in Sect. 4.3.

4.7.2 Approximation Error

Preserving only critical points in trajectory synopses incurs a lossy approximation. To assess the quality of such compressed trajectories, we estimated their deviation from original ones, i.e., without discarding any raw positions (except for those qualified as noise). Such deviation on (x, y) plane (concerning both vessels and aircraft) can be assessed by computing the pairwise Haversine distance H between *synchronized* locations from an original trajectory and its synopsis (consisting only of critical points). If an original location p_i at timestamp τ_i is not critical, then its corresponding time-aligned p_i' in the synopsis is estimated via linear interpolation along the path that connects the two critical points before and after τ_i. A similar assessment is made regarding deviation on the z-dimension for aircraft trajectories. For each original altitude z_i reported at timestamp τ_i from an aircraft, we estimate via linear interpolation a time-aligned altitude z_i' in its trajectory synopsis.

Overall, for each object that reported M raw positions in total, we assessed the *Root Mean Square Error (RMSE)* between the original and synchronized sequences of its locations on the (x, y) plane and over its respective altitudes in the z-dimension:

$$RMSE_{xy} = \sqrt{\frac{1}{M} \cdot \sum_{i=1}^{M}(H(p_i, p_i'))^2}, \quad \text{and} \quad RMSE_z = \sqrt{\frac{1}{M} \cdot \sum_{i=1}^{M}(z_i - z_i')^2}.$$

For each trajectory, $RMSE_{xy}$ is in metres, while $RMSE_z$ is in feet. Table 4.5(a) displays the *average RMSE* over the entire fleet for maritime dataset by varying angle threshold $\Delta\theta$ that is used to recognize significant changes in heading.

The approximation error is almost negligible, mostly due to the increased update frequency in most vessels, but also because movement is constrained in a rather small radius around the port of Brest. Even in the worst case examined with $\Delta\theta = 10°$, average RMSE is only 35 m, which provably indicates the quality of synopses derived from this dataset. In practice, a moderate angle threshold of $5°$ seems adequate for balancing compression efficiency without losing important details in vessel mobility.

However, approximation error is much higher for aircraft trajectory synopses as detailed in Table 4.5(b). This error is not because the approximate spatial path based on critical points deviates significantly from the original. Instead, such an error is caused by the fact that speed of aircraft in the air is orders of magnitude higher compared to that of vessels. Combined with the inherent flaws in timestamp precision and reporting frequency of the original data, this yields a position estimate p' during computation of $RMSE_{xy}$ that deviates a lot from the respective original location p at timestamp τ. Both p and p' are spatially along an almost common path shared between the original and the compressed trajectory, but p' is either ahead or behind p because of speed discrepancies. We deem that this phenomenon could have been alleviated if original data were more precise and had increased arrival rate. Anyway, this error is always less than a nautical mile ($\cong 1852$ m) and certainly can ensure horizontal separation of flights, given that international regulations usually specify a separation of at least 3 miles as a safety precaution.

Quality of the synopses with respect to altitude is far better, as testified in Table 4.5(c). Given that the granularity of reported altitudes is expressed in hundreds of feet, achieving an average $RMSE_z$ up to a few hundred feet is certainly tolerable given the lack of precision in spatiotemporal information as discussed in Sect. 4.3. As international aviation standards specify a vertical separation between commercial flights at 1000 ft, this error magnitude on altitudes is more than acceptable.

4.7.3 Comparison with Trajectory Simplification Methods

Furthermore, we performed a short qualitative comparison regarding the effects of simplification over vessel and aircraft trajectories in terms of compression efficiency and approximation quality. We examined several state-of-the-art algorithms for

Table 4.5 Approximation error of the trajectory synopses on (x, y) and z dimensions

(a) AIS (x, y)		(b) ADS-B (x, y)		(c) ADS-B (z)	
$\Delta\theta$ (°)	RMSE (m)	$\Delta\theta$ (°)	RMSE (m)	$\Delta\gamma$ (feet/sec)	RMSE (in feet)
2.5	6.5	2.5	1056.1	10	288.5
5	16.3	5	1237.5	20	395.4
7.5	26.1	7.5	1449.9	30	533.5
10	35.2	10	1691.0		

trajectory simplification [30] working in *online* mode,[12] but concerning strictly locations on the (x, y) plane (hence, no z-dimensional data is involved). In contrast, our Synopses Generator framework not only identifies which positions matter in terms of simplification but it also labels these points as critical with specific annotations (stop, change in heading, speed, etc.). Since a comprehensive evaluation is beyond the scope of this chapter, we stress that parametrization of each simplification method was not fine-tuned, hence these results can only be seen as indicative.[13]

As detailed in Table 4.6, all methods generally drop many original locations in order to yield a simplified trajectory, although at a different compression ratio depending on the specifications of each algorithm and its objectives. Therefore, they also differ in the resulting approximation quality (i.e., RMSE error). In particular, FBQS [11] is more balanced and yields trajectory approximations with good quality while also drastically reducing the amount of retained locations. SQUISH-E [16] offers supreme quality with minimal error in the resulting simplified trajectories. Yet, it has to retain too many locations (more than 96% of the original *ADS-B* messages in this test) to achieve such accuracy, hence its compression efficiency inevitably worsens. OPERB [10] seems competitive to FBQS in terms of quality and compression, but it apparently needs to keep more original locations to better approximate each trajectory, especially for the aviation dataset, hence its reduced error. STTrace [23] offers high-quality synopses against the maritime dataset when its target compression ratio is set to $\lambda = 80\%$, given that it intends to keep locations that incur minimal error and preserve the shape of the original trajectory. However, quality is much worse against the aviation dataset, most probably because the lack of precision in timestamp values negatively affects good velocity estimations. Compared to these state-of-the-art simplification algorithms, the proposed Synopses Generator framework offers a good balance between approximation quality and compression efficiency. Especially in the maritime use case, proper fine-tuning of the framework parameters has certainly paid off, as it achieves the least approximation error, while also it discards almost 75% of the raw positions. However, it is less successful over the aviation data in terms of approximation quality given the low reporting frequency in the original ADS-B messages, but still it manages to keep only a third of the raw positions.

Overall, these tests indicate that the suggested Synopses Generator can provide quite acceptable accuracy and can annotate online most, if not all, critical changes along each object's course, whereas it also remains competitive to other trajectory simplification techniques.

---

[12]Source code handling moving objects with locations on (x, y) dimensions is available at https://github.com/uestc-db/traj-compression.

[13]Performance efficiency is not examined here, since algorithms are implemented in different languages and frameworks, so a fair comparison would not be possible in terms of execution times.

Table 4.6 Compression efficiency of trajectory simplification techniques on (x, y) dimensions

| | Avg. compression ratio (λ) | | Avg. RMSE (m) | |
Algorithm	AIS	ADS-B	AIS	ADS-B
FBQS [11]	**0.900**	0.521	187.5	373.8
SQUISH-E [16]	0.653	0.035	21.3	**18.4**
OPERB [10]	0.895	0.438	206.4	323.3
STTrace [23]	0.800	**0.803**	31.6	1417.9
Synopses generator	0.746	0.667	**16.3**	1237.5

Values in bold indicate the most efficient algorithm with respect to compression ratio or RMSE

4.8 Towards Cross-Stream Trajectory Maintenance

Next, we introduce our solution for processing surveillance data in order to reconcile and align parts of trajectories from different sources into a unified representation.

4.8.1 Methodology

We apply cross-stream processing to "fill-in" gaps in a trajectory representation from a single source with trajectory segments available from another data source regarding the same moving object. Such correlation of geostreaming data from two sources is particularly challenging, as both streams are of different quality, evolve at diverse and time-varying update frequencies, and present considerable geospatial deviation.

Consider two sources S_1 and S_2 collecting data streams of the same moving objects at different frequencies. These streams pass through the Synopses Generator in order to detect critical points in real time, as described in Sect. 4.4.2. The objective of the cross-stream algorithm is to fill-in the missing information of S_1 by using positions from S_2 concerning a specific object o. The algorithm is invoked once a *communication gap* is identified in S_1, at timestamp τ_i (Fig. 4.6). Then, the method looks at S_2 to identify any appropriate positions that can "fill-in" the missing ones in S_1. Once possible replacements are found, we calculate the spatiotemporal deviation (error) between the "available past points" of the two sources. The "available past points" refer to the locations held in the mobility states of that particular object o in each of the two sources, since these are buffered by the Synopses Generator; Synopses Generator makes use of a small buffer per aircraft holding the latest measurements. If the buffered locations for S_2 include critical points concerning *end of communication gap* or *end of stop* or *landing*, then the "available past points" are those recorded after the detection of these critical points and before τ_i. The "available past points" of the two sources are being used for calculating the spatiotemporal deviation between the two sources. We are able to calculate the

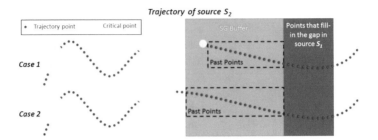

Fig. 4.9 Identifying replacement locations for filling the gaps along a trajectory

deviation of the points between the two sources at each timestep of S_1 by applying an 1-D linear interpolation formula to the "available past points" of S_2. Thus, the produced deviation errors are being used to calculate the mean and median errors. Then, the calculated deviation (either median or mean error) is added to each one of the locations of S_2 that are suitable to "fill-in" the communication gap in S_1. Finally, the "fill-in" stage inserts the calculated points into o's trajectory from source S_1. This procedure results in a "filled-in" S'_1. Figure 4.9 illustrates two trajectories of the same source S_2, which correspond to the above-mentioned possible scenarios of identifying the "available past points". Note that, the past locations, i.e., the positions prior to the identified *communication gap* of S'_1, are collected according to the proposed algorithm. In the first case, the past points include less than those buffered in the mobility state, because of a critical point. In the second case, the past locations include all those currently available in the mobility state.

4.8.2 Proof-of-Concept Evaluation

A challenging case of cross-stream processing would be the correlation of sources of ADS-B messages collected by different providers: (a) FlightAware and (b) ADSBHub. Since ADSBHub covers only Europe and reports positions every second, but with many periods of missing data, there is an urgent need of recovering the missing trajectory information from other sources. In contrast, FlightAware lacks sufficient information only around airports, while the reporting frequency varies across Europe.

We conducted a proof-of-concept evaluation of our cross-stream processing method against data from those two sources regarding all flights of 369 aircrafts that had landed or taken off from Amsterdam on 6 April 2016. For this evaluation, we created artificial gaps in the ADSBHub data, by randomly deleting batches of consecutive positions. The deleted locations are named *real points*. Subsequently, the proposed method was applied to the incomplete ADSBHub, as described in Sect. 4.8.1, i.e., S_1, S_2, and S'_1 stand for the incomplete ADSBHub, the FlightAware, and the "filled-in" ADSBHub, respectively. Finally, the *RMSE* was calculated by

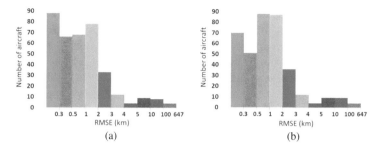

Fig. 4.10 Errors between artificial gaps in ADSBHub "filled-in" with locations from FlightAware. (**a**) Median. (**b**) Mean

using the aforementioned *real points* in conjunction with the "filled-in" points of S'_1.

Figure 4.10 depicts the results generated by the aforementioned evaluation procedure. Considering the speed of an aircraft and the inherent errors in geopositioning and timestamping, a deviation about 300–500 m is quite acceptable. The percentage of aircrafts with RMSE below 500 m is about 60% and 56%, for the median and mean error in calculated deviations, respectively. In both cases, the error is less than 3000 m for 90% of the aircraft. Thus, it can be inferred that the proposed technique advances the real-time integration of data from different sources and allows the successful reconstruction of trajectories.

4.9   Conclusions

In this chapter, we considered concise representations of trajectories of vessels (in the maritime use case) or aircraft (in the aviation use case). We introduced the Synopses Generator framework that can detect and summarize trajectories from a large set of such moving objects. Input is arriving online from surveillance data streams concerning position reports: terrestrial or satellite AIS messages from vessels; ADS-B messages from aircraft or traces from ATC secondary radars in aviation. This methodology is based on a simple idea, yet very effective in practice: instead of retaining every incoming position, we propose to drop any positions that can be predicted with minimal loss in accuracy, and only keep those that may indicate stops, changes in speed or heading or altitude, etc. We have implemented a prototype system that is able: (a) to construct trajectories from input locations in real time and then (b) incrementally maintain trajectory synopses that include only salient mobility features from the incoming surveillance data. We have empirically tested the functionality of this prototype against diverse surveillance data streams in both aviation and maritime domains. Our results confirm that, under careful parametrization and deep knowledge of data characteristics, this software addresses many challenging objectives in a geostreaming context.

Indeed, trajectory detection and summarization is carried out in *online* fashion, thus offering incremental, timely results at *operational latency*. On average, a point can be emitted in the synopses and properly annotated as critical mostly within milliseconds since admission of the corresponding raw message in the input stream.

The framework achieves a particularly *high throughput*, as it is able to consume thousands of messages per second from a large set of moving vessels or aircraft, each streaming its own mobility data over large geographical areas and across varying periods of time. This process is also parallelizable, offering reduced latency in mobility detection and higher throughput and has been confirmed to cope with tens of thousands of incoming messages per second. This confirms the scalability and robustness of this mechanism.

The Synopses Generator can achieve dramatic *compression* over the raw streaming data, effectively digesting massive amounts of position reports into lightweight, annotated trajectory synopses. In case of very frequent position reports, its compression ratio can exceed 95% without harming the quality of trajectory representation. At lower arrival rates in the input as is the case in most real data sources, data reduction is more conservative, but still quite large (around 80% with respect to the input data volumes), as it must not result into distorted compressed representations and lose important mobility features in the data.

Furthermore, compressed representations of the trajectories have more than *tolerable approximation error* with generally small deviations from original traces, also coping with imperfections (such as network delays, noise) inherent in real-world surveillance streams. Even with a first-cut, empirical parametrization, information loss due to summarization is minimized both in terms of location accuracy (in both use cases) as well as with respect to altitude (in aviation), indicating a high spatiotemporal quality in the resulting synopses.

Finally, we have dealt with the issue of *cross-stream* processing by correlating surveillance data from multiple sources in order to "fill-in" trajectories with missing points due to communication gaps and thus provide a coherent trajectory representation. A seamless combination of summarized trajectory segments derived from different input sources is even more demanding to accomplish in an incremental fashion from sparse, time-varying, noisy, and sometimes overlapping motion paths, which are being collected from different data sources in real time. However, the proposed method was tested against real datasets and the results turned out to be very promising in terms of quality.

References

1. Ayhan, S., Samet, H.: Aircraft trajectory prediction made easy with predictive analytics. In: KDD, pp. 21–30 (2016)
2. Cao, H., Wolfson, O., Trajcevski, G.: Spatio-temporal data reduction with deterministic error bounds. VLDB J. **15**(3), 211–228 (2006)
3. Carbone, P., Katsifodimos, A., Ewen, S., Markl, V., Haridi, S., Tzoumas, K.: Apache Flink: stream and batch processing in a single engine. IEEE Data Eng. Bull. **38**, 28–38 (2015)

4. Ding, X., Chen, L., Gao, Y., Jensen, C.S., Bao, H.: UlTraMan: a unified platform for big trajectory data management and analytics. Proc. VLDB Endowment **11**(7), 787–799 (2018)
5. Eldawy, A., Mokbel, M.F.: SpatialHadoop: a MapReduce framework for spatial data. In: ICDE, pp. 1352–1363 (2015)
6. Hagedorn, S., Götze, P., Sattler, K.-U.: Big spatial data processing frameworks: feature and performance evaluation. In: EDBT, pp. 490–493 (2017)
7. Kipf, A., Pandey, V., Böttcher, J., Braun, L., Neumann, T., Kemper, A.: Analytics on fast data: main-memory database systems versus modern streaming systems. In: EDBT, pp. 49–60 (2017)
8. Lange, R., Dürr, F., Rothermel, K.: Efficient real-time trajectory tracking. VLDB J. **20**(5), 671–694 (2011)
9. Li, J., Tufte, K., Shkapenyuk, V., Papadimos, V., Johnson, T., Maier, D.: Out-of-order processing: a new architecture for high-performance stream systems. Proc. VLDB Endowment **1**(1), 274–288 (2008)
10. Lin, X., Ma, S., Zhang, H., Wo, T., Huai, J.: One-pass error bounded trajectory simplification. Proc. VLDB Endowment **10**(7), 841–852 (2017)
11. Liu, J., Zhao, K., Sommer, P., Shang, S., Kusy, B., Jurdak, R.: Bounded quadrant system: error-bounded trajectory compression on the go. In: ICDE, pp. 987–998 (2015)
12. Long, C., Chi-Wing Wong, R., Jagadish, H.V.: Trajectory simplification: on minimizing the direction-based error. Proc. VLDB Endowment **8**(1), 49–60 (2014)
13. Maier, D., Li, J., Tucker, P., Tufte, K., Papadimos, V.: Semantics of data streams and operators. In: ICDT, pp. 37–52 (2005)
14. Meratnia, N., de By, R.A.: Spatiotemporal compression techniques for moving point objects. In: EDBT, pp. 765–782 (2004)
15. Muckell, J., Hwang, J.-H., Patil, V., Lawson, C.T., Ping, F., Ravi, S.S.: SQUISH: an online approach for GPS trajectory compression. In: COM.Geo, pp. 13:1–13:8 (2011)
16. Muckell, J., Olsen, P.W. Jr., Hwang, J.-H., Lawson, C.T., Ravi, S.S.: Compression of trajectory data: a comprehensive evaluation and new approach. Geoinformatica **18**(3), 435–460 (2014)
17. Ozsoyoglu, G., Snodgrass, R.T.: Temporal and real-time databases: a survey. IEEE Trans. Knowl. Data Eng. **7**(4), 513–532 (1995)
18. Pandey, V., Kipf, A., Neumann, T., Kemper, A.: How good are modern spatial analytics systems? Proc. VLDB Endowment **11**(11), 1661–1673 (2018)
19. Patroumpas, K., Sellis, T.: Maintaining consistent results of continuous queries under diverse window specifications. Inf. Syst. **36**(1), 42–61 (2011)
20. Patroumpas, K., Artikis, A., Katzouris, N., Vodas, M., Theodoridis, Y., Pelekis, N.: Event recognition for maritime surveillance. In: EDBT, pp. 629–640 (2015)
21. Patroumpas, K., Alevizos, E., Artikis, A., Vodas, M., Pelekis, N., Theodoridis, Y.: Online event recognition from moving vessel trajectories. GeoInformatica **21**(2), 389–427 (2017)
22. Patroumpas, K., Pelekis, N., Theodoridis, Y.: On-the-fly mobility event detection over aircraft trajectories. In: ACM SIGSPATIAL, pp. 259–268 (2018)
23. Potamias, M., Patroumpas, K., Sellis, T.: Sampling trajectory streams with spatiotemporal criteria. In: SSDBM, pp. 275–284 (2006)
24. Potamias, M., Patroumpas, K., Sellis, T.: Online amnesic summarization of streaming locations. In: SSTD, pp. 148–165 (2007)
25. Terroso-Saenz, F., Valdés-Vela, M., den Breejen, E., Hanckmann, P., Dekker, R., Skarmeta-Gómez, A.F.: CEP-traj: an event-based solution to process trajectory data. Inf. Syst. **52**, 34–54 (2015)
26. Tucker, P.A., Maier, D., Sheard, T., Fegaras, L.: Exploiting punctuation semantics in continuous data streams. Trans. Knowl. Data Eng. **15**(3), 555–568 (2003)
27. Wolfson, O., Sistla, A.P., Chamberlain, S., Yesha, Y.: Updating and querying databases that track mobile units. Distrib. Parallel Databases **7**(3), 257–287 (1999)
28. Xie, D., Li, F., Yao, B., Li, G., Zhou, L., Guo, M.: Simba: efficient in-memory spatial analytics. In: SIGMOD, pp. 1071–1085 (2016)

29. Yu, J., Zhang, Z., Sarwat, M.: Spatial data management in Apache Spark: the GeoSpark perspective and beyond. GeoInformatica **23**(1), 37–78 (2019)
30. Zhang, D., Ding, M., Yang, D., Liu, Y., Fan, J., Shen, H.T.: Trajectory simplification: an experimental study and quality analysis. Proc. VLDB Endowment **11**(9), 934–946 (2018)

Part III
Trajectory Oriented Data Management for Mobility Analytics

The third part of this book specifies solutions towards managing big spatiotemporal data, oriented to the notion of trajectory: The first chapter specifies a generic ontology revolving around the notion of trajectory so as to model data and information that is necessary for trajectory analytics components. This ontology provides a generic model for constructing knowledge graphs integrating data from disparate data sources. In conjunction to this, this chapter describes novel methods for transforming data from archived and streamed data sources to populate the ontology. The second chapter proposes advanced solutions to integrating data. Emphasis is given to enriching data streams and integrating streamed and archival data to provide coherent views of mobility: This is addressed by real-time methods discovering topological and proximity relations among spatiotemporal entities. Finally, the third chapter presents solutions for distributed storage of integrated dynamic and archived mobility RDF data—i.e., large knowledge graphs constructed according to the generic model introduced.

Chapter 5
Modeling Mobility Data and Constructing Large Knowledge Graphs to Support Analytics: The datAcron Ontology

Georgios M. Santipantakis, George A. Vouros, Akrivi Vlachou, and Christos Doulkeridis

Abstract This chapter presents modeling and representation techniques for mobility data, focusing on semantic representations that build around the central concept of semantic trajectory. Moving from mobility data to enriched representations of positional information, associated with contextual data and furthermore with events that occur during the movement of an object, is critical to support advanced mobility analytics. Motivated by these requirements, this chapter describes the datAcron ontology that satisfies these requirements to a larger extent than previous works on semantic representations of trajectories, at multiple, interlinked levels of detail. In addition, we show that this ontology supports data transformations that are required for performing advanced analytics tasks, such as visual analytics, and we present use-case scenarios in the Air Traffic Management and maritime domains.

5.1 Introduction

Analysis of mobility data is often involved in many tasks in critical domains w.r.t. economy and safety. It is often important to combine surveillance data with descriptions of the moving objects, (e.g., geometric information, objects' physical and operational characteristics) and contextual information (e.g., areas and points of interest, weather information, traffic, etc.), originating from disparate and heterogeneous data sources.

Challenging problems include effective information provision for situation awareness, identification of recurrent patterns, decision-making at different scales and levels of abstraction, as well as the prediction of moving objects' behavior under specific circumstances. These challenges are significant, given that their

G. M. Santipantakis · G. A. Vouros · A. Vlachou · C. Doulkeridis (✉)
University of Piraeus, Piraeus, Greece
e-mail: gsant@unipi.gr; georgev@unipi.gr; avlachou@unipi.gr; cdoulk@unipi.gr

© Springer Nature Switzerland AG 2020   123
G. A. Vouros et al. (eds.), *Big Data Analytics for Time-Critical Mobility Forecasting*, https://doi.org/10.1007/978-3-030-45164-6_5

achievement aims to reduce factors of uncertainty regarding operations, enhance punctuality of activities, advance planning efficiency, and reduce operational costs in time critical domains, such as aviation and maritime.

The complexity of these challenges increases significantly to the number of moving objects. Towards reducing this complexity, a shift of operations' paradigm from *location-based*, as it is today, to a *trajectory-based* has been proposed. Trajectories are turned into the main asset and placed in the core of decision-making, assessment of situations, and planning of operations tasks.

Towards addressing these challenges, we need to consider how we represent trajectories to satisfy the data needs and requirements of analysis tasks. Our approach is based on two principles: First, trajectories should reveal objects' behavior in explicit terms, at different levels of abstraction considering their geometric, contextual, and analysis-specific features. In doing so, analysis tasks can retrieve data about trajectories at any level of abstraction that is appropriate for their purposes, switching between abstraction levels, delving into the details of mobility phenomena, and providing overviews in generic terms. Second, data transformations (or conversions) require trajectories to integrate spatial events into temporal sequences, while, on the other hand, these events need to be aggregated into spatial time series, associated with geographic contexts. Combining these abilities allows identifying re-occurring patterns of behavior at varying levels of abstraction, enhancing our understanding of mobility phenomena and thus, decision-making. In this chapter, an "abstraction" is considered any possible combination of aggregation and generalization. When it is necessary, we specify explicitly which kind of abstraction is required.

In this chapter we describe the datAcron ontology for modeling semantic trajectories, integrating spatiotemporal information regarding mobility of objects at multiple, interlinked levels of abstraction, supporting appropriate data transformations, as needed by visual analysis tasks. Visual analytics, as also shown in Chap. 3 of this book, impose specific requirements to support the combination of human and computational data processing through interactive visual interfaces, enabling analysis of spatiotemporal [6] and mobility data [7], sophisticated data analysis, and informed decision-making [5], at varying levels of abstraction.

Existing models and ontologies for the representation of semantic trajectories do not associate data and events at multiple levels of abstraction. They usually specify models for representing trajectories at different levels (from raw to semantic), where each level associates trajectories with a different kind of information. In models where some form of abstractions is supported, these are restricted to specific types and levels. Consequently, switching between levels of abstraction as needed by exploratory analysis tasks is limited; thus, these representation models can hardly serve tasks for visual analytics. These issues are further discussed in detail with explicit references to existing trajectory models, in Sect. 5.3.

The main contributions via this and our previous work on the datAcron ontology [29] are the following:

- We revisit fundamental data types for visual analysis tasks revolving around the notion of semantic trajectory, specifying conversions among these types of data.

These types and conversions provide an in-principle framework for identifying trajectories' constituents, as well as a comprehensive framework for validating ontological specifications towards the provision of appropriately transformed data, satisfying data requirements of visual analysis methods.

- We revisit the notion of "semantic trajectory", as a meaningful sequence of trajectory parts at any level of abstraction. By being meaningful, a semantic trajectory is associated with human-interpretable and machine-processable information, revealing objects behavior in explicit terms. Dealing with multiple levels of abstraction, we support analysis of moving objects' behavior at any scale and/or level of abstraction that is appropriate for analysis tasks.
- We demonstrate the ontology by means of enhanced SPARQL queries, using real-world data from the Air Traffic Management domain and maritime domain.

Section 5.2 of this chapter motivates the need for an ontology, for the representation of semantic trajectories and specifies the requirements. It also outlines the fundamental data types and data transformations for supporting analytics tasks w.r.t. trajectories of moving objects. Section 5.3 briefly reviews weaknesses and limitations of existing proposals for representing semantic trajectories. Section 5.4 presents the datAcron ontology for the representation of semantic trajectories, and Sect. 5.5 demonstrates how data transformations are supported by the ontological specifications, supporting visual analytics tasks for the purposes of Flow Management and maritime cases. A more extensive presentation of the Air Traffic Management cases is also available in a previous work [29]. The chapter concludes with discussion remarks in Sect. 5.6.

5.2 Requirements for Enriched Representation of Mobility Data

In this section, we specify the requirements for the representation of semantic trajectories defining important terms, and then we recall the fundamental types of spatiotemporal mobility data and data transformations/conversions appropriate for supporting visual analytics: Fundamental data types and conversions provide a comprehensive framework for validating ontological specifications.

5.2.1 Requirements for the Representation of Semantic Trajectories

Towards a comprehensive semantic model of trajectories that integrates mobility data, we describe the features that are necessary to the representation of semantic trajectories, including geometric, geographic, and application-specific information [31].

The authors in [31] indicate that geometric information concerns the progression of positions of a moving object during a given time interval. The temporal sequence of raw (surveillance) data specifying the moving object spatiotemporal positions reported from sensing devices defines a *raw trajectory* [19]. The geometric information enables queries like "Return objects which were located at x, y, z at time t." It may be specified at various levels of aggregation, revealing representations regarding the patterns of a moving object at different spatiotemporal scales. For example, computations regarding spatial/topological relations or patterns of movement are often easier when a trajectory is represented as a line, rather than a sequence of positions. Alternatively, a trajectory can be represented as a temporal sequence of lines representing sub-trajectories, each one of special interest on its own (e.g., each one crossing a specific region of interest, or corresponding to a specific phase of movement), or as a sequence of aggregated raw positions with high concentration in spatiotemporal regions or points of interest.

Apparently, the significance of specifying a trajectory at multiple levels of geometric abstraction is strongly related with geographical (e.g., areas or points of interest) and application-specific (e.g., phases of movement) information. The usefulness of having multiple levels of geometric abstractions is that each one serves different purposes towards representing and analyzing the behavior of moving objects. Having these geometric abstractions, we may answer queries like "Return objects that crossed the spatial region X during the time interval [t_begin, t_end]," "Return objects whose trajectories crossed spatial regions that properly include region X during the time interval [t_begin, t_end]," and "Return objects whose trajectories include an aggregation of positions close to a specific point of interest."

Different levels of geometric abstraction provide alternative constituents for structuring trajectories. According to [19] a *structured trajectory* consists of a sequence of *trajectory parts* that can be either *raw positions* reported from any sensing devise, aggregations of raw positions referred as *nodes*, or *trajectory segments*.

A trajectory *segment* is a trajectory itself, which may be part of a whole trajectory. A *node* provides an aggregation of raw positions. Segments and nodes aggregate information that may instantiate a behavior pattern. For example, a sequence of raw positions may instantiate a "turn" or a "stop" event (e.g., the critical points mentioned in Chap. 4). These aggregations can be represented by a single node or segment, associated with an event type (e.g., "turn" or "stop," respectively), and to the corresponding set of raw positions.

It must be noted that events aggregate different types of features, as mentioned in Chap. 9 of this book. An event pattern may comprise contextual features (e.g., crossing a spatial region, or a region with a specific weather condition), features of moving objects (e.g., reaching highest possible altitude), geometric and geographical features, and/or other events regarding the mobility of the object (e.g., moving in low-speed or descending). Events may be *low-level*—associated with basic behavior—or *complex*—associated with complex patterns of behavior.

Segments of trajectories and nodes can be defined with different objectives depending on the application and target analysis and are thus associated with application-specific information. As defined in [19], a maximal sequence of raw data that comply with a given pattern defines an *episode*. In this work we consider *events* as a generalization of episodes. *Events* represent specific or abstract happenings and are associated with trajectory parts, providing application-specific information that is relevant to the trajectory. As a consequence, queries such as "Return objects whose trajectories contributed to congestion events in a specific spatial-temporal region" or "Return objects whose trajectories comprise a segment that is associated with a high-speed event" can be answered.

Geographical features allow turning the geometric information representing the spatial path into a geographical trace [31] which is meaningful for humans and computational processing tasks. This requires associating trajectory parts to (types of) geographic regions, as, for example, link discovery methods that Chap. 6 presents do. These regions of interests may be shops/spots/buildings of different kinds, regions of special interest (e.g., touristic, commercial or industrial), etc. Generalizing geographical features, we can draw semantic associations between trajectory parts, supporting further the abstraction of trajectories (e.g., any trajectory crossing many shops can be a "shopping trajectory," without considering the kind of shops crossed. Specific types of shopping trajectories may indicate specific types of shops crossed). In this work, we generalize geographical features to *contextual*. This comprises features of the moving objects, as well as features of moving objects' environment, considering that these features are associated with objects' movement. These may include weather attributes, space configuration features, as well as aggregated data about co-occurring trajectories—i.e., traffic. This enables answering queries such as "Return trajectories that crossed any region with specific weather conditions [specified as conditions in weather attributes]." A trajectory part may be associated with any event that co-occurs with it spatially and/or temporally: For example, Bad weather conditions, or traffic regulations associated with a spatial region may co-occur with a trajectory crossing-it (thus, related spatially) during a time period (related temporally).

A *semantic trajectory* is a sequence of trajectory parts, associated with contextual information and related events. The association with such information reveals objects' deliberative or accidental behavior in explicit terms, thus contributes to understanding the rationale for that behavior.

It follows that a semantic trajectory can be specified at different levels of abstraction, depending on the geometric features, contextual features, and events considered. Abstraction may happen by means of aggregation, generalization, or both. In doing so, we may retrieve semantically associated trajectories, based on the semantic features they aggregate and information to which they are associated. For instance, we may retrieve "trajectories crossing sensitive areas and associated with suspicious events." Such trajectories may be represented at varying aggregation levels. They may cross areas with different types of sensitivity and they may be associated with different types of suspicious events.

We conjecture that abstractions of a single trajectory should be interlinked, so as any application to be able to get any relevant information that is necessary for its purposes, being able to move in a continuum between specialized/basic information and generalized/aggregated information, through querying and applying data transformations. This supports, for instance, delving into the details regarding a trajectory part associated with a complex event of type "suspicious behavior," by inspecting geometrical, contextual, and application-specific features at the appropriate level of detail.

5.2.2 Fundamental Data Types and Data Transformations for Visual Analytics

Given our aim to represent trajectories towards supporting data-driven approaches to challenging problems in critical domains, this section presents generic spatiotemporal data transformations to serve analysis goals on mobility data. As mentioned in [8], there are three fundamental types of spatiotemporal data associated with mobility: trajectories of moving objects, spatial event data, and spatial time series.

Individual trajectories provide information on the movement of individual objects. Aggregated traffic data are *spatial time series* describing how many moving objects were present in different spatial locations and/or how many objects moved from one location to another during different time intervals. The time series may also include aggregate characteristics of the movement, such as the average speed and travel time. Time series describing the presence of objects are associated with distinct locations, and time series describing aggregated moves (often called fluxes or flows) are associated with directed links between pairs of locations. In both cases *spatial time series* are represented as chronologically ordered sequences of values of time-variant thematic attributes associated with spatial locations or spatial entities (for example, regions of special interest).

Spatial events emerge at spatial locations and exist for a period of time. Spatial events are described by their spatial regions, existence times, and contextual features. Events may occur irrespectively of trajectories, but somehow be related to trajectories (e.g., weather events, regulations imposed in a spatiotemporal region) or may be derived from trajectories (e.g., a turn of a moving object, short distance between a pair of objects, or large number of moving objects in a spatiotemporal region).

Based on these types of spatiotemporal data and following the approach of [20], the fundamental types of queries can be seen as transformations combining three basic components: (a) space (*where*), (b) time (*when*), (c) object or event (*what*). These components can be used in three basic types of queries:

- Retrieve the trajectories/events in a region for a time period (*when&where→ what*).
- Retrieve the region occupied by a trajectory/event or set of trajectories/events, at a given time instant or period (*when&what→where*).
- Retrieve the time periods that a non-empty set of trajectories/events appear in a specific location or area (i.e., *where&what→when*).

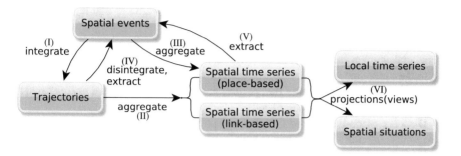

Fig. 5.1 Conversions between different representations

Exploiting these fundamental data types and queries, we aim to support the generic transformations depicted in Fig. 5.1 [8], in support of visual analytics tasks. Briefly, as Fig. 5.1 shows, trajectories integrate spatial events (transformation I), while these events, similarly to trajectories, may be aggregated to spatial time series. These may be either place-based, i.e., associated with a specific spatial region (transformation III), or link-based, such as flows of trajectories between pairs of spatial regions (transformation II). Projections of spatial time series may result in spatially referenced time series or to spatial situations (transformations VI). These transformations impose specific requirements to representations, so as to answer queries regarding trajectories, aggregations of features and events.

More specifically, the left part of the diagram in Fig. 5.1 shows the tight relationships between spatial events and trajectories. In fact, trajectories comprise parts that are associated with spatial events. Even in raw trajectories, each record represents the presence of an object at a specific location at some instant in time. As it is further shown in Fig. 5.1, trajectories are obtained by integrating spatial events. In the simplest case, for each moving object, all (raw) position records are linked in a chronological sequence. Reciprocally, trajectories can be transformed to spatial events either by full disintegration back into the constituent events or by extraction of particular events of interest such as sharp turns, entering/exiting a region, crossing a waypoint, etc. Spatial events that are close in space and time can be united into more complex spatial events. For example, a spatiotemporal concentration of many moving objects entering/crossing a spatial region during a small time window may be treated as a single event of traffic congestion.

Spatial time series can be obtained from spatial events or trajectories through spatiotemporal aggregation. For instance, spatial regions specify *spatial compartments*, and time can be divided into intervals called *time windows*. For each spatial compartment and time window, the spatial events or moving objects that appear in the compartment during the associated time window are binned together and counted. The result is a place-based time series in which temporal sequences of aggregate values are associated with the spatial compartments. From such spatial time series, in turn, it is possible to extract more complex spatial events; for example, events of high traffic density and high demand for a specific spatial region and for specific temporal intervals.

Trajectories can also be aggregated into link-based time series: for each pair of spatial compartments and for a specific time window, the objects that moved from the first to the second compartment during this time interval (specifying a *link* between compartments during that period) are counted. Aggregated characteristics of their movement may be calculated.

Discrete place-based and link-based spatial time series can be viewed in two complementary ways. On the one hand, they consist of temporally ordered sequences of (aggregated) values associated with individual places or links, i.e., *local time series*. On the other hand, a spatial time series is a temporally ordered sequence of the distribution of spatial events, moving objects, or collective moves (flows) of objects over the whole space of interest, together with the spatial variation of various aggregate characteristics. These distributions are called *spatial situations* [7].

Based on the requirements for the representation of semantic trajectories specified in the first part of this chapter and the framework of fundamental types of mobility data and conversions between them presented in this part of the chapter, we proceed in Sect. 5.4 to propose a model for the representation of semantic trajectories, which aims at (a) supporting the representation of semantic trajectories at multiple, interlinked levels of abstraction, (b) structuring trajectories by means of different types of trajectory parts, (c) associating events at varying levels of abstraction with trajectory parts, (d) supporting the transformations needed for visual analysis tasks.

5.3 Semantic Representations of Trajectories: Related Work

Existing approaches for the representation of trajectories can be categorized into those that (a) use plain textual annotations instead of semantic associations to features of interest [3, 9, 10], having limitations towards machine-processable information for the purposes of mobility analysis tasks; (b) constrain the types of events that can be used for structuring a trajectory [3, 9, 26, 31]; or (c) make specific assumptions about the constituents of trajectories [10, 12, 14, 17, 26, 30], thus providing limitations to the specification of trajectories at varying levels of abstraction according to needs.

To a greater extent than previous approaches, we aim to support the representation of trajectories at multiple, interlinked levels of detail [29].

More specifically, although authors in [12] provide a rich set of constructs for the representation of semantic trajectories, these are specified as sequences of episodes, each associated with raw trajectory data, and optionally, with a spatiotemporal model of movement. Beyond representing trajectories only as sequences of episodes, there is no fine association between abstract models of movement and raw data, providing limitations to analysis tasks that need both of them in association.

On the other hand, [10, 26] and [30] provide a two-levels analysis where semantic trajectories are lists of semantic sub-trajectories, and each sub-trajectory is a list of spatial points. The authors in [14], based on the two-levels analysis of trajectory models, introduce an ontological pattern for the specification of trajectories.

Regarding events and episodes, most of the proposed models are based on the "stop-move" model [27, 31], or they are connected to features at specific levels of abstraction: In [10] events—mostly related to the environment rather than to the trajectory itself—are connected to points. This may lead to ambiguities as far as the association of events to trajectories crossing the same points is concerned, especially for the events concerning the trajectory itself rather than the environment. In [12] episodes concern things happening in the trajectory itself and may be associated with specific models of movement. However, it is not clear how multiple models of a single trajectory—each at a different level of analysis—connected to a single episode, are associated. Contextual information in [12] is related to movement models, episodes, or semantic trajectories, which is quite generic as a model, while in [26, 30] and [14] fixes and states represent basic behavioral features of the moving object. These may also represent contextual features and are associated with trajectory points, or in [26] they specify domain-specific features. Finally, in [10] environment attributes are associated with points only and can only be assigned specific values.

As noted in the previous section, the specification of trajectories at various layers, from raw to semantic, depending on the information associated with trajectories (as it is done in [31]) is orthogonal to the goal of providing specifications of trajectories at multiple levels of abstraction. A different approach to that is proposed in [18], where trajectories are associated with qualitative descriptions of movement, at different aggregation levels, much like the distinction between low-level and complex events made above. However, trajectories are specified as sequences of segments associated with at least two key points providing quantitative information on movement, with no association to any type of events or activities.

This lack of flexibility to specify semantic trajectories at multiple levels of abstraction regarding geometric and contextual information, as well as events, and the lack of the capability to link these specifications so as to be able to switch between abstractions flexibly, is a common feature among previous efforts. In addition to that, to the best of our knowledge, there is no work that considers the requirements of analysis tasks in structuring trajectories, so as to support fundamental types of data and transformations between them.

Specifically, considering data transformations for analysis tasks, apart from the structural transformations between or within the different types of spatiotemporal data specified in Sect. 5.2, there exist transformations that change the scale, or level of detail, which may be beneficial for particular tasks. For example, Chu et al. [11] transform trajectories into sequences of traversed map regions (e.g., streets) and apply text mining methods for discovery of "topics," i.e., combinations of regions that have a high probability of co-occurrence in one trip. The extraction of "topics" is done for different time intervals. By investigating the temporal evolution of the topics, it is possible to understand where objects travel in different times of the day and days of the week. Al-Dohuki et al. [1] transform trajectories into texts consisting of region names and text labels denoting speeds (low, medium, and high). Furthermore, a discrete representation of aggregated movements between places can be treated as a graph, to which graph analysis methods can be applied [13, 16]. As such, these various transformations enable the comprehensive analysis of traffic data from multiple complementary perspectives [4].

To the best of our knowledge, the ontology presented in this chapter, namely the *datAcron ontology*, is the first one to provide the flexibility needed to represent trajectories at multiple, interlinked levels of abstractions. Furthermore, and to a greater extent than other models and ontologies proposed, it is validated in the context of data transformations needed by analysis tasks, in highly complex problem cases in the aviation and maritime domains.

The datAcron ontology has been succinctly presented in [21, 22, 29]. In this chapter we provide the details of the specifications, while we show how the datAcron ontology supports a range of generic data transformations that are required by analysis tasks in air traffic management and maritime domains, supporting the provision of information at various levels of analysis and form.

5.4 The datAcron Ontology

The datAcron ontology[1] was developed by group consensus over a period of 12 months following a data-driven approach according to the HCOME methodology [15]. It has been designed to be used as a core ontology towards integrating data from heterogeneous data sources of surveillance and contextual data, in association to recognized (low-level and high-level) events, towards supporting analysis tasks exploiting semantic trajectories. It has been designed and implemented as a generic ontology, to satisfy needs for the representation of trajectories across domains, supporting a wide range of generic data transformations that are required by analysis tasks. As a proof of concept, the ontology has been successfully used in both aviation and maritime domains. For this reason, the ontology also describes aviation and maritime specific concepts and relations.

Following the HCOME methodology, the following specific phases of engineering have been followed:

Specification of Aim, Scope, Requirements, and Identification of Collaborators In this initial phase, we had to be acquainted with terminology regarding semantic trajectories and with analysis goals related to mobility data in several scenarios in two critical domains: Air Traffic Management and Maritime Situation Awareness (cf. Chaps. 1 and 2). Thus, we had to identify the data requirements of analysis tasks and specify the queries to be answered from the ontology. The fundamental

---

[1] http://ai-group.ds.unipi.gr/datacron_ontology/.

data types specified in Sect. 5.2.2 provide the basic framework for representing and exploiting mobility data through transformations.

Knowledge Acquisition, Development, and Ontology Maintenance The development of the datAcron ontology has been driven by ontologies related to our objectives: DUL, SimpleFeature, NASA Sweet and SSN, as well as schemes and specifications regarding data from different data sources. These ontologies served as top ontologies, whose specifications are further refined to the specification of datAcron and domain-specific classes/properties. Standard ontology development and maintenance tasks (e.g., improvisation, versioning, documentation), together with consultation from experts on data analysis and domain-specific tasks took place. It must be pointed out that following a data-driven approach, the major goal was to provide "interfaces" with computational and analysis tasks that either provide data to populate the ontology or fetch data to be exploited for analysis purposes. Thus, ontological specifications should support ontology population and querying in adequate and lossless ways. That is, annotating, representing, and associating data using the appropriate terms, adequately, and without losing any valuable bit of information that would affect analysis results.

Exploitation and Validation During this phase, the ontological specifications have been validated in (a) populating the ontology by means of RDF generators, and in (b) providing data in appropriate forms for data analysis tasks. Refinements of ontological specifications proposed during this phase, or changes in the required features to be exploited, had to be incorporated in the ontology.

It must be pointed out that these phases happened iteratively, for example, the specification of a new data source providing any kind of features in different forms, trigger the first phase, with potential consequent activities in the other phases.

5.4.1 Core Vocabulary and Overall Structure

As explained in Sect. 2.1 and illustrated in Fig. 5.2, a trajectory (`Trajectory`) can be segmented to several trajectory parts (`TrajectoryParts`). Each trajectory part can be a trajectory segment, a trajectory node, or a position provided by a raw surveillance data source. Segments and nodes can be further analyzed iteratively to other, less abstract trajectory parts.

The generic pattern of specifying structured trajectories is presented in Sect. 5.4.2.

Trajectories and trajectory parts can be associated with geometric and contextual information, as well as with events represented by the class `dul:Event`. As already pointed out, events are important happenings associated with the mobility of objects. These may occur in the environment of moving objects and affect their mobility, or may be derived from trajectories. Ontology patterns for associating contextual information and events to trajectory parts are presented in Sect. 5.4.2.

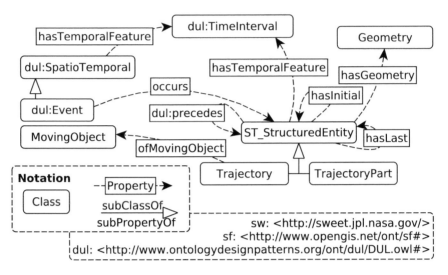

Fig. 5.2 The main concepts and relations of the proposed ontology

5.4.2 Patterns for the Representation of Semantic Trajectories

Figure 5.3 illustrates the generic pattern of structured trajectories. The main concept in this pattern is the `Trajectory`, which is a subclass of Spatiotemporal Structured Entities represented by the class `ST_StructuredEntity`. This, being a subclass of `dul:Region` represents a region in a dimensional space and time, used as a value for a quality of an entity (e.g., a storm covering an area), while it also represents (structured) trajectories and their parts. A structured trajectory, as well as any of its parts of type `TrajectoryPart`, can be a temporal sequence of `TrajectoryPart` entities.

Direct subclasses of `Trajectory` are the

- `IntendedTrajectory`: planned trajectories specified by an `dul:Infor-mationEntity`. These are different from actual trajectories, since they may not be realized. They specify the intention of a moving object. A specific example from the FM domain is a `FlightPlan`,
- `ActualTrajectory`: trajectories constructed from actual positioning data[2] and associated with low-level events representing important trajectory changes (e.g., turns, increase/decrease of speed, change of altitude, etc.),
- `RegulatedTrajectory`: trajectories that have been modified by an operational event, such as a regulation,

---

[2]In datAcron we construct actual trajectories after compression of the raw data, as Chap. 4 presents. In general, different applications may have different requirements in aggregating raw data.

- `RawTrajectory`: trajectories constructed by the raw unprocessed sequence of positional data of moving objects.

An `ActualTrajectory` can be further distinguished to a `Closed Trajectory` (i.e., a trajectory that has reached its destination) and to an `OpenTrajectory` (i.e., a trajectory in progress).

The `TrajectoryPart` class is further refined to the following subclasses:

- `Segment`: associated with a spatial region and a time proper interval.
- `Node`: associated with a point in space and a time instant, or time interval. The latter holds in case the node aggregates several raw positions. A `Node` can be the result of a data processing component computing aggregations and abstractions of raw positional data.
- `RawPosition`: represents the raw (unprocessed) positional data. Each raw position instance is associated with a point in space and a time instant.

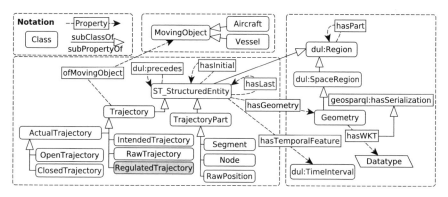

Fig. 5.3 The pattern of structured trajectories. Domain-specific concepts in gray

A specific trajectory, as well as any of its trajectory parts, being instances of `dul:Region` can be associated with its parts via the `dul:hasPart` property or via the subproperties `hasInitial`, `hasLast` which indicate the first and last part of the `ST_StructuredEntity`, respectively. For instance, a trajectory may comprise a sequence of trajectory segments (e.g., segments within sectors), who on their own turn comprise other segments (e.g., segments within air blocks), nodes (e.g., entering or exiting any airspace compartment), raw positions, and so on. The temporal sequence of structured entities is specified by means of the property `dul:precedes`. Trajectories related via the property `dul:precedes` represent subsequent trajectories of a specific object, and thus, we can keep a long history of its movement. It must be noted that this combination of properties supports sharing trajectory parts between trajectories even of the same object with no ambiguity: For instance, a trajectory node or segment can be shared between the actual and the intended trajectory of an aircraft, without mixing the trajectories.

Each structured entity (i.e., trajectory or trajectory part) can be associated with a specific *geometry* (sf:Geometry), representing a point or region of occurrence, and a *temporal entity* (dul:TimeInterval) specifying a time interval of occurrence. The geometries of structured entities can be serialized into Well-Known-Text (WKT) and asserted as values to the data property hasWKT, which is sub-property of geosparql:hasSerialization.

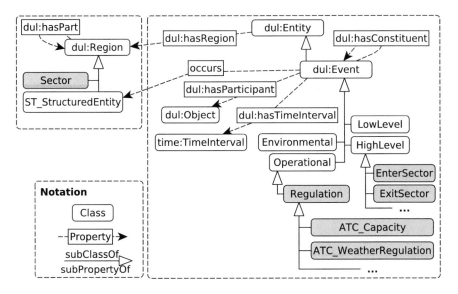

Fig. 5.4 The pattern of trajectories linked with events. Domain-specific concepts in gray

Trajectories and trajectory parts can be associated with events and contextual features of importance. Specifically, events can be associated with any ST_Structu-redEntity (i.e., with any trajectory and trajectory part), via the property occurs. This is illustrated in Fig. 5.4. An event can be associated with other events via the properties dul:hasConstituent or dul:hasPart. This is the case for high-level (complex) events (e.g., hotspot occurrence in the FM domain) associated with other high-level (e.g., regulation imposed to a sector and events signifying individual flights entering a sector) or low-level events. An event may involve participants (associated via the property dul:hasParticipant) and it holds for a specific TimeInterval specified by the property dul:hasTimeInterval. An event can be a:

– Low-Level event, in case its detection requires data from a single trajectory: For instance, a TopOfClimb is such an event. These events are detected by low-level event detectors, such as those incorporated in the synopses generator presented in Chap. 4 of this book, or by link discovery methods, such as those presented in Chap. 6.

- High-Level event, in case its detection requires contextual data and maybe data from multiple trajectories. For example, events of type EnterSector involve information about sectors crossed by a trajectory. As another example, the occurrence of hotspots requires data about sectors and multiple trajectories. These events are detected by complex events processing engines, such as the RTEC presented in Chap. 9 of this book.

Orthogonal to the classification between low-level and high-level events, we also have the following classes of events:

- Operational event, if it is issued by operators, affecting regions or groups of entities for a specific time interval. For example, a regulation (Regulation) is applied on a sector and remains active for a time interval and indirectly affects all the trajectories crossing the sector.
- Environmental event, if it happens in the environment and affects the mobility of moving objects. Extreme weather conditions are such events.

It must be noted that associating events to trajectory parts satisfies the requirement to associate events at varying levels of trajectory aggregation. For instance, a low-level event associated with a node (e.g., a "turn" event) is associated with any trajectory part (e.g., trajectory segment) that comprises that node. Also, each trajectory part may be associated with multiple events, and thus, provide rich information about objects' behavior. For example, a low-level "turn" event associated with a node may co-occur with a low-level "descend" event associated with a trajectory segment comprising that node. In addition to that, the trajectory segment can be further associated with other types of events (e.g., events of type "CrossingSector").

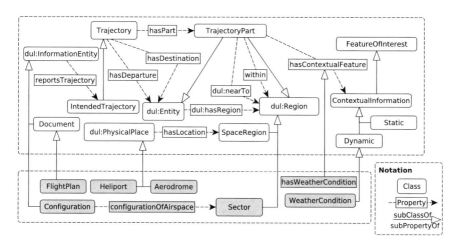

Fig. 5.5 The pattern of trajectories linked with contextual information. Domain-specific concepts in gray

In addition to events, trajectory parts can be linked to contextual information. Such information may concern static aspects of the environment (e.g., airports, airspaces, etc.), dynamic aspects (e.g., changing sector configurations, opening schemes, forecasts of weather conditions). The pattern for linking trajectory parts with contextual information is illustrated in Fig. 5.5. Without loss of generality, subsequent paragraphs and Fig. 5.5 provide examples of associating trajectories to contextual entities of interest for the FM cases.

In general, each `TrajectoryPart` can be associated with entities of type `ssn:FeatureOfInterest`, providing contextual information. For example, given the importance of weather conditions in the FM domain, each `TrajectoryPart` can be associated with entities of type `WeatherCondition`, which is defined as a subclass of `ssn:FeatureOfInterest`. This represents any entity whose properties are being estimated or calculated in the course of an observation.

Additionally, as in many domains where specific regions and places are of importance, airspace regions are of major importance in the FM domain. In general, structured entities can be linked to spatial regions (instances of `dul:Region`) of particular interest through the properties `within` and `dul:nearTo`. Also, although any trajectory part can be associated with an entity, the departure and destination of a trajectory can be considered as contextual information, linked to trajectories via the properties `hasDeparture` and `hasDestination`, respectively. These properties range to the class `dul:Physical-Place`. These in the case of the aviation (maritime) domain can be further refined to domain-specific classes such as `Airport` or `Heliport` (or port, fishing area for the maritime domain, respectively).

Finally, an `IntendedTrajectory` is associated via the property `reportsTrajectory` with an entity of type `dul:InformationEntity`, specifying the details of the intended trajectory. For example, flight plans in the FM domain provide information on the intended trajectory and, in case a regulation has affected the trajectory, report the regulated intended trajectory.

As a concrete and simple example of a trajectory specified at multiple levels of abstraction, Fig. 5.6 shows the representation of a trajectory crossing an airspace compartment: The trajectory is represented both as a geometry projected in two dimensions, and as a temporal sequence of trajectory segments, which are indicated in different color, depending on whether each segment occurs within the compartment or not. This structure results through a topological link discovery process where the trajectory geometry is used as a first indication of the potential fact that the trajectory crosses the air compartment (filtering step). This is further verified by exploiting the raw trajectory positional data and identifying the trajectory segments that spatially occur within the compartment. Chapter 6 of this book provides further details on the link discovery process. Additional information to trajectory segments is provided by associated events that are not shown in the figure, to keep it simple. Hence, beyond the representation of the trajectory as a sequence of trajectory segments, at a second level of abstraction, the trajectory is represented as a temporal sequence of semantic nodes, each one signifying an important event occurring across the trajectory. For instance, trajectory nodes H, L, M, and K are associated with entry/exit events, representing the relation of raw positions with the airspace compartment. Trajectory segments and nodes are further associated with positional raw data.

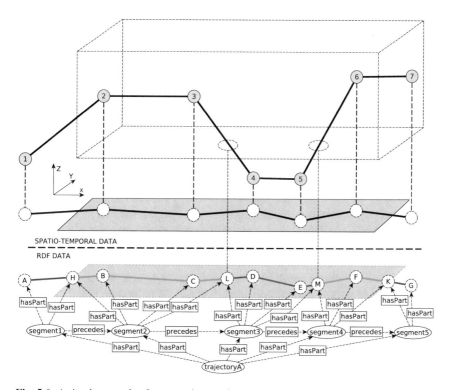

Fig. 5.6 A simple example of representing a trajectory crossing an airspace compartment

As a furthermore elaborated example, Fig. 5.7 shows an example of associating trajectories with information about events and contextual information. The two maps in the upper part show the trajectories of the flights performed between Paris Orly and Lisbon (left) and between London Heathrow and Madrid (right) during April 2016. Information about crossing sectors in which various types of regulations were applied has been attached to the points of the trajectories, denoting the regulation reason codes. In the map, the trajectories are represented by segmented lines; the segments are colored according to the regulation reasons of their starting points. For the segments that were not in regulated sectors, the regulation reason code is empty. These segments are represented by thin dashed lines.

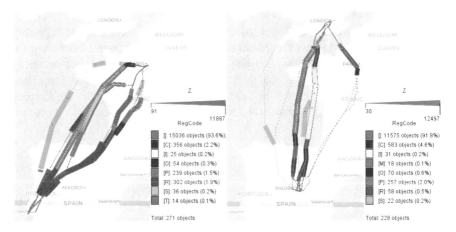

Fig. 5.7 Examples of trajectories enriched with information about crossing sectors in which regulations were applied (Figure provided by G.Andrienko, IAIS Fraunhoffer)

5.5   Supporting Data Transformations

In this section we show how the proposed ontology supports the data transformations, regarding the needs of visual analysis tasks in two major FM cases in the aviation domain: Case FM01, aiming to the discovery of patterns of regulations, and case FM02, aiming to the analysis of hotspots occurrences. Regarding the maritime domain, we demonstrate aggregations of cargo and fishing vessels' trajectories intersecting with high risk areas, by transforming these to spatial time series (place based). Cases specify scenarios with specific analysis objectives and data needs. Appropriate visualizations show data-driven exploratory analysis results towards identifying patterns of behavior and supporting decision-making.

5.5.1   Data Sets

To explore the capacities of the ontology to support visual analysis tasks we exploit the following data sets for aviation cases, as also described Chap. 2 of this book:

- Network Manager Regulations: This data set provides historical data of regulations applied by the Network Manager on sectors in the European airspace, during April 2016.
- Sector Configuration: This data set describes the structure of sector configurations for specific periods of time within April 2016.

- Flight Plans: This data set contains the submitted flight plans prior to the takeoff for the flights operated during April 2016, to/from airports worldwide. However, only a few flights have destination/origin a non-European airport.
- Entry/Exit points: This data set is derived from the combination of sector configurations and flight plans. A spatiotemporal link discovery task [23] interpolates the altitude, latitude, longitude, and time an aircraft enters/exits each air block (and sector), as also described in Chap. 6 of this book. Having these entry/exit points we can specify trajectories as sequences of trajectory segments, each one topologically being "within" a crossed airspace compartment (shown in Fig. 5.6).
- NOAA grib binary files: This data set is a collection of 96 binary files reporting 3-h weather forecasts, starting from April 1st to April 24th, 2016.

For the maritime case we exploit the following data sets,[3] as these have been described also in Chap. 1 of this book:

- Maritime surveillance data in Europe during January 2016.
- World Port Index, reporting 3684 major ports worldwide.
- Vessel types as reported in AIS messages during 2016.

These data sets are provided by heterogeneous (and often voluminous) data sources. We have introduced the RDF-Gen [24, 25] method which converts data into triples with low latency, w.r.t. a given ontology (in our case, the datAcron ontology). The main idea of RDF-Gen is to use a SPARQL-like triple template for each data source, to convert raw data from the source to RDF triples. RDF-Gen templates allow the use of custom functions for cleaning and converting data values, generating URIs, and generating triples populating the ontology. This ontology population task by means of the appropriate RDF generator, is an ontology validation task performed during ontology development.

Among the data sets listed, the flight plans data set is the most voluminous. Specifically, this data set reports 958,288 flight plans (please recall that flight plan updates are possible, and flight plans can report at most three trajectory types), which are converted to 1,548,628,183 RDF triples. The link discovery task for interpolating entry/exit positions for air blocks and constructing the corresponding trajectory segments for each trajectory generates 283,906,720 additional RDF triples, resulting in a total of 1,832,534,903 triples.

The absence of an information document to report the indented trajectory in the maritime domain restricts the identification of trajectories to be only on runtime,

---

[3] Also available online at https://zenodo.org/record/1167595.

i.e., there is no trivial method to construct intended/planned trajectories of moving objects in this domain. The method that computes the (actual) trajectories from reported positions reconstructs each trajectory in memory, appending a list of aggregated nodes from reported positions for each moving object. The method takes into consideration the annotations attached to aggregated nodes and applies three basic criteria, when the last node in the list of aggregated nodes of some moving object reports zero speed:

1. If the current node reports non-zero speed, then a new trajectory starts on the current node,
2. if the current node has temporal distance from the last node greater than t_{th}, then a new trajectory starts on current node,
3. if the current node has spatial distance from the last node greater than s_{th}, a new trajectory starts.

An important difference between aviation and maritime trajectories is that while the first have exactly one and defined origin and destination places, the latter have not a defined origin/destination and these have to be identified. This also indicates that there can be cases of trajectories in the maritime domain that do not end (or start) within some port. When a trajectory ends (respectively, starts) near to some port, we associate the trajectory with the port as its destination (respectively, origin). If both origin and destination of a trajectory have been assigned to some port, the trajectory is classified as "closed" (otherwise it is considered to be "open"). The threshold t_{th} expresses the amount of time a vessel has to spend at some place in order to be considered to be stopped (in our experiments we set this threshold to 1 h). The threshold s_{th} aims to distinguish trajectory termination from communication gaps within the trajectory.

5.5.2   datAcron Namespaces for Functions

Most of the queries involve spatial and temporal functions, thus data transformations cannot be fully supported by standard SPARQL 1.1. For this reason we have extended standard SPARQL 1.1 with the following namespaces regarding functions:

- SPARQL_functions.converters: These include functions for converting given values to a specific format, e.g., the conversion of latitude, longitude, altitude, and time values into a single string representation for each 4D point. An important function in this namespace is the getWeatherAVG(), which given the name of a weather variable, a geometry, an altitude range, and a timestamp retrieves the average value for the weather variable within the airspace volume defined from the geometry and the altitude range.

- SPARQL_functions.distance: These are various distance functions between geometries. For cases where high performance is preferred over accuracy, the GeoEllipticDistance() function (based on *Vincenty's formulae* [28]) can be used in the computations. For all the cases where accuracy is important, this namespace provides the function geodesicDistance() which is implemented on top of geographicLib.[4] This function computes the distance between the centroids of given geometries in meters and provides accuracy up to 10^{-9} m.
- SPARQL_functions.spatial: These are functions implementing all the OGC topological relations between pairs of geometries. Each function accepts WKT representations of geometries as arguments and returns Boolean true if the topological relation holds or false otherwise.
- SPARQL_functions.temporal: These are functions implementing all the temporal relations described in Allen's interval algebra [2]. Each function returns true if the corresponding temporal relation holds, or false otherwise. For example, the function *during_sf()* returns true if the temporal interval defined by the first two arguments (start and end time instants), is during, starts or finishes within the interval specified by the third and fourth arguments.

5.5.3 Validation Setup

We have set up a SPARQL 1.1 endpoint, on top of which we have developed procedures for producing the required time series spanning within specific time periods. These procedures take as input duration and time step of a shifting time window, instantiate query parameters (e.g., moving object types), and pose the queries. Subsequently, placeholders of parameters in queries are identified by "$." The results are returned in tabular form, such that any visualization tool can be used.

For instance, in cases where we need to generate time series of counts of entities, the corresponding procedure uses a parameterized SPARQL query, where the time period of interest, the time window, and the time step for shifting the window are parameters to be instantiated. The procedure builds a sequence of queries for subsequent time windows of a given duration. The starting points of subsequent windows differ by a number of minutes equal to the time step specified.

Specifically, given a time step Δt, a time window duration wd and a period $[TimeStart, TimeEnd]$, the i-th query of n iterations, where $n = \frac{(TimeEnd - wd - TimeStart)}{\Delta t}$, concerns the time interval $[TimeStart + (i * \Delta t), TimeStart + wd + (i * \Delta t)]$.

---

[4]Publicly available online at https://geographiclib.sourceforge.io/.

5.5.4   Pre-processing Steps and Auxiliary Structures

To increase the efficiency of query answering, we pre-compute intermediate results and store these in auxiliary structures. This method is by analogy to the spatial databases which rely on specialized indices (e.g., spatial indices such as R-Tree) to improve query answering performance. This is an additional way of exploiting data fetched via the SPARQL endpoint. Furthermore, the auxiliary structures (in addition to custom made functions) overcome limitations of SPARQL (such as forming iterative queries), and in the same time simplify the SPARQL queries used in the end (e.g., to increase the computational efficiency of query answering, no nested queries are used for the use cases) without affecting the validation of ontological specifications.

As already specified above, the link discovery process segments a trajectory to those parts that are within air blocks for the aviation domain (or regions in general), by computing the spatiotemporal entry/exit points per trajectory and air block. Given that sectors comprise air blocks we can represent trajectories at different aggregation levels, depending on whether we are focusing on air blocks or sectors, according to the ontology specifications. The additional triples computed by the link discovery process are of the form (?x :within ?y.) representing trajectory segments ?x that occur spatially in air blocks ?y. Similar relations can of course be computed for the maritime domain, representing trajectories crossing, for instance, fishing or protected areas.

To further increase efficiency we use an in-memory HashMap relating sectors with sets of airblocks. For the cases where a sector comprises another sector, we associate the former with the set of airblocks composing the latter. The HashMap is constructed using the query:

```
PREFIX : <http://www.datacron-project.eu/datAcron#>
PREFIX dul: <http://www.ontologydesignpatterns.org/ont/dul/DUL.owl#>
PREFIX sp: <java:SPARQL_functions.spatial.>
SELECT ?s ?airblock_wkt (str(?lower) as ?lowerLevel)
(str(?upper) as ?upperLevel) WHERE {
?s dul:hasPart+ ?airblock .
?airblock :hasGeometry ?g ;
:hasLowerLevel ?lower ;
:hasUpperLevel ?upper .
BIND(sp:getGeom(?g) as ?airblock_wkt) .
}
```

where (`?s dul:hasPart+ ?airblock.`) traverses the property path built from one or more occurrences of `dul:hasPart`, specifying the structure of sectors in terms of constituent air blocks and sectors. The above query reports the URIs of sectors, as well as the air block projection geometry in WKT and the lower/upper flight levels for each air block that a sector comprises.

Furthermore, the ontology is populated with triples stating regulations imposed on sectors (i.e., regulation events) for specific time intervals, with a potential cancellation time per regulation. The duration of a regulation is the time interval between the starting time and the earliest time instant between regulation cancellation (if it is specified) and ending time.

5.5.5 *Visual Analytics Enhanced Via Data Transformations on Aviation Use Cases*

As a use case demonstrating data transformations supported by the proposed ontology in the aviation domain, we discuss the rationale for the choice of sector configurations, based on the expected evolution of demand. In doing so, we first retrieve all sectors (active or not) crossed by any trajectory and then, we provide a time series of the number of trajectories intended to cross any sector, providing the evolution of demand per sector.

To compute the evolution of demand we aggregate the trajectories specified by flight plans into spatial time series by sectors and time windows. Two time-dependent attributes may be computed for any sector: *entry count* (how many flights enter the sector during each time interval) or *occupancy* (how many flights are present in that sector during each time interval). These may be counted in overlapping time windows, depending on the step used for shifting the time window. As usually, to produce spatial time series we use a time window of specific duration and a time step, which specifies the time difference between the starting points of two consecutive windows.

These cases require in the first place transforming trajectories (as specified by flight plans) into time series of spatial events, and then, transforming trajectories into spatial time series of demands by aggregating them by (active) sectors (aggregation II in Fig. 5.1). A more detailed analysis and additional transformations have been presented in [29].

(a) This case first requires for a given intended trajectory specified by a flight plan to retrieve the series of sectors S (active and inactive) crossed by that trajectory, and the trajectory segments crossing each sector in S. For example, the following query returns the sectors crossed by the trajectory of a given flight plan, e.g., `:flight_plan_AA51147955`:

```
PREFIX : <http://www.datacron-project.eu/datAcron#>
PREFIX dul: <http://www.ontologydesignpatterns.org/ont/dul/DUL.owl#>
SELECT ?sector (min(?start) as ?timeEnter) (max(?end) as ?timeExit)
WHERE {
:flight_plan_AA51147955 :reportsTrajectory ?t .
?t a :IntendedTrajectory ;
dul:hasPart ?segment .
?segment a :Segment ;
:within ?airblock ;
:hasTemporalFeature ?time .
?time :TimeStart ?start ;
:TimeEnd ?end .
?sector dul:hasPart+ ?airblock .
} Group By ?sector
Order By ?timeEnter
```

A more restricted version of the above query concerns only the active sectors during the time period of the flight defined by the first and last node of the trajectory reported by the given flight plan, according to the active sector configurations. The query is as follows:

```
PREFIX : <http://www.datacron-project.eu/datAcron#>
PREFIX dul: <http://www.ontologydesignpatterns.org/ont/dul/DUL.owl#>
PREFIX tmp: <java://SPARQL_functions.temporal.>
SELECT ?sector (min(?start) as ?timeEnter) (max(?end) as ?timeExit)
WHERE {
:flight_plan_AA51147955 :reportsTrajectory ?t .
?t a :IntendedTrajectory ;
dul:hasPart ?segment .
?segment a :Segment ;
:within ?airblock ;
:hasTemporalFeature ?time .
?time :TimeStart ?start ;
:TimeEnd ?end .
?sector dul:hasPart+ ?airblock .
?f a :FM_Configuration ; :configurationOfAirspace ?airspace ;
:hasTemporalFeature ?time.
?airspace dul:hasPart ?sector.
?time :TimeStart ?ts ; :TimeEnd ?te.
FILTER(tmp:overlap(?start,?end,?ts,?te))
} Group By ?sector
Order By ?timeEnter
```

Finally, we use the following query to compute per sector and time window, the demand for that sector, i.e., the number of trajectories intended to cross that sector during the corresponding period specified by the temporal window. The time window shifts with a step of Δt minutes.

```
PREFIX : <http://www.datacron-project.eu/datAcron#>
PREFIX dul: <http://www.ontologydesignpatterns.org/ont/dul/DUL.owl#>
PREFIX tmp: <java://SPARQL_functions.temporal.>
SELECT (count(DISTINCT ?tr) as ?demand) WHERE
{
?flightPlan :reportsTrajectory ?tr .
?tr a :IntendedTrajectory ;
dul:hasPart ?segment .
?segment :within ?airblock ;
:hasTemporalFeature ?time .
$Sector$ dul:hasPart+ ?airblock .
:entersRegion :occurs ?segment .
?time :TimeStart ?s .
 FILTER(tmp:during_sf(?s,?s,$t+k*Δt$,$t+wd+k*Δt$))
}
```

We can restrict this query to the number of trajectories crossing active sectors (i.e., considering the periods in which each sector is active):

```
PREFIX : <http://www.datacron-project.eu/datAcron#>
PREFIX dul: <http://www.ontologydesignpatterns.org/ont/dul/DUL.owl#>
PREFIX xsd: <http://www.w3.org/2001/XMLSchema#>
PREFIX tmp: <java://SPARQL_functions.temporal.>
PREFIX sp: <java://SPARQL_functions.spatial.>
SELECT (count(DISTINCT ?t) as ?demand) WHERE
{
?f a :FM_Configuration ;
:configurationOfAirspace ?airspace ;
:hasCapacity ?capacity ;
:hasTemporalFeature ?time .
?airspace dul:hasPart $Sector$ .
?sector dul:hasPart+ ?airblock .
?time :TimeStart ?ts ; :TimeEnd ?te.
?t a :IntendedTrajectory ;
dul:hasPart ?segment .
?segment :within ?airblock ;
:hasTemporalFeature ?tn .
?tn :TimeStart ?s ;
:TimeEnd ?e .
FILTER(sp:overlaps(?s,?e,?ts,?te) &&
 tmp:during_sf(?s,?s,$t+k*Δt$,$t+wd+k*Δt$))
}
```

5.5.6 Visual Analytics Enhanced Via Data Transformations on Maritime Use Cases

As a use case demonstrating data transformations and visual analytics enabled by the proposed ontology in the maritime domain, we make the hypothesis that regions at sea with high potential of a polluting incident are those that are crossed by trajectories of vessels carrying hazardous cargo, and fishing vessels during the same time period. This hypothesis requires aggregating trajectories to spatial time series (place based) and providing their projection to spatial situations, indicating high risk in areas where trajectories of cargo and fishing vessels intersect. We partition the space into cells of 0.5° (latitude/longitude) and we assume that the risk of pollution incident is higher in those cells that trajectories of cargo and fishing vessels frequently intersect. The query is similar to computing the demand of airblocks in the air traffic management domain: we aggregate the trajectories of specific vessel types that intersect in a given cells, for a sliding window.

It is important to highlight at this point that the proposed ontology enables the trajectory reconstruction method from spatial events, which subsequently allows the computation of intersecting points from these trajectories simply by their geometries. Indeed, the SPARQL query to retrieve all the trajectories within a predefined sliding window (using also as a parameter $vesselType for the vessel type), is the following:

```
PREFIX : <http://www.datacron-project.eu/datAcron#>
PREFIX dul: <http://www.ontologydesignpatterns.org/ont/dul/DUL.owl#>
PREFIX xsd: <http://www.w3.org/2001/XMLSchema#>
PREFIX tmp: <java://SPARQL_functions.temporal.>
SELECT ?tr WHERE
{
?v a $vesselType .
?tr :ofMovingObject ?v ;
:hasGeometry ?g ;
:hasInitial ?start ;
:hasLast ?end .
?g :hasWKT ?wkt .
?start :hasTemporalFeature/:TimeStart ?s .
?end :hasTemporalFeature/:TimeStart ?e .
FILTER(
 tmp:during_sf(?s,?s,$t+k*Δt$,$t+wd+k*Δt$))
}
```

Each trajectory of vessels type {Tanker, Cargo, TankerHazard} and {Fishing} is partitioned into its constituent trajectory parts within each constructed cell. Each trajectory is associated with its trajectory parts via the dul:hasPart relation. The trajectory parts are also associated with the temporal feature that describes the time

that the vessel enters and exits each cell. We can therefore retrieve and count the
trajectory parts that intersect within each cell specified, using the query:

```
PREFIX  : <http://www.datacron-project.eu/datAcron#>
PREFIX dul: <http://www.ontologydesignpatterns.org/ont/dul/DUL.owl#>
PREFIX tmp: <java://SPARQL_functions.temporal.>
PREFIX sp: <java://SPARQL_functions.spatial.>
SELECT count(distinct ?tr1) as ?intersections WHERE
{
?v1 a $vesselTypeA .
?v2 a $vesselTypeB .

?tr1 :ofMovingObject ?v1 ;
dul:hasPart ?p1 .
?tr2 :ofMovingObject ?v2 ;
dul:hasPart ?p2 .
?p1 :hasTemporalFeature/:TimeStart ?s1 .
?p1 :hasTemporalFeature/:TimeEnd ?e1 .
?p2 :hasTemporalFeature/:TimeStart ?s2 .
?p2 :hasTemporalFeature/:TimeEnd ?e2 .
?p1 :hasGeometry/:hasWKT ?g1 .
?p2 :hasGeometry/:hasWKT ?g2 .

FILTER(
 tmp:during_sf(?s1,?s1,$t+k*Δt$,$t+wd+k*Δt$) &&
 tmp:during_sf(?e1,?e1,$t+k*Δt$,$t+wd+k*Δt$) &&
 tmp:during_sf(?s2,?s2,$t+k*Δt$,$t+wd+k*Δt$) &&
 tmp:during_sf(?e2,?e2,$t+k*Δt$,$t+wd+k*Δt$) &&
sp:intersects(?g1,?g2) &&
sp:within(?g1, $cellWKT) &&
sp:within(?g2, $cellWKT)
)
}
```

The arguments $vesselTypeA and $vesselTypeB in the above query correspond
to one of the types {Tanker, Cargo, TankerHazard} and {Fishing}, respectively, and
$cellWKT defines the WKT of each cell to be evaluated.

Please notice that all of s1, e1, s2, e2 need to be within the sliding time window.
Thus, the sliding window has to be sufficiently wide, so that any of the vessels
observed can cover the distance of $\frac{\sqrt{2}}{2}$ deg, which is the hypotenuse of a 0.5×0.5
cell. Also, the above query counts the distinct intersections between trajectories, i.e.,
multiple intersections between the same trajectories are count as one.

Figure 5.8 illustrates the computed regions for the entire surveillance data set of
January 2016. We observe that the reported positions of antipollution vessels in the
surveillance data set (as indicated by the red dots in this figure) validate the initial
hypothesis.

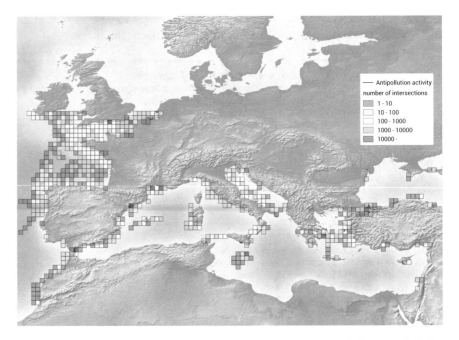

Fig. 5.8 Regions where trajectories of Tanker, Cargo, and Fishing vessels intersect. Red dots indicate antipollution vessel activity

5.6 Conclusions

This work contributes a generic ontology for the representation of semantic trajectories of moving objects at varying levels of spatiotemporal analysis. This representation improves our understanding of moving phenomena and of significant events that affect entities' mobility, through analysis tasks. Trajectories can be seen as temporal sequences of moving objects' positional data, aggregations of positional data signifying meaningful events, as temporal sequences of trajectories segments, or as geometries.

Delving into these specifications, we have shown how the different levels of trajectory enable data transformations required for visual analysis tasks at query time, in both aviation and maritime domains. We also provide the SPARQL queries executed in the populated ontology for the purposes of the demonstrated real-world cases. Generic data transformations shown in the complex Air Traffic Management and maritime examples adapt available data to the analysis goals, or to specific requirements of the methods that the analyst wants to apply.

In the process of developing this ontology, several lessons have been learned. The ontology can be re-used in different domains where trajectories are important in the analysis of behavior of the subjects. Each domain can enrich the ontology with domain-specific events that can be described by any combination of spatial, temporal, or other domain-specific relations with trajectory parts of moving objects. Furthermore, having in mind the requirements of the analytical tasks to be supported can guide the decision of the level of abstraction to be applied on the trajectory. For example, if some analytical task requires the identification of those trajectories that cross a specific region of interest, it usually suffices to use trajectory segments instead of positional data. Finally, the presence of multiple levels of abstraction for the representation of trajectories enables more efficient processing of queries regarding trajectories. The more abstract representation of a trajectory is usually less complex to be evaluated against a set of regions or points of interest, compared to the more detailed representation of the same trajectory. Thus, having multiple levels of abstraction for the same trajectory allows the application of a "filter and refine" approach, where during "filtering" the more abstract representation of a trajectory is evaluated, and during the "refinement," the appropriate representation of the same trajectory (as required by the analytical task) is evaluated against detailed criteria. This approach will eliminate a large number of trajectories that definitely are not related by any means with the regions or points of interest.

As future work we aim to re-use and expand this ontology in different domains where trajectories play important role in analysis of behavior: Either for traffic analysis in cities, or for human behavior analysis in crowded places (e.g., buildings, touristic places, festivals, etc.), under normal or emergency circumstances, or even in domains where trajectories do not involve spatiotemporal entities, but space-temporal entities, where space is any n-dimensional space where information entities (e.g., images) do exist.

Acknowledgments The research work has been supported by the datAcron project, which has received funding from the European Union's Horizon 2020 research and innovation program under grant agreement No 687591.

References

1. Al-Dohuki, S., Wu, Y., Kamw, F., Yang, J., Li, X., Zhao, Y., Ye, X., Chen, W., Ma, C., Wang, F.: SemanticTraj: a new approach to interacting with massive taxi trajectories. IEEE Trans. Vis. Comput. Graph. **23**(1), 11–20 (2017). http://doi.org/10.1109/TVCG.2016.2598416. http://doi. ieeecomputersociety.org/10.1109/TVCG.2016.2598416
2. Allen, J.F.: Maintaining knowledge about temporal intervals. Commun. ACM **26**(11), 832–843 (1983). http://doi.org/10.1145/182.358434. http://doi.acm.org/10.1145/182.358434
3. Alvares, L.O., Bogorny, V., Kuijpers, B., de Macêdo, J.A.F., Moelans, B., Vaisman, A.A.: A model for enriching trajectories with semantic geographical information. In: GIS, p. 22 (2007)
4. Andrienko, N., Andrienko, G.: Visual analytics of movement: an overview of methods, tools and procedures. Inf. Vis. **12**(1), 3–24 (2013). https://doi.org/10.1177/1473871612457601
5. Andrienko, G., Andrienko, N., Jankowski, P., Keim, D., Kraak, M., MacEachren, A., Wrobel, S.: Geovisual analytics for spatial decision support: setting the research agenda. Int. J. Geogr. Inf. Sci. **21**(8), 839–857 (2007). https://doi.org/10.1080/13658810701349011
6. Andrienko, G., Andrienko, N., Demsar, U., Dransch, D., Dykes, J., Fabrikant, S.I., Jern, M., Kraak, M.J., Schumann, H., Tominski, C.: Space, time and visual analytics. Int. J. Geogr. Inf. Sci. **24**(10), 1577–1600 (2010). https://doi.org/10.1080/13658816.2010.508043

7. Andrienko, G., Andrienko, N., Bak, P., Keim, D., Wrobel, S.: Visual Analytics of Movement. Springer Publishing Company, Incorporated, Berlin (2013)
8. Andrienko, G., Andrienko, N., Chen, W., Maciejewski, R., Zhao, Y.: Visual analytics of mobility and transportation: state of the art and further research directions. IEEE Trans. Intell. Transp. Syst. **18**(8), 2232–2249 (2017). http://doi.org/10.1109/TITS.2017.2683539
9. Baglioni, M., de Macêdo, J.A.F., Renso, C., Trasarti, R., Wachowicz, M.: Towards semantic interpretation of movement behavior. In: Advances in GIScience, pp. 271–288. Springer, Berlin (2009)
10. Bogorny, V., Renso, C., de Aquino, A.R., de Lucca Siqueira, F., Alvares, L.O.: Constant - a conceptual data model for semantic trajectories of moving objects. Trans. GIS **18**(1), 66–88 (2014)
11. Chu, D., Sheets, D.A., Zhao, Y., Wu, Y., Yang, J., Zheng, M., Chen, G.: Visualizing hidden themes of taxi movement with semantic transformation. In: Proceedings of the 2014 IEEE Pacific Visualization Symposium, PACIFICVIS '14, pp. 137–144. IEEE Computer Society, Washington, DC (2014). http://dx.doi.org/10.1109/PacificVis.2014.50
12. Fileto, R., May, C., Renso, C., Pelekis, N., Klein, D., Theodoridis, Y.: The baquara^2 knowledge-based framework for semantic enrichment and analysis of movement data. Data Knowl. Eng. **98**, 104–122 (2015)
13. Hamad, K., Quiroga, C.: Geovisualization of archived ITS data-case studies. IEEE Trans. Intell. Transp. Syst. **17**(1), 104–112 (2016). https://doi.org/10.1109/TITS.2015.2460995
14. Hu, Y., Janowicz, K., Carral, D., Scheider, S., Kuhn, W., Berg-Cross, G., Hitzler, P., Dean, M., Kolas, D.: A geo-ontology design pattern for semantic trajectories. In: Tenbrink, T., Stell, J., Galton, A., Wood, Z. (eds.) Spatial Information Theory, pp. 438–456. Springer International Publishing, Cham (2013)
15. Kotis, K., Vouros, G.A.: Human-centered ontology engineering: the HCOME methodology. Knowl. Inf. Syst. **10**(1), 109–131 (2006)
16. Kraak, M., Ormeling, F.: Cartography: Visualization of Spatial Data, 3 edn. Guilford Publications, New York (2010)
17. Nogueira, T.P., Martin, H.: Querying semantic trajectory episodes. In: Proc. of MobiGIS, pp. 23–30 (2015)
18. Paiva Nogueira, T., Bezerra Braga, R., Martin, H.: An ontology-based approach to represent trajectory characteristics. In: Fifth International Conference on Computing for Geospatial Research and Application. Washington, DC (2014). https://hal.archives-ouvertes.fr/hal-01058269
19. Parent, C., Spaccapietra, S., Renso, C., Andrienko, G.L., Andrienko, N.V., Bogorny, V., Damiani, M.L., Gkoulalas-Divanis, A., de Macêdo, J.A.F., Pelekis, N., Theodoridis, Y., Yan, Z.: Semantic trajectories modeling and analysis. ACM Comput. Surv. **45**(4), 42 (2013)
20. Peuquet, D.J.: It's about time: a conceptual framework for the representation of temporal dynamics in geographic information systems. Ann. Assoc. Am. Geogr. **84**(3), 441–461 (1994)
21. Santipantakis, G., Vouros, G., Glenis, A., Doulkeridis, C., Vlachou, A.: The datAcron ontology for semantic trajectories. In: ESWC-Poster Session (2017)
22. Santipantakis, G.M., Vouros, G.A., Doulkeridis, C., Vlachou, A., Andrienko, G.L., Andrienko, N.V., Fuchs, G., Garcia, J.M.C., Martinez, M.G.: Specification of semantic trajectories supporting data transformations for analytics: The datacron ontology. In: Proceedings of the 13th International Conference on Semantic Systems, SEMANTICS 2017, Amsterdam, 11–14 Sept 2017, pp. 17–24 (2017). https://doi.org/10.1145/3132218.3132225
23. Santipantakis, G., Doulkeridis, C., Vouros, G.A., Vlachou, A.: Masklink: Efficient link discovery for spatial relations via masking areas, arXiv:1803.01135v1 (2018)
24. Santipantakis, G.M., Glenis, A., Kalaitzian, N., Vlachou, A., Doulkeridis, C., Vouros, G.A.: FAIMUSS: flexible data transformation to RDF from multiple streaming sources. In: Proceedings of the 21th International Conference on Extending Database Technology, EDBT 2018, Vienna, 26–29 March 2018, pp. 662–665 (2018). https://doi.org/10.5441/002/edbt.2018.79

25. Santipantakis, G.M., Kotis, K.I., Vouros, G.A., Doulkeridis, C.: RDF-gen: Generating RDF from streaming and archival data. In: Proceedings of the 8th International Conference on Web Intelligence, Mining and Semantics, WIMS '18, pp. 28:1–28:10. ACM, New York (2018). http://doi.org/10.1145/3227609.3227658. http://doi.acm.org/10.1145/3227609.3227658

26. Soltan Mohammadi, M., Mougenot, I., Thérèse, L., Christophe, F.: A semantic modeling of moving objects data to detect the remarkable behavior. In: AGILE 2017. Wageningen University, Chair group GIS & Remote Sensing (WUR-GRS), Wageningen (2017). https://hal.archives-ouvertes.fr/hal-01577679

27. Spaccapietra, S., Parent, C., Damiani, M.L., de Macêdo, J.A.F., Porto, F., Vangenot, C.: A conceptual view on trajectories. Data Knowl. Eng. **65**(1), 126–146 (2008)

28. Vincenty, T.: Direct and inverse solutions of geodesics on the ellipsoid with application of nested equations. In: Survey Review XXII, pp. 88–93 (1975). https://doi.org/10.1179%2Fsre.1975.23.176.88. https://www.ngs.noaa.gov/PUBS_LIB/inverse.pdf

29. Vouros, G., Santipantakis, G., Doulkeridis, C., Vlachou, A., Andrienko, G., Andrienko, N., Fuchs, G., Martinez, M.G., Cordero, J.M.G.: The datacron ontology for the specification of semantic trajectories: specification of semantic trajectories for data transformations supporting visual analytics. J. Data Semant. **8** (2019). http://doi.org/10.1007/s13740-019-00108-0. http://link.springer.com/article/10.1007/s13740-019-00108-0

30. Wen, Y., Zhang, Y., Huang, L., Zhou, C., Xiao, C., Zhang, F., Peng, X., Zhan, W., Sui, Z.: Semantic modelling of ship behavior in harbor based on ontology and dynamic Bayesian network. ISPRS Int. J. Geo-Inf. **8**(3) (2019). http://doi.org/10.3390/ijgi8030107. http://www.mdpi.com/2220-9964/8/3/107

31. Yan, Z., Macedo, J., Parent, C., Spaccapietra, S.: Trajectory ontologies and queries. Trans. GIS **12**(s1), 75–91 (2008). http://doi.org/10.1111/j.1467-9671.2008.01137.x. https://onlinelibrary.wiley.com/doi/abs/10.1111/j.1467-9671.2008.01137.x

Chapter 6
Integrating Data by Discovering Topological and Proximity Relations Among Spatiotemporal Entities

Georgios M. Santipantakis, Christos Doulkeridis, Akrivi Vlachou, and George A. Vouros

Abstract Link discovery (LD) is the process of identifying relations (links) between entities that originate from different data sources, thereby facilitating several tasks, such as data deduplication, record linkage, and data integration. Existing LD frameworks facilitate data integration tasks over multidimensional data. However, limited work has focused on spatial or spatiotemporal LD, which is typically much more processing-intensive due to the complexity of spatial relations. This chapter targets spatiotemporal link discovery, focusing on topological and proximity relations, proposing a framework with several salient features: support both for streaming and archival data, support of spatial relations in 2D and 3D, flexibility in terms of input consumption, improved filtering techniques, use of blocking techniques, proximity-based LD instead of merely topological LD, and a data-parallel design and implementation. The efficiency of the proposed spatiotemporal LD framework is demonstrated by means of experiments on real-life data from the maritime and aviation domains.

6.1 The Role of Link Discovery in Data Integration

Data integration is a critical task for applications managing data that originates from different and often heterogeneous data sources. In the current era of big data [9], where data is of massive volume, generated at unprecedented rates in the form of data streams, and in different representation models, formats and modalities, big data integration [6] raises additional challenges.

One significant step to facilitate data integration is *link discovery* (LD) [11], which is defined as the process of identifying relations (links) between entities/objects that originate from different data sources. In the following, we present different

G. M. Santipantakis · C. Doulkeridis (✉) · A. Vlachou · G. A. Vouros
University of Piraeus, Piraeus, Greece
e-mail: gsant@unipi.gr; cdoulk@unipi.gr; avlachou@unipi.gr; georgev@unipi.gr

© Springer Nature Switzerland AG 2020   155
G. A. Vouros et al. (eds.), *Big Data Analytics for Time-Critical Mobility Forecasting*, https://doi.org/10.1007/978-3-030-45164-6_6

applications of link discovery tasks that are meaningful in real-life scenarios: record linkage [4], entity resolution [5, 16], and data deduplication [7].

Example 6.1 (Record Linkage Between Data Sources) Consider a merge taking place between two companies, which requires that their customer databases need to be reconciled. Customers are described by their names and identifying common customers requires joining records based on alphanumeric values (names). However, in practice, using exact matching does not work sufficiently well, due to various reasons, such as spelling errors (e.g., "Jon Smith" vs. "John Smith"), use of middle name (e.g., "George Vouros" vs. "George A. Vouros"), etc. Therefore, approximate matching techniques (a.k.a. approximate or fuzzy joins) are required, which constitute a special case of link discovery between entities.

Example 6.2 (Entity Resolution) Consider two different data representations corresponding to two entities, where the problem is to determine if the two entities are identical. For instance, a Wikipedia page describing an athlete and a record in a table of a relational database of athletes. The problem is to discover that two entities refer to the same real-world object.

Example 6.3 (Deduplication) Consider the case of two persons that perform data entry using different naming conventions. In this case, we may encounter multiple records that are not identical, yet they refer to the same object. As a representative example one can think of "Los Angeles" vs. "LA", "fifth avenue" vs "5th avenue", etc.

In order to support application scenarios such as the above, efficient and effective link discovery techniques are sought. This book chapter focuses on a special case of link discovery, where the underlying data sets are of spatiotemporal nature and the relations to be discovered are also spatial or spatiotemporal.

The remainder of this book chapter is structured as follows: Section 6.2 provides background knowledge on link discovery and explains the problem of spatiotemporal link discovery. Section 6.3 presents the design and implementation of a spatiotemporal link discovery framework. Then, Sect. 6.4 focuses on scalability aspects of link discovery, under the prism of streaming data and also for voluminous data sources. Section 6.5 presents the results of an empirical study of link discovery over real-world spatiotemporal data sets. Finally, Sect. 6.6 summarizes the most important findings regarding spatiotemporal link discovery.

6.2 Background on Spatiotemporal Link Discovery

In this section, we briefly introduce basic concepts of link discovery, followed by the problem of spatiotemporal link discovery, and an overview of existing systems and frameworks.

6.2.1   Principles of Link Discovery

Link discovery is a challenging topic which relates to record linkage [4], deduplication [7], and data fusion [2].

The problem of link discovery can be formalized as follows. Consider two data sets T and S, which are usually called *target* and *source*, respectively. Also, consider a relation r that may hold between entities of the two data sets, i.e., $r(\tau, \sigma)$, such that: $\tau \in T$, $\sigma \in S$, and the pair (τ, σ) satisfies the relation r. We use $r(\tau, \sigma)$ to denote that relation r holds over entities τ and σ.

A brute force link discovery algorithm would have to evaluate all entities in T against all entities in S, thereby producing the result using $O(n \cdot m)$ comparisons, where $n = |T|, m = |S|$. However, this cost may be prohibitively expensive in practice for large data sets.

Hence, *blocking* techniques [16] are typically employed to reduce this cost. Essentially, entities are grouped in *blocks* in such a way that only entities within the same block need to be compared against each other for potential satisfaction of the given relation. In turn, this drastically reduces the processing cost of link discovery in practice.

Two of the most popular link discovery frameworks that have appeared in the literature are LIMES [15] and SILK [8]. LIMES is a generic link discovery framework which facilitates different approximation techniques to compute estimates of the similarity between instances. Specifically, the approaches implemented in LIMES include the original LIMES [15], HR^3 [13] (which aims to further reduce the number of comparisons), HYPPO [12] (a hyperspace approximation algorithm), and ORCHID [14] (a combination of the Hausdorff and orthodromic metrics to compute the distance between geo-spatial objects). The original LIMES algorithm uses the triangular inequality in order to avoid processing all possible pairs of objects. For this purpose, it employs the concept of exemplars, which are used to represent areas in the multidimensional space, and tries to prune entire areas (and the respective enclosed entities) from consideration during the refinement step. Another link discovery framework is SILK, proposing a novel blocking method called *MultiBlock*, which uses a multidimensional index in which similar objects are located near to each other. In each dimension the entities are indexed by a different property or different similarity measure. Then, the indices are combined together to form a multidimensional index, which is able to prune more entities by taking into account the combination of dimensions.

6.2.2   The Problem of Spatiotemporal Link Discovery

Spatiotemporal link discovery is a subarea of link discovery where the underlying data sets are spatial or spatiotemporal and the relations of interest are also of spatiotemporal nature. In this chapter, we focus mainly on two types of spatial relations: topological and proximity relations. However, we also consider the extension of these relations for spatiotemporal data, rather than just spatial data.

6.2.2.1 Topological Relations

Topological relations are defined for spatial objects, such as points, polylines, and polygons, in order to express the location of one object with respect to the other [10]. One of the most popular models for topological relations between spatial objects is the Dimensionally Extended nine-Intersection Model (DE-9IM). For any pair of spatial objects (e.g., a and b), the following nine relations are defined:

- equals: a and b are topologically equal, i.e., $a \cap b = a$ and $a \cap b = b$.
- disjoint: a and b have no common point, i.e., $a \cap b = \emptyset$.
- intersects: a and b have at least one common point, i.e., $a \cap b \neq \emptyset$.
- touches: a and b have at least one common boundary point, but no interior point.
- contains: a contains b if it holds that $a \cap b = a$.
- covers: a covers b if every point of b lies in the interior of a.
- coveredBy: this is symmetrical to covers, i.e., a is covered by b, if b covers a.
- within: a lies in the interior of b.

Some additional properties can be derived based on the definitions above. For instance, equals(a, b), if within(a, b) and contains(a, b). Also, disjoint(a, b), if intersects(a, b)=false. Also, covers(a, b) is equivalent to within(b, a).

6.2.2.2 Proximity Relations

Proximity relations are also known as distance relations and practically determine distances between spatial objects. The most common proximity relation is nearby, which uses a spatial distance threshold θ to determine pairs of spatial objects a, b whose distance is at most θ. Obviously, this requires the use of a distance function, and the definition of distance between spatial objects (e.g., point to polygon or polyline to polygon). Another proximity relation is nearest (and its generalization k-nearest) which identifies the nearest object b (the k nearest objects $\{b_1, \ldots, b_k\}$, respectively) to a given object a.

6.2.2.3 Temporal Relations

Temporal relations between temporal intervals are based on Allen's interval algebra, a calculus for temporal reasoning [1]. Practically, the calculus defines a set of 13 possible relations between time intervals. The following list describes 7 relations, while for the first 6 there exists the inverse relation too. Consider two temporal intervals a, b, each associated with *begin* and *end* timestamps, then the temporal relations can be defined as follows:

- `before`: a is before b, if a ends prior to the begin of b.
- `meets`: a meets b, if the end of a coincides with the begin of b.
- `overlaps`: a overlaps b, if a begins before b and a ends after the begin of b.
- `starts`: a starts b, if their begins coincide.
- `during`: a begin is after b begin, and a end is prior to b end.
- `finishes`: a finishes b, if their ends coincide.
- `equals`: both begin and end of a coincide with begin and end of b, respectively.

6.2.2.4 Motivating Real-Life Applications and Examples

Figure 6.1 depicts two real-life data sets, namely spatiotemporal positions of vessels and protected Natura2000 areas (represented as spatial regions) in the wider area of Spain. Maritime surveillance authorities as well as environmental agencies are interested in monitoring obedience to rules regarding crossing such protected areas. This can be seen as a link discovery task, which aims at identifying a spatial relation such as `within` between the position of a vessel and a polygon representing the protected area.

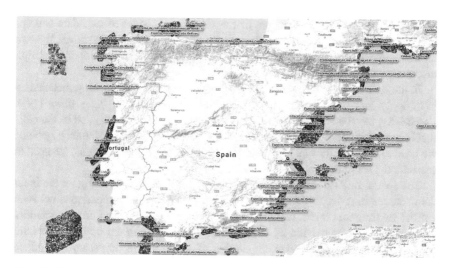

Fig. 6.1 Natura2000 regions and vessels recorded to be within these regions (different colours represent different vessels)

Another application of spatiotemporal link discovery is complex activity recognition. For example, in the maritime domain vessels are required to report their positions using systems such as AIS,[1] and authorities are interested in the detection of different (complex) patterns, such as trawling, loitering, potential collision, approaching a point/region of interest (cf. Chap. 1 of this book). For the detection of such complex patterns, especially at large scale or in real time, complex event recognition software can be greatly assisted by spatiotemporal link discovery, which detects spatiotemporal relations between moving and static objects [18].

6.2.3 Existing Spatial Link Discovery Systems and Frameworks

Despite the wide applications of link discovery and the interest that it has attracted lately (see [11] for a survey), there is not much work on the challenging topic spatiotemporal link discovery, and practically no work on spatial link discovery for big data. In this section, we provide an overview of existing spatial link discovery systems that target relatively small and static data sets.

HR^3 [13] and HYPPO [12] address link discovery tasks when the property values that are to be compared are expressed in an affine space with a Minkowski distance. Both approaches are designed with main objectives to be efficient and lossless. In addition, HR^3 [13] comes with theoretical guarantees on reduction ratio, a metric that corresponds to the percentage of the Cartesian product of two data sources that was not explored before reporting the link discovery results. However, the two aforementioned approaches for LD do not explicitly focus on spatiotemporal link discovery, nor do they tackle the streaming nature of data sources.

In the case of link discovery for spatial data, the prevalent blocking mechanism is to apply *grid partitioning* of the 2D space (also known as *space tiling* [14]). Essentially, the space is partitioned to cells, and any entity is assigned to the cells that include this or any of its parts (e.g., in the case of entities represented by points, we need to consider the cell that includes the point, and in case of areas, we need to consider the set of cells overlapping with the area).

The spatial link discovery methods [14, 20] apply grid partitioning on the input data sources in order to create blocks of entities and avoid comparisons between entities of different blocks, which are guaranteed not to result to a link association. Then, during the processing of each individual block, different optimizations are employed in order to minimize the number of computations necessary to produce the correct result set.

RADON [20] is the most recent approach for discovering topological relations between data sources of areas, and can discover efficiently multiple relations using space tiling. One of its main techniques for efficiency relies on the use of caching

---

[1]The automatic identification system (AIS) is an automatic tracking system that uses transponders on ships and is used by vessel traffic services (VTS).

to avoid recomputing distances. However this imposes non-negligible requirements for main memory, especially for large data sources. Furthermore, RADON employs techniques that achieve efficiency for data sets that consist of polygons rather than points. In this sense, it addresses a specific case of spatial LD.

ORCHID [14] is another grid partitioning method, which studies the problem of discovering all pairs of polygons, whose Hausdorff distance (practically Max–Min distance) is below a given threshold. As most spatial link discovery approaches, it also employs space tiling to improve the filtering step, and bounding circles are used as approximations of polygons. Moreover, already computed distances are maintained in order to avoid (re-)computing new distances, when possible. Finally, the triangular inequality is used for pruning areas without distance computations.

Smeros et al. [21] study link discovery on spatiotemporal RDF data. The authors study several topological relations that are defined on polygons. As usual, the algorithm relies on a grid to filter out cells that contain polygons which cannot satisfy the relation. Unfortunately, the topological relations do not take into account proximity nor distance between polygons, and their approach primarily targets polygons (rather than points).

6.3 A New Spatiotemporal LD Framework

In this section, we present the design and implementation of a generic link discovery framework [19] that is tailored for spatiotemporal data. We present its extensible design and flexibility in terms of supported spatiotemporal data types and relations (Sect. 6.3.1). Then, we outline a state-of-the-art method for link discovery of topological relations (Sect. 6.3.2). To complement this approach, we present improvements that apply to certain cases of spatiotemporal data and may lead to performance gains (Sect. 6.3.3). We then present methods for proximity-based link discovery (Sect. 6.3.4). Last, but not least, we consider the case of more complex geometries, such as polylines (Sect. 6.3.5).

6.3.1 The Architecture of a Spatiotemporal Link Discovery Framework

The generic architecture of the spatiotemporal link discovery stLD framework is illustrated in Fig. 6.2. Its input is two data sources represented in RDF format.[2] RDF is the de-facto standard for semantic data representation, and it makes possible to express the discovered relations (links) also in RDF. Hence, the output of stLD is

---

[2] Any other data source can be transformed in RDF using an appropriate data transformation tool, such as RDF-Gen [17].

also provided in RDF and can be either: (a) only the linked entities discovered or (b) the linked entities concatenated with the input RDF fragment. The first option allows to decouple the processing of input data sets from the processing of discovered links and reduces the overall link discovery time. The second option provides synchronized and sequential RDF fragments to the output, which is beneficial for certain applications that need to process enriched input entities (i.e., with additional spatiotemporal links and properties).

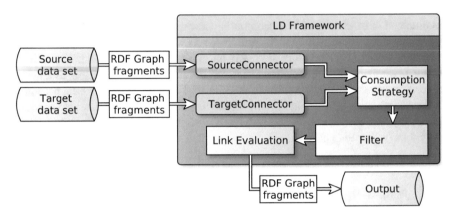

Fig. 6.2 High-level architecture of the spatiotemporal link discovery framework stLD

The stLD framework follows the *filter-and-refine methodology* for performing the link discovery task. In the filtering step, a blocking method is employed to drastically reduce the number of candidate pairs of entities, whereas in the refinement step the candidate pairs need to be examined to check if they satisfy the relation of interest. In brief, the main components of the stLD framework (also presented in Fig. 6.2) include:

- *Connectors*: responsible to parse and validate the input RDF fragments, and select only those entities and their properties that are necessary for the link discovery tasks. We distinguish the two connectors by their name (SourceConnector and TargetConnector), because the two sources may in principle represent different types of spatiotemporal data.
- *Consumption Strategy*: specifies the order in which the target (T) and source (S) data sets are consumed by the framework. As will be explained shortly, this offers flexibility and makes the framework suitable for a variety of link discovery tasks.
- *Blocking Methods*: determine the type of blocking employed by the stLD framework from a pool of available blocking methods, and practically correspond to the *filtering* part of the filter-and-refine methodology.
- *Link Evaluation*: performs the actual evaluation of the specified relation $r \in R$ over candidate pairs (τ, σ), with $\tau \in T'$ and $\sigma \in S'$, where T' and S' are subsets

of T and S, respectively ($T' \subseteq T$ and $S' \subseteq S$), determined by the blocking method. Essentially, this corresponds to the *refinement* part of the methodology.

6.3.1.1 Connectors and Configurations

The Connectors implement access to the input RDF fragments based on a Configuration, which is also expressed in RDF format. Table 6.1 reports the core terms of the configuration namespace. A complete example of a configuration file is also available in Fig. 6.3. The core part of a configuration is the link specification. It states the conditions that need to be satisfied by properties of entities, for a given link to be established between these entities. The conditions are implemented as functions (called "evaluation functions"), executed during the refinement part of the link discovery task. The evaluation functions vary from arbitrary combinations of simple or weighted distance/similarity functions to combinations of Boolean spatial relations, as specified by the Open Geospatial Consortium (OGC).

Table 6.1 The core terms of the configuration namespace

Name	Description
dcf:prefix	Specifies the prefices used in the SPARQLfilter part. Zero or more prefices can be used.
dcf:source	Specifies the path to the data source (it can be local or remote).
dcf:SPARQLfilter	Provides a SPARQL query as a filter to the triples provided by the data source. The results of the SPARQL query are used for the compilation of spatiotemporal entities evaluated in the link discovery tasks.
dcf:compiler	Specifies the implemented function to be used to compile entities in the source from the query results of dcf:SPARQLfilter.
dcf:EvaluationFunction	Specifies the implemented function applied to evaluate if a relation holds.
dcf:Link	Defines the term to be used for associating entities that satisfy a relation. This option affects the triples generated for the linked entities.
dcf:output	Configures the framework to either report (a) only the computed links (using the value "links_only") or (b) the consumed data enriched with computed links (using the value "all").

```
...
< dcf : Prefix rdf : about ="http ://www. datacron −project . eu / config#prefix1">
< dcf : PrefixKey >:</dcf : PrefixKey >
< dcf : PrefixValue >
http ://www. datacron −project . eu / datAcron#
</dcf : PrefixValue >
</dcf : Prefix >
< dcf : Prefix rdf : about ="http ://www. datacron −project . eu / config#prefix2">
< dcf : PrefixKey >dul :</dcf : PrefixKey >
< dcf : PrefixValue >
http ://www. ontologydesignpatterns . org / ont / dul /DUL. owl#
</dcf : PrefixValue >
</dcf : Prefix >

< dcf : LD_Configuration
        rdf : about ="http :// test /20170329/ source / ais_brest_source ">
< dcf : source >../ ais_data_20170329 . ttl </dcf : source >
< dcf : SPARQLfilter >SELECT ?vesselID ?nodeID ?heading ?time ?wkt WHERE{
   ?event : occurs ?nodeID .
   ?nodeID : ofMovingObject ?vesselID ;
     : hasHeading ?heading ;
     dul : hasConstituent ?t ;
     dul : hasConstituent ?g .
   ?t : TimeStart ?time .
   ?g : hasWKT ?wkt .
}
</dcf : SPARQLfilter >
< dcf : compiler >gr . unipi . ailab . datacron ... movingVessel </dcf : compiler >
< dcf : EvaluationFunction >gr . unipi . ailab . datacron ... within
</dcf : EvaluationFunction >
< dcf : Link >: within </dcf : Link >

< dcf : output >all </dcf : output >
...
```

Fig. 6.3 Example of a link discovery task configuration

6.3.1.2   Consumption Strategy

The consumption strategy specifies how input data should be processed, and it is specified in the configuration. The stLD framework supports a variety of strategies:

- *target-first*: This strategy processes first the target data set in its entirety, organizes its entities using a blocking method, and then proceeds to the source data set. This strategy enables tuning for better performance, such as swapping target-source data sources, asserting in the grid the smaller data set, and reducing the overall processing time.
- *target-first-n*: This strategy is applicable on data sources where a single entity is referenced on multiple spatiotemporal positions. For instance, a trajectory is represented as a sequence of spatiotemporal positions, so it is practically a list

of records. Thus, a trajectory needs to be reconstructed as an object, prior to identifying links with entities of the other data source.

- *jointly*: This strategy processes and consumes the two data sources together. This method is quite different from strategies used by state-of-the-art link discovery frameworks, which typically first consume and organize one data source and then process the other. Instead, it can be applied for the consumption of two streaming data sources.

6.3.1.3   Blocking Methods (Filter)

As already mentioned, LD tasks rely on blocking techniques [11] to split the space in blocks (cells in geographical terms)—a process known as *grid partitioning* or *space tiling*—assign spatial objects to cells based on overlap, and eventually compare only pairs of objects in each cell. This avoids the cost of exhaustive comparison ($O(|T| \cdot |S|)$) between each pair of objects. Essentially, the grid cells help in filtering the majority of pairs of objects, thus retaining only few candidate pairs (typically those within a cell), which need to be evaluated in a subsequent refinement step.

The proposed framework supports different blocking methods, from grid structures to spatial indexes (such as R-trees). In the following, for clarity of presentation, we restrict the discussion to a grid structure that consists of cells, but the ideas are applicable to other indexing structures.

The grid specifies how the spatiotemporal entities of the target data source should be organized. In its general form, it consists of a set of cells and the functions to add and retrieve entities to and from the cells, respectively. In the implementation of stLD, two different types of grid structures are provided: (a) EquiGrid, which given the granularity of each dimension, constructs equally sized cells, (b) Hierarchical grid, where multiple EquiGrids are employed with different granularities organized in a hierarchy. The functions that add and retrieve entities to/from cells are very efficient. Specifically, for a given entity we retrieve the cells in $O(D)$, where D is the dimensionality of entities. For low dimensionality values,[3] this is practically performed in constant time.

6.3.1.4   Link Evaluation

This component practically corresponds to the refinement phase of link discovery. Its input are pairs of entities (τ, σ) that have been output from the filtering step, and are *candidate pairs* for producing a link r, i.e., a valid relation $r(\tau, \sigma)$. The link evaluation implements the necessary algorithms for checking the condition for a link to hold, according to the specifications. The link specifications are set in

---

[3]Only 2 dimensions are needed to represent spatial data, whereas 3 dimensions are needed for spatiotemporal data.

the configuration of the source data set. Commonly used links include `nearby`, `within`, `equals`, and depend on proximity and topological relations, between any combination of entities.

In the implementation, all relations between any type of spatial objects, as specified by the Open Geospatial Consortium (OGC),[4] are supported: points, polygons, and polylines. In addition, proximity relations have been implemented and the framework has been generalized to be applicable in 3D geographical space.

6.3.2 Link Discovery of Topological Relations

In this section, we present solutions for link discovery of topological relations between areas, as the ones mentioned above. At first, the focus is on a baseline approach, followed by the state-of-the-art algorithm called RADON [20]. Finally, we present the proposed solution MaskLink.

6.3.2.1 A Baseline Link Discovery Algorithm

Given the *target* T, and *source* S data sources, and a relation r in the set of topological relations R defined above, the goal of link discovery is to detect the pairs $(\sigma, \tau) \subseteq S \times T$, where $\sigma \in S$ and $\tau \in T$, such that (σ, τ) satisfies r.

As already mentioned, a brute force link discovery algorithm would have to perform the geometrical test between all pairs of entities. To avoid this excessive cost, blocking techniques are typically employed in order to prune the candidate pairs of entities and reduce the number of candidates considered during the refinement step. In the case of link discovery for spatial data, the prevalent blocking mechanism is to apply *grid partitioning* of the 2D space (also known as *space tiling*).

Essentially, the space is partitioned to cells, and any entity is assigned to the cells that include this or any of its parts (e.g., in the case of entities represented by points, we need to consider the cell that includes the point, and in case of areas, we need to consider the set of overlapping cells).

Grid Construction For the construction of the grid and its constituent cells, the first step is to derive the bounding box of the two data sets S and T. Essentially, this bounding box is the 2D data space, which is of interest for the current LD task. The bounding box is defined by two points: its lower left corner $(x_L, y_L) = (\min_{\forall i} x_i, \min_{\forall i} y_i)$ and its upper right corner $(x_U, y_U) = (\max_{\forall i} x_i, \max_{\forall i} y_i)$, where x_i and y_i represent the X- and Y-coordinate value of the i-th point of a spatial object. Let a function MBR(.) s.t. the bounding box of a given geometry g is denoted as MBR$(g) = \langle (x_L, y_L), (x_U, y_U) \rangle$.

---

[4]Definition and requirements of relations are available online at https://www.opengeospatial.org/standards/geosparql.

The second step is to construct the grid cells. For this purpose, two parameters m_x and m_y need to be set, corresponding to the number of splits in the X-axis and Y-axis, respectively. As a result, the number of grid cells is $m_x \cdot m_y$. These parameters also define the *granularity* of the grid, since the sides of a cell are $\frac{x_U - x_L}{m_x}$ and $\frac{y_U - y_L}{m_y}$, respectively. For example, a 2D grid with 0.5×1.5 granularity is built by cells of size $0.5°$ by $1.5°$.[5]

Assignment to Grid Cells After the grid has been constructed, the next step is to organize the entities of the given data sets in grid cells. This step is performed using a function that assigns each entity to one or more cells of the grid. We say that a spatial representation $\sigma \in S$ is *assigned* to a cell c, if c intersects with σ. Then, to compute each relation $r \in R$, the spatial representation σ is compared only against those entities in T assigned to the cells where σ has been assigned. This approach results to much fewer comparisons of candidate pairs, since only pairs corresponding to the same cell need to be examined. We refer to this link discovery technique as *baseline*.

6.3.2.2 The RADON Algorithm

RADON [20] is an efficient algorithm for topological link discovery, which is optimized for polygons and static data sources. Its basic operation relies on the use of an equi-grid partitioning of the 2D space, where an arbitrary number of rectangular cells are created. As a result, RADON relies on the same baseline (admittedly with optimizations): first, it organizes the target data source T in the grid, by assigning minimum bounding boxes (MBBs) of areas to cells, and then it processes each polygon of S by assigning also to overlapping cells.

To achieve improved performance, RADON utilizes the following techniques. First, it selects the data set that will be organized in the cells, based on a heuristic, thus offering performance gain at runtime (this is called *swapping*). Second, it only assigns areas $\sigma \in S$ to a cell that already contains areas from data set T. This is mentioned as sparse space tiling. Third, it applies a caching mechanism in order to avoid re-computing relations for pairs (σ, τ) that have been previously computed, e.g., due to σ, τ spanning multiple cells.

Despite the benefits in efficiency offered by these techniques, their adoption also limits the applicability of RADON. For instance, swapping requires that both data sets are available beforehand, thereby rendering this technique (and consequently RADON) inapplicable in the case of streaming data sources. Also, RADON requires both data sets to be stored in-memory, and in combination with the caching mechanism, it adds a significant overhead in main memory.

---

[5]Other parameters for the construction of the grid include (but are not limited to) the precision of the spatiotemporal representations and the Coordinate Reference System (CRS) to be used.

In contrast, in the following, the MaskLink approach is presented where only the target data source T is organized in memory, thus: (a) offering scalability regardless of the size of S and (b) making it applicable in the case of a streaming data source.

6.3.3   The MaskLink Technique for Link Discovery of Topological Relations

In this section, we present the MaskLink technique in the case where at least the target data source includes entities whose spatial representation are polygons. The MaskLink technique is in the core of the proposed Link Discovery framework.

The basic idea exploited by MaskLink is that in several cases of spatial link discovery, grid cells contain a significant amount of "empty space", namely the part of the cell that overlaps with no areas of the target data set T. Consider a spatial representation $\sigma \in S$ of an entity that overlaps with a cell, and k spatial representations $\{\tau_1, \ldots, \tau_k\} \in T$ that also overlap with the given cell. Our observation is that if σ is disjoint to all the spatial representations in $\{\tau_1, \ldots, \tau_k\}$ in this cell, then we can safely infer that there are no other (except "disjointness") topological relations to be discovered in this cell between σ and any area in $\{\tau_1, \ldots, \tau_k\}$.

Motivated by this observation, we propose the masking technique to explicitly represent the empty space within cells as yet another area. Thus, for each grid cell c, we construct an artificial polygon called *mask* of c, which is defined as the difference between the cell and the union of areas overlapping with the cell, i.e.,

$$mask(c) = area(c) - (area(c) \cap \bigcup_i area(\tau_i)).$$

Figure 6.4 shows an example of the mask of a cell; the middle cell overlaps with areas in $\{\tau_1, \ldots, \tau_5\}$, and the mask of the cell is the area represented in black colour.

Having the mask of a cell as yet another area, we can devise an efficient algorithm for link discovery that eagerly avoids comparisons to geometries for spatial representations enclosed in the mask of a cell. In practice, first we identify a cell c to which the spatial representation σ is assigned. On case σ is a polygon, we first construct the $Overlap\_S(c, \sigma)$. We then compare $Overlap\_S(c, \sigma)$ to the mask of the cell, to check if it is enclosed in the empty space. If this single comparison returns true, we stop processing this entity, thereby saving k comparisons. For the typical case where a cell contains several areas, this technique prunes several candidate pairs of entities, saving computational time in the refinement step of the LD process.

Algorithm 6.1 presents the pseudo-code for discovering any topological relation link between the spatial representation σ and the spatial representations of T. As a prerequisite, the grid has already been constructed and the spatial representations of T have been assigned to cells. This is essentially a pre-processing step. In

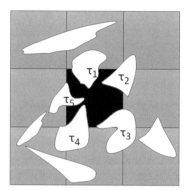

Fig. 6.4 Illustration of the mask of a grid cell that overlaps with five areas $\tau_1, \ldots, \tau_5$

Algorithm 6.1 Spatial LD algorithm for topological relations using mask

---

1: **Input:** Grid cells $C = \{c_1, \ldots, c_m\}$, spatial representations $T = \{\tau_1, \ldots, \tau_n\}$, σ
2: **Output:** Set of relations $r(\sigma, \tau_i)$ s.t. r is a topological relations and $\tau_i \in T$
3: **Requires:** Grid has been constructed, and spatial representations in T have been assigned to overlapping cells
4: $T^w \leftarrow \emptyset$
5: locate cells $\Psi \subseteq C$ that enclose or overlap with σ
6: **for** each $c_i \in \Psi$ **do**
7:     **if** within$(Overlap\_S(c_i, \sigma), mask(c_i))$ **then**
8:         **for** each $\tau_j \in c_i$ **do**
9:             $T^w \leftarrow T^w \cup \{disjoint(\sigma, \tau_j)\}$
10:        **return** T^w
11:    **else**
12:        **for** each τ_j assigned to c_i **do**
13:            **for** each relation r **do**
14:                **if** (σ, τ) **then**
15:                    $T^w \leftarrow T^w \cup \{r(\sigma, \tau_j)\}$
16: **return** T^w

---

the first step, the cells Ψ to which σ is assigned are determined (line 5). This operation is performed in constant time $O(1)$ in the case of equi-grids. Then, for each $c_i \in \Psi$, the algorithm checks if $Overlap\_S(c_i, \sigma)$ is enclosed in the mask $mask(c_i)$ of cell c_i (line 7). In the latter case, no further processing is required, and the algorithm terminates returning the inferred set of disjoint relations to the spatial representations assigned to c_i. If it is not contained, then each relation against all areas τ_j in cell c_i (line 12) is checked. For those areas τ_j that satisfy a relation $r(\sigma, \tau_j)$, we append the discovered links in the result set T^w (line 15), and return T^w.

The lines 12–15 of Algorithm 6.1 can be processed in parallel, i.e., each iteration in the for loop is carried out by a different thread ("worker"). The number of concurrent workers is usually a predefined constant w.r.t. system configuration, to allow uninterrupted system operation (in the experiments, 4 workers are used).

Multi-thread processing is enabled using a *pool of tasks*, populated with the refinement tasks of $r(\sigma, \tau_j)$. As soon as a worker is available and the pool contains tasks, the next task is selected and assigned to the worker for processing.

6.3.4 Link Discovery of Proximity Relations

Interestingly, the MaskLink technique is applicable also for link discovery of proximity relations between spatial representations of entities in S and T. More concretely, let us consider the `nearby` relation. Recall that the `nearby` relation is defined using a spatial threshold θ and returns true when it holds that $dist(\sigma, \tau) \leq \theta$, for two spatial representations σ, τ.

Proximity link discovery concerns the identification of all such relations between entities in data sets S and T. In more detail, given the spatial representation of any entity (either a point or a polygon) $\sigma \in S$, we wish to discover the subset of spatial representations in T that are located at most at distance θ from σ.

Let $\tau' = buff(\tau, \theta)$ denote the expanded area of τ that contains all points in space located in distance less or equal to θ from any point in τ, and can be computed using a standard library for computational geometry. Any such area, depending on threshold θ, is called θ-buffered area.

Assuming that σ is an area, we make the following basic observation: if $\sigma, buff(\tau, \theta)$ are disjoint, then `nearby`$(\sigma, \tau, \theta) = false$. We extend this observation for more than one area τ, to make it applicable for cells.

More formally, given a cell c that overlaps with the area $\sigma \in S$ and k areas $\{\tau_1, \ldots, \tau_k\} \subseteq T$, if $Overlap\_S(c, \sigma)$ is disjoint to any $buff(\tau, \theta)$, then we can infer that `nearby`$(\sigma, \tau, \theta) = false$ for all τ_i, $1 \leq i \leq k$. A similar case holds in case σ is a point: if σ is not enclosed in any $buff(\tau, \theta)$, then we can infer that `nearby`$(\sigma, \tau, \theta) = false$ for all τ_i, $1 \leq i \leq k$.

Based on these observations, MaskLink can be slightly adjusted to support the relation `nearby`. The main adjustment concerns the way the mask of each cell is computed. First, we expand each area τ_i by θ, and then the mask of the cell is computed as previously, using the θ-buffered areas instead of the actual areas τ_i. To differentiate this mask of a cell c from the one used in the previous algorithm, we denote it by $mask^\theta(c)$.

Algorithm 6.2 presents the pseudo-code for LD of relation `nearby`. Notice that the grid is constructed exactly as before, using the original areas in T. In the pre-processing stage, the algorithm computes the mask $mask^\theta(c_i)$ for each cell using the θ-buffered areas τ_j assigned to c_i. First, the cells Ψ that enclose or overlap with σ (i.e., the cells to which σ is assigned) are located (line 5). For each cell $c_i \in \Psi$, if $mask^\theta(c_i)$ encloses $Overlap\_S(c_i, \sigma)$ (line 7), then only disjoint relations to τ_j areas assigned to c_i can be inferred. This is because of the way $mask^\theta(c_i)$ has been constructed. If this is not true, all areas τ_j assigned to c_i need to be examined, and if $buff(\tau, \theta)$ overlaps or encloses $Overlap\_S(c_i, \sigma)$, then we append with the discovered link `nearby`(σ, τ, θ) the result set T^n (lines 12–14).

Algorithm 6.2 Spatial link discovery algorithm for relation `nearby` using mask

---

1: **Input:** Grid cells $C = \{c_1, \ldots, c_m\}$, Areas $T = \{\tau_1, \ldots, \tau_n\}$, σ, threshold θ
2: **Output:** Set of relations $r(\sigma, \tau_j)$ s.t. $r \in \{nearby, disjoint\}$ and $\tau_j \in T$
3: **Requires:** Grid has been constructed and areas have been assigned to cells
4: $\quad T^n \leftarrow \emptyset$
5: locate cells $\Psi \subseteq C$ to which σ is assigned
6: **for** each $c_i \in \Psi$ **do**
7: $\quad$ **if** $within(Overlap\_S(c_i, \sigma), mask^\theta(c_i))$ **then**
8: $\quad\quad$ **for** each $\tau_j \in c_i$ **do**
9: $\quad\quad\quad T^n \leftarrow T^n \cup \{disjoint(\sigma, \tau_j)\}$
10: $\quad\quad$ **return** T^n
11: $\quad$ **else**
12: $\quad\quad$ **for** each τ_j assigned to c_i **do**
13: $\quad\quad\quad$ **if** $buff(\tau, \theta)$ *overlaps or encloses* σ **then**
14: $\quad\quad\quad\quad T^n \leftarrow T^n \cup \{nearby(\sigma, \tau_j)\}$
15: **return** T^n

6.3.5 Refined Blocking Method

An important part of the blocking-based link discovery is the process followed for deciding the cells of the grid for a given spatial representation. A fast approach that can be used efficiently for most of the spatial representations (polygons or points) is to compute the MBR of the given geometry, and assign the cells for the MBR instead of the given geometry. Since usually a geometry has more points than its MBR, this method can be efficient in terms of performance for most of the cases, i.e., finding the cells overlapping with MBR is trivial, while deciding the cells for the actual geometry may require considerably more evaluations w.r.t. the cell size and geometry type.

However, the MBR-based blocking cannot be as efficient when blocking polylines. This is particularly true when polylines span large geographic areas as it is the case for trajectories of moving objects, such as aircraft or vessels. For example, the trajectory of a flight from London to Moscow will produce an MBR that will cover a large part of the northern Europe, thus filtering will include a considerable amount of cells, far away from the trajectory, wasting computational resources on comparisons that will not lead to links.

To address this limitation, an improvement on the blocking method can be applied. In this method, the minimum necessary set of cells is computed that are needed to cover any given polyline, w.r.t. the granularity of the grid. The general idea of the method is to compute the cells for each point of the polyline, and for the case of closed geometries, the cells for all the interior points. This approach will provide the minimum necessary cells to cover the polyline (the proof is trivial). A naive implementation would be to iterate all the points of the given polyline (including the interpolated points w.r.t. grid granularity) and compute the corresponding cells. This method however would result to linear complexity to the number of points in a polyline, which will be expensive, for polylines consisting of many points.

On the other hand, the proposed "refined blocking" method recursively segments the polyline and terminates when both ends of the segment are within the same cell. This method has linear complexity to the number of cells that will be used for the geometry, w.r.t. the grid granularity. The cost of refined blocking is not higher than the cost of using simply the MBR of the geometry, since the computation of MBR iterates through all the points of the geometry to decide the minimum and maximum latitude/longitude values of the MBR. On the other hand, the proposed method has a worst-case scenario of iterating all the points (i.e., the case where each point of the geometry should be placed in a separate cell w.r.t. the grid granularity).

In addition to that, selecting the minimum number of cells is an important benefit for the refinement task. For example, consider the case of Fig. 6.5, where the blue line illustrates the trajectory of a flight, the red cells are those that will be selected using an MBR blocking method, and the green cells are those selected by our proposed method. We observe that for a trajectory with $(\max(latitude) - \min(latitude)) = m$ cells and $(\max(longitude) - \min(longitude)) = n$ cells w.r.t. the granularity of the grid, the MBR method will return $m \times n$ cells, while the "refined blocking" returns considerably fewer number of cells. In the example of Fig. 6.5, the MBR returns 1170 cells, while the refined blocking method returns 86 cells. If we consider that each cell contributes with a number of candidate entities to be compared with the trajectory for discovering a link, the benefit from refined blocking method is clear. Obviously this blocking method can be applied on buffered trajectories as well, to include adjacent cells when computing proximity relations.

6.4 Scalable LD for Spatiotemporal Data

In this section, we show how the link discovery framework can be adapted to support scalable link discovery, by exploiting big data technologies. Consequently, we focus on parallel data processing methods, which aim at sharing computation with multiple computers (workers), in order to complete the link discovery task, while satisfying real-life performance requirements, focusing on low-latency processing.

6.4.1 Setup for Scalable Link Discovery

Consider a cluster of commodity machines that is available for the task of link discovery in the context of big data. We consider two aspects of big data: volume and velocity. By volume, we refer to data sets of such size that cannot be handled by traditional database systems. By velocity, we refer to streaming data sources of high input rate. In both cases, our problem is how to exploit the available cluster infrastructure in order to minimize the completion time for link discovery.

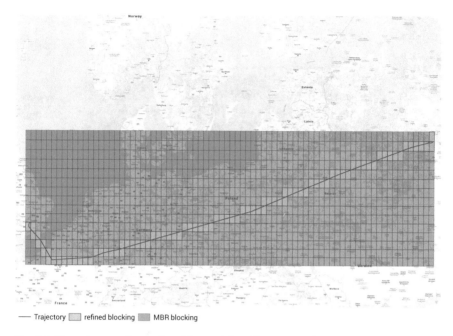

— Trajectory ☐ refined blocking ▨ MBR blocking

Fig. 6.5 The trajectory of a moving object and the cells computed by MBR-based blocking (red cells), and by refined blocking (green cells) on a grid of $0.5° \times 0.5°$ granularity

Our premise is to exploit big data technologies that come with various salient features, such as scalability, fault-tolerance, ease of programming, etc., in order to devise a solution for parallel link discovery. We choose Apache Flink [3] as the underlying big data framework, which provides a high-throughput, low-latency streaming engine that fits our requirements, even though our techniques can be developed on top of other frameworks as well.

6.4.2   Stream-Based Link Discovery

In the following, we present our techniques for link discovery having as inputs a streaming data source and a static data source (stream-static). Then, we discuss other potential setups, such as static-static or stream-stream.

We demonstrate two variants for parallelization, which are applicable in different setups. The first setup concerns a streaming data source and a static data set, which is additionally not extremely large. The second setup is applicable for two streaming data sources, but also for the case of extremely large static data sets.

Stream-Static   In the case that the target data set is static and not extremely large, we parallelize the LD task by broadcasting the blocks of this data set to all worker

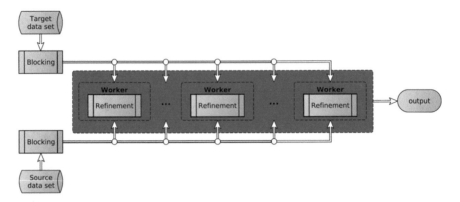

Fig. 6.6 Parallel LD for one streaming data source and a static data set

nodes, as depicted in Fig. 6.6. Then, parallelism is achieved by having each entity of the source data set sent to a worker using either a round-robin strategy or by hash-based partitioning. In this way, each worker can discover links between entities of the target data set and the portion of the source data set that was assigned to it. Moreover, this is performed independently of other workers, thus it is expected that throughput can increase linearly when adding more workers, as long as the partitioning strategy distributes the source data set fairly to workers.

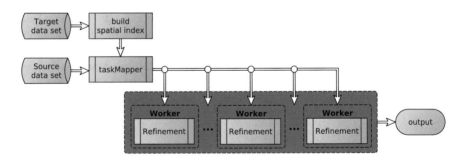

Fig. 6.7 Parallel LD for two streaming data sources

Stream-Stream In the more general case, both data sets may be stream based or too large to be broadcast. In this case, the idea is to split the blocking structure to the workers, by having each worker being responsible for a subset of blocks (i.e., grid cells).

Figure 6.7 depicts this case, where both data sources are blocked in the same way and subsets of the same blocks are distributed to worker nodes. Essentially, all entities of the target and source data sets that are located in the same block are assigned to the same worker, which can then proceed with the discovery of the requested relations.

One subtle issue is how to guarantee correctness of link discovery, i.e., avoid the case that two entities $\tau \in T$ and $\sigma \in S$ that satisfy the relation of interest end up in different workers. This largely depends on the link discovery task at hand, and it has to be carefully addressed when assigning entities to blocks.

In more concrete terms, consider the case of proximity-based LD, where we are interested in finding pairs of entities with distance below a given application-defined threshold. In this case, the mapping of entities to grid cells is not necessarily one-to-one. For example, consider the case of vessel positions and ports, and the link discovery task is to identify vessels that have distance to a port below a threshold value. In this case, the target data set (ports) is assigned to grid cells based on containment, but a vessel position may have to be assigned to multiple cells, which are determined based on overlap with a circle centered at the vessel's position and radius equal to the distance threshold. In turn, this guarantees correctness of the LD task.

6.5 Empirical Evaluation

This section presents the results of an empirical study using real-life data sets, in order to demonstrate the efficiency of spatiotemporal link discovery. The goal of the empirical evaluation is twofold:

- To demonstrate the efficiency of MaskLink compared to the state of the art for topological link discovery (RADON [20]), while being more generally applicable.
- To quantify the performance gain achieved by the extension of MaskLink for proximity-based link discovery compared to the baseline approach.

6.5.1 Experimental Setup

For the first experiment on topological link discovery, we employ the data sets used in the evaluation of RADON [20], namely CLC and NUTS:

- S=CORINE Land Cover[6] (CLC) is provided by the European Environment Agency, which collects data regarding the land cover of European countries.
- T=NUTS,[7] provided by Eurostat group of the European Commission, contains a detailed hierarchical description of statistical areas for the whole Europe.

---

[6]See also https://datahub.io/dataset/corine-land-cover.

[7]Version 0.91 (http://nuts.geovocab.org/data/0.91/).

Since CLC contains 44 data sets varying in size (from a few hundreds to hundreds of thousands), we merged all data sets into one big data set. For testing scalability, we exported from CLC data sets of varying size {500K, 1M, 1.5M, 2M, 2.5M} and evaluated MaskLink against RADON[8] and baseline. We also preprocessed the NUTS data sets, to convert the ngeo:posList serialization to Well-Known-Text format. All the reported experimental results do not include the preprocessing step. RADON has been configured using the default settings, as in [20]. The MaskLink and the baseline technique use a 2.5° granularity grid.

For the second experiment on proximity link discovery, we evaluate the proximity relation nearby using real-world data sets compiled from positions of vessels and fishing areas. Specifically, we use as S a data set that contains kinematic messages of vessels in the Mediterranean Sea spanning between 01-11-2016 and 31-01-2016, whereas T is a data set of fishing areas that contains 5076 polygons generated from raster images depicting the fishing intensity in European waters (reported by European Union). The goal is to identify links between vessel positions and fishing areas that these vessels approach. We report results for different numbers of position messages {500K, 1M, 1.5M, 2M, 2.5M}, while using the complete data set of fishing areas in the Mediterranean Sea. Since RADON does not support point-to-region relations (the heuristic for deciding the size of cells based to the size of areas in input, cannot be applied), the MaskLink technique has been evaluated only against the baseline approach.

Table 6.2 Comparison of total execution time (in sec) of RADON, MaskLink, and baseline for all topological relations

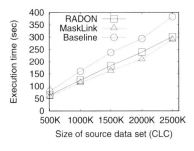

#Entries in CLC	RADON	MaskLink	Baseline
500K	63.81	59.05	79.65
1000K	124.55	114.05	160.28
1500K	183.36	162.99	237.89
2000K	239.31	210.00	293.84
2500K	301.29	289.97	382.84

6.5.2 Results for Region-to-Region Topological Relations

We compare the MaskLink technique for all topological relations to RADON using the data sets CLC and NUTS. RADON requires that the entire data sets are loaded

---

[8]Downloaded from https://github.com/dice-group/LIMES, accessed on December 2017.

in memory, which is not possible in our experimental setting. We overcome this memory limitation, by loading NUTS (as the smaller data set) in memory, and accessing CLC in batches of lines. For a fair comparison of techniques, we repeat the same procedure for MaskLink, although it can be directly applied on the given data sources.

Table 6.2 and the corresponding chart on its left report the total execution time of MaskLink for different number of CLC entries, in comparison to RADON and baseline. Specifically, the first column of the table indicates the size of CLC data set, and the next columns present the total processing time for RADON, MaskLink, and baseline for computing all topological relations.

We observe that the MaskLink consistently outperforms RADON by approximately 10%. This is an important finding, as the main interest is the scalability of LD methods with input size. Also, recall that MaskLink achieves this performance using less memory and even when one data source is streaming. When comparing MaskLink to baseline, we observe that the gain achieved by MaskLink increases with the size of the data source, indicating that large instances of the problem cannot be efficiently solved by the baseline algorithm.

As regards the overhead imposed due to the computation of the geometry of a mask, we observed that the time needed to compute the mask areas of all cells for the given target data set was approximately 3 s. This is deemed tolerable, since it is a one-time cost and provides a considerable gain as shown by comparing the processing time of MaskLink to that of baseline.

Table 6.3 Total execution time (in sec) for topological and proximity relations between points and areas using MaskLink and baseline

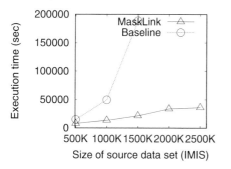

#Entries in IMIS	MaskLink	Baseline
500K	7,769.05	15,064.83
1000K	13,494.58	49,630.11
1500K	21,528.29	188,763.63
2000K	33,801.71	timeout
2500K	36,275.21	timeout

6.5.3 Results for Point-to-Region Proximity-Based Relations

In this experiment, we report results for different number of position messages and using the complete data set of fishing areas in the Mediterranean Sea. Table 6.3 and the chart on its left report the total processing time for all topological and proximity relations that computed between points and areas. The first column indicates the size of source data set, while the second and third columns report the total processing time for MaskLink and baseline, respectively.

Again, MaskLink outperforms the baseline consistently, but this time by a much larger margin, since the problem of proximity link discovery is harder in general. It is very significant to notice that the baseline is not scalable for this problem, as it does not terminate in reasonable time when the input size is larger than 1500K. In contrast, MaskLink scales gracefully with the size of input data.

In the case of very small input sizes, the baseline is faster than MaskLink. This can be explained by the fact that for small data sets the cost of computing the mask of each cell becomes comparable to the total processing time. In any case, our focus is to improve the performance of LD for large data sets.

Furthermore, as a side-note, we observed in this experiment that the time required for preparing the target data set (i.e., populating the grid and computing the mask) is considerably larger than the time needed for the first experiment. This is explained by the fact that the geometries used in the second experiment are produced from raster images (vectorized), thus the geometries have more fine-grained detail and the mask computation requires more processing time.

6.6   Conclusions

Spatiotemporal link discovery is a challenging topic with numerous applications in mobility analytics. This chapter provided an overview of the current landscape in link discovery focusing primarily on spatial and spatiotemporal data.

Limitations of existing works in this field include handling the temporal dimension together with spatial data as well as the lack of scalable link discovery frameworks targeting big spatial and spatiotemporal data.

Motivated by such shortcomings, this chapter presents a framework for spatiotemporal link discovery, along with its constituent modules, which covers several link discovery tasks, both for the detection of topological as well as proximity relations. In addition, the proposed framework targets spatiotemporal data, rather than spatial only. Last, but not least, a scalable, data-parallel version of the spatiotemporal link discovery framework has been demonstrated, aiming at link discovery over big data, such as streaming data sources or voluminous data that cannot be processed on a single computer.

Acknowledgments The research work has been supported by the datAcron project, which has received funding from the European Union's Horizon 2020 research and innovation programme under grant agreement No 687591 and by the Hellenic Foundation for Research and Innovation (H.F.R.I.) under the "First Call for H.F.R.I. Research Projects to support Faculty members and Researchers and the procurement of high-cost research equipment grant" (Project Number: HFRI-FM17-81).

References

1. Allen, J.F.: Maintaining knowledge about temporal intervals. Commun. ACM **26**(11), 832–843 (1983)
2. Bleiholder, J., Naumann, F.: Data fusion. ACM Comput. Surv. **41**(1), 1:1–1:41 (2008)
3. Carbone, P., Katsifodimos, A., Ewen, S., Markl, V., Haridi, S., Tzoumas, K.: Apache flink™: stream and batch processing in a single engine. IEEE Data Eng. Bull. **38**(4), 28–38 (2015)
4. Christen, P.: A survey of indexing techniques for scalable record linkage and deduplication. IEEE Trans. Knowl. Data Eng. **24**(9), 1537–1555 (2012)
5. Christophides, V., Efthymiou, V., Stefanidis, K.: Entity Resolution in the Web of Data. Synthesis Lectures on the Semantic Web: Theory and Technology. Morgan & Claypool Publishers, San Rafael (2015)
6. Dong, X.L., Srivastava, D.: Big Data Integration. Synthesis Lectures on Data Management. Morgan & Claypool Publishers, San Rafael (2015)
7. Elmagarmid, A.K., Ipeirotis, P.G., Verykios, V.S.: Duplicate record detection: a survey. IEEE Trans. Knowl. Data Eng. **19**(1), 1–16 (2007)
8. Isele, R., Jentzsch, A., Bizer, C.: Efficient multidimensional blocking for link discovery without losing recall. In: Proceedings of WebDB (2011)
9. Jagadish, H.V., Gehrke, J., Labrinidis, A., Papakonstantinou, Y., Patel, J.M., Ramakrishnan, R., Shahabi, C.: Big data and its technical challenges. Commun. ACM **57**(7), 86–94 (2014)
10. Mamoulis, N.: Spatial Data Management. Synthesis Lectures on Data Management. Morgan & Claypool Publishers, San Rafael (2011)
11. Nentwig, M., Hartung, M., Ngomo, A.N., Rahm, E.: A survey of current link discovery frameworks. Semant. Web **8**(3), 419–436 (2017)
12. Ngomo, A.N.: A time-efficient hybrid approach to link discovery. In: Proceedings of OM (2011)
13. Ngomo, A.N.: Link discovery with guaranteed reduction ratio in affine spaces with Minkowski measures. In: Proceedings of ISWC, pp. 378–393 (2012)
14. Ngomo, A.N.: ORCHID - reduction-ratio-optimal computation of geo-spatial distances for link discovery. In: Proceedings of ISWC, pp. 395–410 (2013)
15. Ngomo, A.N., Auer, S.: LIMES - a time-efficient approach for large-scale link discovery on the web of data. In: Proceedings of IJCAI, pp. 2312–2317 (2011)
16. Papadakis, G., Skoutas, D., Thanos, E., Palpanas, T.: A survey of blocking and filtering techniques for entity resolution. CoRR abs/1905.06167 (2019)
17. Santipantakis, G.M., Kotis, K.I., Vouros, G.A., Doulkeridis, C.: RDF-gen: Generating RDF from streaming and archival data. In: Proceedings of the 8th International Conference on Web Intelligence, Mining and Semantics, WIMS 2018, Novi Sad, 25–27 June 2018, pp. 28:1–28:10 (2018)
18. Santipantakis, G.M., Vlachou, A., Doulkeridis, C., Artikis, A., Kontopoulos, I., Vouros, G.A.: A stream reasoning system for maritime monitoring. In: 25th International Symposium on Temporal Representation and Reasoning, TIME 2018, Warsaw, 15–17 October 2018, pp. 20:1–20:17 (2018)
19. Santipantakis, G.M., Glenis, A., Doulkeridis, C., Vlachou, A., Vouros, G.A.: stLD: towards a spatio-temporal link discovery framework. In: Proceedings of the International Workshop on Semantic Big Data, SBD@SIGMOD 2019, Amsterdam, 5 July 2019, pp. 4:1–4:6 (2019)
20. Sherif, M.A., Dreßler, K., Smeros, P., Ngomo, A.N.: Radon - rapid discovery of topological relations. In: Proceedings of AAAI, pp. 175–181 (2017)
21. Smeros, P., Koubarakis, M.: Discovering spatial and temporal links among RDF data. In: Proceedings of LDOW (2016)

Chapter 7
Distributed Storage of Large Knowledge Graphs with Mobility Data

Panagiotis Nikitopoulos, Nikolaos Koutroumanis, Akrivi Vlachou, Christos Doulkeridis, and George A. Vouros

Abstract This chapter presents novel solutions for storage and querying of large knowledge graphs, represented in RDF, which consist of mobility data. Such knowledge graphs are generated and updated daily based on incoming positional information of moving entities, possibly linked with contextual information and weather data. To cope with the massive size of knowledge graphs, several challenges need to be addressed related to distributed storage and parallel query processing. This chapter presents the design and implementation of a parallel processing engine for spatiotemporal RDF data built on top of Apache Spark. The engine is comprised of a storage layer, which stores deliberately encoded spatiotemporal RDF triples and a dictionary of mappings between integer identifiers and RDF resources, and also uses Property tables and columnar storage layout for improved performance. Also, the engine uses a processing layer, which is comprised by a query parsing component, a logical query builder, and a physical query constructor in order to produce execution plans that efficiently handle spatiotemporal constraints along with SPARQL processing. The performance of our engine is demonstrated by means of experiments over large knowledge graphs of real-life mobility data.

7.1 Introduction

Knowledge graphs of mobility data represented in RDF are produced from data integration of positional information of moving objects with other external data sources: static databases, weather and contextual data. Efficient management of large knowledge graphs is of utmost importance, since it may assist in the advanced data analysis of integrated data for the discovery of hidden patterns of movement. For example, visual analytics (cf. Chap. 3) and trajectory analytics (cf. Chap. 10) can

P. Nikitopoulos · N. Koutroumanis · A. Vlachou · C. Doulkeridis (✉) · G. A. Vouros
University of Piraeus, Piraeus, Greece
e-mail: nikp@unipi.gr; koutroumanis@unipi.gr; avlachou@unipi.gr; cdoulk@unipi.gr; georgev@unipi.gr

© Springer Nature Switzerland AG 2020   181
G. A. Vouros et al. (eds.), *Big Data Analytics for Time-Critical Mobility Forecasting*, https://doi.org/10.1007/978-3-030-45164-6_7

exploit scalable querying of large knowledge graphs in order to efficiently retrieve integrated data for further analysis.

Even though scalable management and querying of large RDF graphs has been extensively studied recently [1, 7], one differentiating factor of knowledge graphs of mobility data is the spatiotemporal dimension of data, which requires special treatment. The reason is that queries to such knowledge graphs typically entail both constraints imposed to the structure of the graph as well as spatiotemporal constraints on some entities. Practically, when confronted with such spatiotemporal RDF data, existing distributed RDF engines would have to process the spatiotemporal constraint in a post-processing step, after having evaluated the RDF part of the query. However, this incurs significant overhead, because large portions of data that could be filtered based on the spatiotemporal constraint need to be processed. In turn, this deteriorates the performance of query processing.

Motivated by these shortcomings, in this chapter, the design and implementation of a parallel/distributed RDF engine capable to store spatiotemporal RDF data is presented. The prototype engine, called DiStRDF, consists of a storage layer that keeps spatiotemporal data using an encoding scheme and a processing layer implemented in Apache Spark. By means of a detailed experimental evaluation, the performance of DiStRDF is studied for various queries over large knowledge graphs constructed from real-life mobility data.

The rest of the chapter is organized as follows. In Sect. 7.2, we familiarize the reader with related work on the field of scalable querying on large knowledge graphs. Section 7.3 presents the encoding scheme for spatiotemporal RDF data. Section 7.4 provides a high-level overview of DiStRDF, whereas Sects. 7.5 and 7.6 present its storage and processing layers, respectively. The results of the empirical evaluation are discussed in Sect. 7.7, while Sect. 7.8 provides concluding remarks.

7.2 Background

The Resource Description Framework (RDF) is a specification recommended by W3C[1] for modeling and interchanging data over the Web. It specifies a structured way for storing data from various data sources, while providing several techniques for managing semantic heterogeneity, namely schema matching, ontologies, and schema repositories. RDF data is represented as a set of triples (*subject*, *property*, *object*) or $\langle s, p, o \rangle$, also known as *statements*.

W3C recommends the use of the declarative SPARQL language for querying RDF datasets. SPARQL relies on graph pattern matching queries to extract relevant data. A *triple pattern* (*tp*) is an RDF triple where variables may occur in subject, predicate, or object position. Conjunctive queries are expressed by using shared

---

[1] https://www.w3.org/.

variables across different triple patterns, also called *basic graph patterns*. Thus, a basic graph pattern is evaluated to a set of RDF triples.

In terms of geospatial and spatiotemporal standardization, the Open Geospatial Consortium (OGC)[2] has proposed standards for modeling and querying spatial data and queries in a unified manner. OGC standardizes the representation of geometries by using a text representation, called Well Known Text (WKT).[3] WKT can represent various geometry types in 2D or 3D space (e.g., line, point, polygon, etc.), by referencing the respective coordinates. Moreover, OGC has proposed a standardized language for querying geospatial data, namely the GeoSPARQL language.[4] This is an extension to the standard SPARQL query language, which provides additional features (vocabulary) for querying this type of data. The study in [8] has proposed the use of stRDF model and stSPARQL language for modeling and querying spatiotemporal data, respectively. The work of [9] describes the various models and query languages that exist in more details.

7.2.1 Distributed Processing Frameworks

Apache Hadoop[5] is a popular open source fault-tolerant distributed processing framework. It uses the MapReduce [4] processing model to operate on a distributed set of data. The Hadoop project is comprised of two major components: the processing framework, namely YARN (Yet Another Resource Negotiator) and the storage component, namely HDFS (Hadoop Distributed File System).

YARN is the computing resource management and job scheduling service of Hadoop; it manages the execution of Hadoop jobs by allocating CPU cores and RAM on every node of the Hadoop cluster. It also schedules tasks to be executed on different nodes of the Hadoop cluster. Unexpected disruptions of nodes availability are handled by restarting failed or timed-out tasks on other available nodes. YARN is comprised by two major services: the ResourceManager and the Node-Manager service. Typically, a Hadoop cluster consists of one ResourceManager acting as the master node and several NodeManagers acting as slave nodes. The ResourceManager service collects information from all NodeManagers (e.g., their availability, number of CPU cores, amount of memory, etc.) and coordinates the execution of Hadoop jobs. A Hadoop job is split to tasks, which are assigned to the NodeManagers by the ResourceManager service. The ResourceManager may reschedule a failed task to other NodeManagers. The NodeManager services are responsible for executing the tasks assigned to them and reporting their status back to the ResourceManager.

---

[2]http://www.opengeospatial.org/.

[3]https://www.opengeospatial.org/standards/wkt-crs.

[4]https://www.opengeospatial.org/standards/geosparql.

[5]https://hadoop.apache.org/.

HDFS is a generic distributed file system which is optimized for storing large sets of data on a set of computing nodes. It is fault tolerant and designed to be run on low-cost hardware. The files written to HDFS are split to blocks of user-defined size (the default size of an HDFS block is 128MB). Every block is stored on one of the available nodes and optionally replicated to other nodes to increase the availability of the stored data. Hence, every file stored in HDFS is distributed and replicated among the available nodes. Typically, the user needs to define a *replication factor* which indicates the number of times a block will be replicated to other nodes. This setting can be defined system-wide, folder-wide, or for individual files. HDFS is comprised by two major services: the Namenode and the Datanode. The Namenode service is responsible for: (a) keeping the metadata of the file system (i.e., file names and directory structure) and (b) managing the Datanodes by assigning blocks to them and monitoring their availability. The Datanode is responsible for storing the data assigned by the Namenode service, into the node's secondary memory. Typically, an HDFS cluster hosts one Namenode service acting as the master node and several Datanode services acting as slave nodes of the cluster.

Apache Spark[6] is an open source distributed processing engine that operates on a cluster of commodity computing nodes. It provides a set of abstractions which facilitate working seamlessly with distributed data, while ensuring that the processing will not be affected by any hardware failure. Spark is able to operate either standalone, or by utilizing other resource managers such as YARN. It can also read and write data from various data sources and destinations, including popular databases and regular files on local disk and HDFS.

Spark offers several APIs for querying distributed data: (a) the Spark Core API which is the foundation of the overall Spark project, (b) the Spark SQL API for querying data using SQL language, (c) the Spark Streaming API for monitoring streaming data, (d) the Spark MLlib API for applying machine learning techniques, and (e) the Spark GraphX API for distributed processing of graph data. The DiStRDF project relies on Spark SQL API, since it provides a high-level API for executing query operations on distributed data. The main abstraction of Spark SQL is the DataFrame which is an immutable data structure containing distributed data residing in the available nodes. It is comprised by a fixed set of named columns and provides several operations to be executed on the distributed data, such as filtering, sorting, grouping, etc. Every operation on a DataFrame produces a new immutable DataFrame containing the processed data.

7.2.2 Scalable Querying of Large Knowledge Graphs

The topic of parallel and scalable SPARQL query processing has been extensively studied in literature (cf. [1, 7] for related surveys). However, most of these works do

---

[6]https://spark.apache.org/.

not deal with spatial or spatiotemporal query processing. For example, the studies in [13, 16, 17] propose innovative techniques for in-memory, distributed query processing, but do not handle efficiently the special case of spatiotemporal data. Such systems need an additional step prior or after query execution to apply the spatial or spatiotemporal filters on the input data. Instead, a system supporting natively the management of spatiotemporal data must be able to incorporate spatiotemporal processing during query execution. Hence, the aforementioned techniques lead to reduced performance when dealing with spatial or spatiotemporal data, since they are inclined to add an expensive processing step.

Other studies [2, 6, 8, 10] focus on centralized processing of spatiotemporal RDF data. However, these proposals cannot handle the increased complexity of vast-sized distributed data. Novel parallel data processing techniques need to be employed to cope with data that do not fit in main memory of a single node. Furthermore, distributed systems need to minimize the interaction between the nodes to reduce the overhead incurred by network latency.

Motivated by these limitations, we focus on distributed systems for parallel processing of spatial and spatiotemporal RDF data. The requirements include: handling RDF data, supporting spatial and spatiotemporal representations and querying, as well as scalable processing of complex queries that combine triple patterns and spatiotemporal constraints.

7.3   One-Dimensional Encoding

To encode the spatiotemporal information of a moving entity, we map its spatial and temporal constituents to a single integer value. The spatial location can be in 2D (for cars, vessels, etc.) or 3D (for aircraft). However, only the 2D case is presented in the following descriptions for simplicity of the presentation. The mapping of the spatial location is based on the grid partitioning method, where space is split into $2^m = (2^{m/2} * 2^{m/2})$ equi-sized cells. Also, a space-filling curve is used, providing an ordering for the spatial cells of the grid. Figure 7.1 shows an example for the 2D case for $m = 4$, where the Hilbert and Z-order curves are depicted together, with an integer value (ID) assigned to each cell. Space-filling curves aim to preserve the *spatial locality* by having nearby cells be also close on the curve. Obviously, some information loss is inevitable when going from 2D to 1D, but the spatial locality is mostly preserved. In principle, both curves can be used in the proposed encoding, albeit Z-order is easier to extend for higher dimensions (e.g., 3D).

Every entity whose spatial location is enclosed in a spatial cell is assigned with a unique identifier. Therefore, k bits are reserved for assigning unique IDs to different entities in the same spatial cell. As such, the maximum number of entities that fit in a spatial cell is 2^k.

The temporal dimension is handled as follows; a temporal partitioning $\mathcal{T} = \{T_0, T_1, \ldots\}$ of the time domain is considered, where T_i represents a temporal interval. No assumptions on specific properties of the partitioning are made, i.e.,

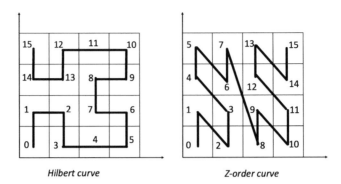

Fig. 7.1 Space-filling curves (Hilbert and Z-order) used for ordering the spatial cells [14]

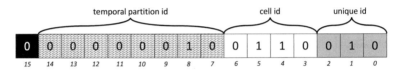

Fig. 7.2 IDs encoding using bits: b total bits, m bits for spatial part (cell id), k bits for uniqueness, $b - (m + k + 1)$ bits for the temporal partition [14]

the length (or duration) of temporal partitions can vary, apart from the fact that the partitions are disjoint, covering the entire time domain ($\bigcup T_i = \mathcal{T}$), and that T_i precedes T_{i+1} in the temporal order. Every temporal partition T_i is associated with a spatial grid, as depicted in Fig. 7.1. The only restriction is that the identical grid structure (i.e., 2^m equi-sized cells) is used for all temporal partitions T_i.

Figure 7.2 shows how the information is combined together, resulting to a *unique* identifier for any spatiotemporal position of a moving entity. The figure depicts the binary representation of the identifier, consisting of b bits. The k least significant bits are used to encode the identifier of an entity in a specific spatiotemporal cell. The ID of the cell is recorded in the next m bits. The most-significant bit is reserved so as to discern between spatiotemporal RDF entities and other RDF entities. By convention, the most-significant bit is set to 0 for all IDs of spatiotemporal RDF entities, while for IDs of all other RDF entities is set to 1. The remaining $b - (m + k + 1)$ bits are used for encoding the time, thus $2^{b-(m+k+1)}$ temporal partitions can be stored in total. The following example explains how the one-dimensional (1D) identifier of an entity is generated.

Example 7.1 In Fig. 7.2, the case of $b = 16$, $m = 4$, and $k = 3$ is considered, and the depicted identifier is $2^8 + 2^5 + 2^4 + 2 = 306$. The spatial cell in which it belongs is 6 (=0110), and the spatial grid contains $2^4 = 16$ cells in total. This encoding can accommodate at most $2^{b-(m+k+1)} = 2^8 = 256$ temporal partitions. It can be observed that this ID corresponds to the second entity found in the given spatiotemporal cell and that the entity is located in the second temporal partition.

Given the ID of a spatiotemporal entity, the 3D spatiotemporal cell where the entity belongs to can be identified. Also, given a 3D cell, a range of IDs that correspond to any entity belonging to the cell can be computed. The encoding scheme guarantees that entities with similar spatiotemporal representations are assigned nearby IDs (that belong to small ranges of values), thus preserving data locality. For example, given a time partition T_i, all entities in T_i belong to the interval $[2^i \cdot (2^{m+k}), 2^{i+1} \cdot (2^{m+k})]$, where 2^m is the number of spatial cells, and 2^k is the maximum number of objects within each spatial cell. Essentially, 2^i is used to shift the time intervals, thus different temporal partitions can be mapped to different 1D intervals of identifiers.

In summary, the encoding scheme (a) allows to identify a spatiotemporal approximation of a position in the spatiotemporal domain given an ID, and (b) achieves to reflect the spatiotemporal locality of spatiotemporal positions in the 1D integer domain, by assigning nearby integer values to positions which are close to each other in the spatiotemporal space. More details considering the computation of the identifier as well as various strategies for partitioning the temporal domain dynamically, are provided in [19].

7.4   Overview of the DiStRDF Engine

The DiStRDF engine is a distributed query processing engine able to process spatiotemporal SPARQL queries in parallel. Since it focuses on analyzing massive datasets, it uses Apache Spark as an execution engine to evaluate the queries on the stored data. At a high abstraction level, the DiStRDF engine consists of two layers: the storage layer, described in more details in Sect. 7.5 and the processing layer, described in more details in Sect. 7.6.

The storage layer interacts with the query execution engine to provide the query-relevant data when needed. Essentially, it is a scalable parallel storage solution which supports replication, in-memory lookups, indexing, compression, and efficient query execution. It consists of two data stores: (a) the dictionary which stores the mappings between RDF triples and their encoded values and (b) the RDF data store which contains the encoded RDF triples stored in HDFS. The storage layer is designed as a distributed storage solution. Hence, even in the case of hardware failures on some nodes of the cluster, the availability of the stored data will be unaffected.

The processing layer takes as input a SPARQL query and outputs the result of the query. Since the result might be relatively large, we opt to store the result on HDFS and report a sample of the result to the user's screen. The processing layer consists of the following components:

- The *SPARQL Query Parsing* component, which transforms the input SPARQL query into an internal representation that facilitates the process of building a logical plan.

- The *Logical Plan Builder* component, which builds the logical plan: a tree representation of the SPARQL query consisting of a set of logical operators.
- The *Logical Plan Optimizer* component, which optimizes the previously built logical plan to make its execution more efficient.
- The *Physical Plan Constructor* component, which composes a query execution plan by selecting an implementation algorithm for each operator of the logical plan.
- The *Execution Engine* component, which executes the query on the data.

7.5   The Storage Layer

The storage layer is a scalable storage solution of high availability, which stores RDF data in a set of cluster computing machines (nodes). Its high availability feature is supported by replicating the stored data in the available nodes. Hence, if a node goes offline, the remaining nodes will have the replicated data to answer the queries.

Since the data of interest are of spatiotemporal nature, and the goal of DiStRDF engine is to support efficient execution of spatiotemporal queries, we employ the encoding technique described in Sect. 7.3, to encode the RDF triples having spatiotemporal information. By using this technique, our DiStRDF engine benefits by increased compression of the stored data and increased efficiency in querying the stored data with spatiotemporal criteria. However, the encoding technique requires the storage of the dictionary—a mapping table between the encoded and decoded values. Hence, the DiStRDF storage layer stores the dictionary and the encoded RDF triples in distributed and highly available data stores.

7.5.1   Storing the Dictionary

The DiStRDF dictionary data store needs to support efficient lookups, and horizontal scalability. Hence, we turn our attention to use an in-memory, distributed key-value store, such as Memcached,[7] MICA [11], or Redis.[8] A key-value store is a NoSQL database that is designed for storing, retrieving, and managing dictionaries or hash tables. A dictionary in a NoSQL store is a data structure that contains a set of keys associated with their respective values. The store is responsible for storing those pairs and retrieving the corresponding value when a key is provided. Any of the aforementioned key-value stores is a good candidate for building the dictionary, but in DiStRDF we opt to use Redis to store the dictionary, since it is one of the most popular distributed key-value stores [12].

---

[7]https://memcached.org/.

[8]https://redis.io/.

Redis answers the queries by looking up data stored in main memory; the secondary memory is only used for persisting snapshots of the data in case a restart is needed. Redis also supports replication to enable high availability of the stored data and partitioning for horizontal scaling. It also uses hash-based indices which enable lookups in constant time for most of the queries. It supports several types of data types, such as strings, hashes, lists, sets, sorted sets with range queries, bitmaps, hyperloglogs, and even geospatial indexes with radius queries. Notice that in the DiStRDF engine we use the 1D encoding to support spatiotemporal queries rather than the native spatial-only feature of Redis.

We also have implemented a data loading mechanism, which takes as input the source RDF data and transforms it into integer values as follows: if the subject/object concerns spatiotemporal information, then the 1D encoding scheme is used; otherwise, a random unique negative integer value is provided[9] to a non-spatiotemporal subject/object or property. Then, the dictionary stores the mapping between the integer (key) and the RDF representation (value, of type string). In addition, we also keep a reverse dictionary mapping between the RDF representation (key) and the integer identifier (value), since we need both dictionaries to efficiently compute the SPARQL queries and translate results from and back to the application.

7.5.2 Storing RDF Triples

There are several works that study the problem of distributed storage of RDF data (cf. [1, 7] for related surveys). In DiStRDF, the RDF triples are transformed to encoded integer values and stored in an HDFS cluster. Our design covers several aspects of the data store, such as compression, file layout, data organization, indexing, and partitioning.

Table 7.1 RDF property table example

Node	ofMovingObject	hasHeading	hasGeometry	hasTemporalFeature
node15	ves376609000	15	15.3W 47.8N	2017-01-03 02:20:05
node22	ves369715600	0	19.4E 35.9N	2017-01-30 17:20:00
node58	ves376609000	3	23.2E 35.7N	2017-02-05 10:42:08

The RDF nodes containing spatiotemporal information are organized in *Property tables* as demonstrated in Table 7.1. Property tables show good performance when a group of properties always exists for a given resource, thereby avoiding the need of costly joins to reassemble this information. In our case, the majority of the query workload targets nodes representing the position of moving objects (a.k.a. *semantic*

---

[9]Negative integer values are preferred to avoid collisions with integer values generated from the 1D encoding.

nodes), therefore we build a property table that maintains information related to the positions of moving objects. In Table 7.1, the nodes 15, 22, and 58 correspond to traced positions of moving vessels. These nodes contain spatial and temporal information (hasGeometry and hasTemporalFeature predicates, respectively). Thus, these RDF nodes are stored as property tables, along with other useful predicates, such as the moving object's identifier. A spatiotemporal query benefits from such data organization, since it will probably require fewer joins to process upon answering a query.

The RDF nodes containing no spatiotemporal information are stored separately in a table called *leftover triples*. This table is used as a regular triples data store, which stores subject, predicate, and object in three different columns. An example of a leftover triples table is depicted in Table 7.2, where non spatiotemporal information is stored, such as event occurrence and static information about the moving vessels.

Table 7.2 RDF leftover triples example

Subject	Predicate	Object
turnInit	occurs	node22
stoppedInit	occurs	node15
ves376609000	hasBuildDate	2009-05-31

These tables are stored in Parquet formatted files; Apache Parquet[10] is a column-based data layout with native support for several compression codecs (lzo, gzip, snappy). Parquet formatted datasets are usually split into several files; every file contains a set of metadata such as ranges of values for every column of the file. This metadata can significantly boost query performance, e.g., in case a query retrieves few columns of a property table. Predicate and projection pushdown are also features supported by Parquet, which avoids the cost of reading all data from disk at query time.

The partitioning mechanism used in DiStRDF storage layer partitions the data according their spatiotemporal similarity. More specifically, we use the 1D encoded integers computed for the spatiotemporal data to range-partition them across the available nodes. Notice that since in the 1D encoding scheme the temporal value has greater impact on the encoding, the similarity between the spatiotemporal nodes is considered to be on the temporal dimension first; a range will contain similar nodes on their temporal dimension, which may not be that similar in their spatial dimension. Leftover triples which do not have spatiotemporal information to be encoded are distributed evenly among the cluster nodes.

---

[10]https://parquet.apache.org/.

7.6   The Processing Layer

The DiStRDF processing layer is a SPARQL query engine that processes batch queries over huge amounts of spatiotemporal RDF data. The processing layer is implemented on Apache Spark, a popular engine for parallel in-memory data processing based on the MapReduce model. Apache Spark addresses many of the limitation of Hadoop [5] and can achieve significant performance gains to competitor systems, such as Hadoop [18].

The processing layer is comprised of the following components (a) the SPARQL Query Parsing, (b) the Logical Plan Builder, (c) the Logical Plan Optimizer, and (d) the Physical Plan Constructor.

7.6.1   The SPARQL Query Parsing Component

When a SPARQL query is declared, the first task that is performed prior to its execution is the *query parsing task*, which:

- checks the correctness of syntax and ensures that the query is specified correctly, and
- transforms the query into an internal representation, used by the other modules of the processing engine.

Assuming that the syntax is correct, the SPARQL query is translated into a set of *basic graph patterns* (BGP). These graph patterns are given to the logical planner so as to construct a *logical query plan* (depicted in Fig. 7.3). The query parsing component is built using the functionality of the Apache Jena software so as to obtain the BGPs from the SPARQL query.

7.6.2   The Logical Plan Builder

After query parsing has been completed successfully, the logical query builder is assigned with the task of constructing a logical query plan, which consists of logical operators ordered in hierarchical form. Specifically, the logical plan represents a way to execute the respective SPARQL query.

Five types of logical operators are used for processing the BGP part of a SPARQL query. Obviously, these operators do not cover the entire SPARQL specification (such as grouping, sorting, etc.), but they cover a wide variety of SPARQL queries most commonly encountered in practice. The operators constitute a fundamental and challenging part of a distributed RDF processing engine. Nominally, the operators are described as follows:

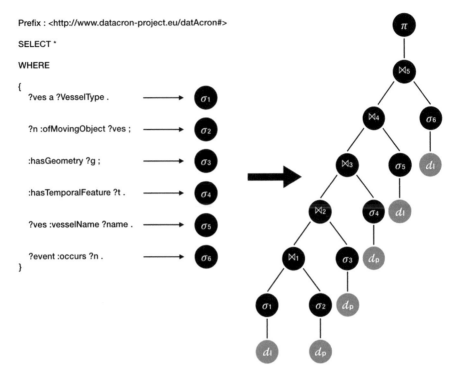

Fig. 7.3 Example of a SPARQL query (left) that is translated into BGPs and then provided to the logical planner in order to construct the query plan (right)

- *Join operator* (⋈): A binary operator which takes as input two sets of data, performing join based on the following triple pattern fields; Subject-Object, Object-Object, Object-Subject, and Subject-Subject.
- *Projection operator* (π): A unary operator that keeps only a subset of the available fields of the input set of data.
- *Selection operator* (σ): A unary operator that takes as input a set of data and a triple pattern and returns the data that match the pattern.
- *Distinct operator*: A unary operator that takes as input a set of data and selects only the distinct values of a field.
- *Sort operator*: A unary operator that takes as input a set of data and performs a sorting operation on the records, based on a field.
- *Limit operator*: A unary operator that takes as input a set of data and limits the records (results) to a number.
- *Datasource operator* (d): A unary operator that is used to access a set of data stored at a data source.
- *Spatiotemporal filter estimation operator* (σ_ϵ): This operator is provided with a spatiotemporal box constraint and applies *approximate filtering* by exploiting the 1D encoding scheme introduced in Sect. 7.3. It inspects only the subject field

to extract the embedded approximate spatiotemporal information (i.e., cell and temporal partition), keeping only the data which satisfy the range constraint. Due to the fact that the extracted information is approximate, a refinement is needed to discard the false positives from the result set. To this end, a spatiotemporal refinement operator is introduced.

• *Spatiotemporal filter refinement operator* (σ_ρ): This operator is provided with a spatiotemporal box constraint and performs *exact filtering*, based on the exact spatiotemporal information provided in the data, namely in the corresponding spatial and temporal fields.

Commonly, a query plan is represented as a tree that consists of a set of connected nodes corresponding to *logical operators*. The nodes take as input the results of their children nodes and calculate the result of the operation defined by their Logical Operator; these results (a node's output) are given to their parent nodes for continuing the query execution. If a node has zero children (equivalently, if it is a leaf node), then its input data is a physical source of data. If a node has no parent (equivalently, if it is a root node), then its input is considered to be the query result.

Algorithm 7.1 Logical Plan Builder

1: **Input:** SPARQL query
2: **Output:** Logical Plan Tree
3: $LP \leftarrow \{\}$
4: **for** each triple pattern $t_i \in query.BGP$ **do**
5:    $so_i \leftarrow new\,Select\,Operator(t_i)$
6:    **for** each field $f_{ij} \in t_i$ ($j \in \{Subject, Predicate, Object\}$) **do**
7:      **if** f_{ij} is value **then**
8:        so_i.addFilter(f_{ij})
9:    LP.add(so_i)
10: **while** LP.size > 1 **do**
11:    $jo \leftarrow$ new JoinOperator(LP_1, LP_2)
12:    LP.remove(LP_1, LP_2)
13:    LP.add(jo)
14: $operator \leftarrow$ getTreeRoot(LP)
15: **if** FILTER exists in query **then**
16:    $operator \leftarrow$ new SelectOperator($operator$)
17: **if** $DISTINCT$ exists in query **then**
18:    $operator \leftarrow$ new DistinctOperator($operator$)
19: **if** $ORDERBY$ exists in query **then**
20:    $operator \leftarrow$ new SortOperator($operator$)
21: **if** $LIMIT$ exists in query **then**
22:    $operator \leftarrow$ new LimitOperator($operator$)
23: $operator \leftarrow$ new ProjectOperator($operator$)
24: **return** $operator$

The pseudocode of the logical plan builder is depicted in Algorithm 7.1. Getting the BGP as an input (which is provided by the query parsing component), it outputs the root node of the logical plan tree. Initially, it iterates through the triple patterns of the BGP (lines 4–9) and creates a select operator for every triple pattern (line 5). For every value specified in the three fields of the triple pattern (subject, predicate, or object), the select operator is assigned with a filter value (lines 6–8). The Selection operator is added to the LP list (line 9) in order to be used later in the construction of the plan. Adopting this approach, the Selection operators are pushed down (as low as possible in the plan), thus minimizing the amount of data provided to the parent nodes. For this reason, the leaf nodes of the query plan are usually Selection operators.

In the second step, the logical plan builder iterates through the operators contained in the LP list in order to form Join operators. In every execution of the loop (lines 10–13), the first two elements of the LP are combined under a Join operator. Then, the combined operators are removed from the LP list before adding the new Join operator in it. The loop ends when one operator is left in the LP list. Such an example is illustrated on Fig. 7.3. In case that two Select operators with common variable names between their corresponding filters are combined under a Join operator, the Join corresponds to an inner join between the underlying datasets. Otherwise, the Join corresponds to a cross join between them.

The third step integrates the procedure of transforming the elements (operators) of the LP list (line 14) into a tree (hierarchical) structure and then forming its root node (lines 15–24). It is certain that the type of root node of the final tree will be a Projection operator which is the output of the Logical Plan Builder component. By forming the tree structure of the LP list, the tree may obtain a new root node (operator) whose type will be one of: Select, Distinct, Sort, or Limit.

7.6.3 The Logical Plan Optimizer

Optimizations of the logical plan are classified in two main categories:

* *Rule-based optimizations*: based on a set of pre-defined rules that are known to result in improved performance.
* *Cost-based optimizations*: based on estimates of the output size of operators, where the estimates are derived from the maintained statistics of the underlying data.

In *rule-based optimization*, the joins that are performed on the subject node are discerned from other joins. Essentially, join-subject operators are identified, in order to be differentiated from typical join operators. The rationale behind this differentiation is that they can be exploited during physical planning to perform in a more efficient way.

In technical terms, the procedure performing this task attempts to form star joins (as subqueries of the initial query) among the triple patters, whenever possible. If a set of triple patterns (corresponding to Selection operators) form a star join, a join-subject operator is created and becomes the parent node of these Selection operators. Put differently, these triple patterns share a common variable in the subject field. The remaining triple patterns that do not form a star join remain as Selection operators

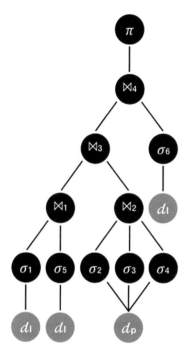

Fig. 7.4 The resulting logical query plan of rule-based optimization of SPARQL query depicted in Fig. 7.3

in the logical plan. Then, Join operators are formed by taking into account the remaining Selection operators and the join-subject operators (see Fig. 7.4), where the SPARQL query of Fig. 7.3 is used.

In *cost-based optimization*, the formation and the ordering of Join operators is determined, as this is well-known to have a significant impact on the performance of query execution.

- *Join operators Formation.* By having the Select operators added in the LP list (lines 4–9 of Algorithm 7.1), the optimizer iterates the list to find common variable names between their corresponding filters. The matching operators are combined together by a Join operator, along with their matching fields which are used as the join condition. The optimizer prioritizes the formation of Joins that are to be performed on the datasource that constitutes a property table (d_p).
- *Operators ordering.* After the formation of the Join operators, the optimizer exploits statistics (in the form of histograms) that are constructed during the data loading phase. Specifically, for each operator, the cardinality of its output size is estimated. The optimizer that determines the ordering of nodes prioritizes the execution of operators that return the fewest results. We refer to Fig. 7.5 for an illustrative example. This final logical query plan corresponds to the initial query plan that was formed.

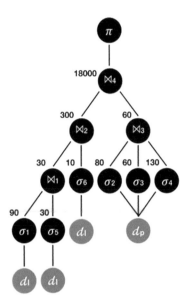

Fig. 7.5 The resulting logical query plan of cost-based optimization of SPARQL query depicted in Fig. 7.3. The numbers indicate estimations of output size

7.6.4　The Physical Plan Constructor

The physical plan constructor component takes as input the optimized logical plan and transforms it to a physical query plan. A physical query plan is comprised of a set of physical operators, which represent an algorithm for every stage (operation) of the query execution. Hence, the physical plan constructor aims to choose the best performing algorithm for every stage of the query execution. In this section we present the various physical operators supported by DiStRDF engine.

All operators are designed to operate by utilizing the features of Apache Spark SQL API. They process a distributed set of data (i.e., a Spark DataFrame) by performing parallel operations on it. Therefore, the workload of any DiStRDF physical operator is in fact distributed among the available computing nodes to be computed in parallel.

7.6.4.1　Datasource Operators

In DiStRDF there are two variants of data sources: the property table data source and the leftover triples data source. The implementation of the physical Datasource operator is straightforward for both of these data sources: it creates a new Spark SQL DataFrame from the specific Parquet file data source. Notice that, in DiStRDF, typically the next operation, right after a Datasource operator, is to filter the data needed by the query (Selection operation). As already discussed, Spark supports filter push-down, thus the first selection operator will be pushed down to the data source level. This is handled automatically by Spark; however, we need to emphasize that caching the DataFrame before applying the selection operation will break the filter push-down mechanism. Hence we opt to not cache any data source, to benefit by the improved performance of predicate push-down.

In the case of a property table data source, the resulting DataFrame contains a table similar to Table 7.1. The first column contains the subject of all the triples that are joined together, and all other columns correspond to the individual predicates found in the RDF triples; the values inside the cells are the values found in the object position of the underlying triples.

On the other hand, the leftover triples data source results to a DataFrame containing a table similar to Table 7.2. It contains three columns, one for each position of an RDF triple: subject, predicate, and object. Notice that this data representation is different from the one used for the property table data. This anomaly could result to handling increased complexity during query evaluation, since each representation needs different processing algorithms. To overcome this issue, we transform the current one-triples representation to the property table one, during the next selection operation applied on this DataFrame.

7.6.4.2   Selection Operators

The physical Selection operators take as input a DataFrame and apply one or more filters on it, based on the parameters defined in the corresponding logical Selection operator. Since in DiStRDF the Selection operators are created by both BGP and FILTER parts of the SPARQL query, the physical selection operators can handle any of these cases. Two physical selection operators are supported in DiStRDF: the *IndexScan* and *TableScan* operations.

The IndexScan operation is applied only in the case the Selection operation is placed immediately after a Datasource operator. Basically, it applies a filter on the DataFrame which is pushed-down on the data source level. The IndexScan operator is very efficient, since it reads only the data that are relevant to the query, decreasing the overhead posed by having to read the whole dataset from disk. Since both data sources are stored as Parquet files, the IndexScan operator can be used on DataFrames from both of them. The TableScan operator is applied whenever an index is not provided (i.e., in the case the Selection operator is not immediately after a Datasource operator). It filters a DataFrame by scanning all the containing data and keeping those that satisfy the filtering criteria.

If the Selection operator is performed on top of a leftover triples Datasource operator, then before applying the IndexScan operation, it transforms the DataFrame as shown in Table 7.3. In order to make this transformation, the Selection operator first filters the DataFrame to keep only the records containing the correct value of the "Predicate" field. Hence, the resulting DataFrame contains only records of a single value on the Predicate column, as shown in Table 7.3(a). Then, the Predicate

Table 7.3 Example of transforming leftover triples to property table format

(a) Leftover triples format   →   (b) Property table format

Subject	Predicate	Object		Subject	hasLength
ves376609000	hasLength	52		ves376609000	52
ves376609005	hasLength	97		ves376609005	97
ves376609008	hasLength	65		ves376609008	65

column is deleted, and the Object column is renamed to match the value of the fields on the Predicate column. The example depicted in Table 7.3(b) uses the name "hasLength" on the Object column, since all Predicate field values are filtered to have the value "hasLength." After applying this transformation, the DataFrame is filtered by applying the IndexScan or TableScan filters, as described earlier.

7.6.4.3 Join Operators

The physical Join operator takes as input two tables and combines their tuples based on a related column between them. Notice that in some cases, the join operation is pre-computed by the stored property tables. The DiStRDF processing layer is designed to exploit such information and avoids evaluating a pre-computed join operation.

Given the fact that data is distributed, processing a join operator requires distributed join processing. This is a challenging operation because it usually results in transferring large amounts of data from one node to another, and this cost can dominate the entire execution cost. As a result, optimizing the processing of joins is a critical factor for a distributed RDF processing engine.

By exploiting the Spark SQL API, we can utilize two physical join operators: Broadcast Hash Join and Sort-merge Join. The details of these algorithms can be found in [3]. However, they are briefly described in the following, assuming that *datasetA* and *datasetB* are joined together, when the size of *datasetB* is estimated to be smaller than the size of *datasetA*, based on the available statistics:

- **Broadcast Hash Join.** This algorithm is typically more efficient for smaller sizes of *datasetB*. It broadcasts *datasetB* to all nodes available in the cluster. Then, each node performs a join operation, using the portion of *datasetA* available locally. The execution steps of this algorithm are described below:

  1. *datasetB* is collected at a single node of the cluster (also called driver node).
  2. A hashed structure of *datasetB* is built locally on the driver node.
  3. Hashed *datasetB* is broadcast to all the nodes.
  4. The broadcast *datasetB* is joined with local portions of *datasetA* in parallel, using the hash join algorithm.

- **Sort-merge Join.** This algorithm performs better for larger sizes of *datasetB*. It can also be used when the actual size of *datasetB* is unknown and cannot be estimated accurately. It performs a shuffling (i.e., repartitioning) of both *datasetA* and *datasetB* on all nodes of the cluster and then joins together the local subsets. Sort-merge Join is a more decentralized algorithm compared to Broadcast Hash Join, at the cost of potentially higher network bandwidth consumption.

1. *datasetA* and *datasetB* are repartitioned (shuffled) using the same *partitioner*[11] on their respective join keys. Thus, records from both datasets will reside on the same node if and only if these records share the same join keys.
2. Each local subset of *datasetA* is sorted in parallel on all nodes.
3. The Sort-merge Join algorithm is applied on the subsets of sorted *datasetA* and *datasetB*.

7.6.4.4 Projection Operator

The Projection operator takes as input a Spark DataFrame and adjusts the number, names, and order of columns according to the parameters defined in the logical Projection Operator. Its primary purpose in DiStRDF is to select, rename, and order the columns that will be delivered to the user as the final result set. Hence, the users will be able to recognize the resulting set of columns, according to the query they have provided.

7.6.4.5 Spatiotemporal Operators

The spatiotemporal operators are special cases of Selection operators with spatiotemporal criteria. Since DiStRDF engine is designed to specifically to address efficient execution of SPARQL queries with spatiotemporal criteria, these operators are essential towards achieving the DiStRDF's goal. When the physical plan constructor identifies spatiotemporal criteria in the logical plan, it intervenes in the execution plan, by adding two operators: the approximate spatiotemporal operator and the exact spatiotemporal operator.

Essentially, the approximate spatiotemporal operator is always pushed to the bottom of the execution plan tree, to prune early and efficiently as many records as possible. It utilizes the 1D spatiotemporal encoding, described in Sect. 7.3, to prune all records that are not enclosed in a spatiotemporal box determined by the spatial and temporal constraints. To accomplish this, two numbers are computed corresponding to the lower and upper cells that enclose the spatiotemporal box, respectively. Since these cells are a superset of the spatiotemporal box, these numbers represent the approximate location of the spatiotemporal box, in the spatiotemporal grid constructed by the 1D encoding. This filter returns only an approximate result, thus an additional step is required to prune the records

---

[11] A partitioner is a mechanism that determines the location (i.e., node) of each record, on the repartitioning process.

that do not belong in the final result set (false positives). To this end, the exact spatiotemporal filter is added later in the execution plan, to refine the results based on the actual spatiotemporal filter defined in the SPARQL query.

7.6.4.6 Other Operators

All other operators, such as Distinct, Limit, and Sort, are designed to exploit the corresponding native Spark SQL DataFrame operations and their description is out of scope of this book chapter.

7.7 Experimental Evaluation

In this section, we present the results of our experimental study. Our algorithms are implemented using Scala 2.11 and Apache Spark 2.1. We deployed our code on a proprietary cluster of 10 physical nodes, each having 64GB RAM, a 6-core 1.7GHz processor and running Ubuntu 16.04.

7.7.1 Experimental Setup

DataSets The datasets that are used in the experiments contain surveillance and static information from the maritime and aviation domains. The maritime surveillance data cover the Mediterranean Sea and part of the Atlantic Ocean for the entire month of January 2016. The aviation surveillance data cover the entire space of Europe for a week of April 2016. In order to represent all data in RDF format, the datAcron ontology is used (cf. Chap. 5), described in [15].

The total size of the dataset is about 400 million triples: 300 and 100 million triples compose the maritime and aviation domain, respectively. These triples are encoded to integer values using the encoding scheme described in Sect. 7.3 to form the one-triples table, which is approximately 9GB in text format, and 4GB in Parquet format using snappy compression. Also, property tables are built for the *Semantic Node* and *Vessel* entities of datAcron Ontology so as to enable efficient access to their corresponding properties during query execution. In the context of the experiments data is stored in HDFS using Parquet file format, to enable efficient access for the Spark applications and benefit from columnar storage, compression, and predicate push-down.

Furthermore, a dictionary is used for the mapping of the encoded and decoded values. It is stored in a Redis cluster instance, running on the cluster, with no replication enabled. The total number of records (key-value pairs) stored in the Redis dictionary is roughly 106 million.

Configuration Apache Spark was configured on YARN, using Hadoop 2.7.2. One node was set to be the driver node, while the others contained the Spark Executors. For all of the conducted experiments, 9 Spark Executors are used, with 5 executor cores and 4GB memory each. HDFS was configured with replication factor of 3. Also, Jedis[12] library was used in order to establish the communication with the Redis cluster.

Data Statistics The Statistics Manager component is configured to build equi-width histograms with 1000 buckets per dimension. This resulted in 10^3 buckets for the Predicate histogram, and 10^6 buckets for Subject-Predicate and Object-Predicate histograms. The max frequency values held in the buckets of the histogram are used during query planning; hence, the logical query optimizer is based on worst-case estimations for calculating the occurrence frequency of any field value.

Metrics The execution time needed for a SPARQL query to be performed on the Spark cluster, is the main evaluation metric used in the experiments. Specifically, the actual execution time needed for the queries to be evaluated is measured, by omitting (a) any overhead caused by Spark initialization processes and (b) the time needed to store the result set in HDFS. In technical terms, only the time needed to calculate the result set in main memory is measured. This is feasible by executing a call to the count method of the Spark DataFrame containing this result.

Moreover, each experiment is performed for 10 times before reporting any time needed for query's execution; only the time needed for the 11th execution is mentioned, as a warm-up procedure takes place. This ensures that the cost of the Java JIT compiler and the overhead for establishing connections to the Redis cluster from all YARN containers is also omitted. The output size of our experiments in terms of number of records is also reported, as an indication of query selectivity (i.e., the number of records that matches the query's basic graph pattern).

Methodology The experiments are presented with 2 real-world SPARQL queries, one for each domain (maritime and aviation). The logical and physical optimizers are enabled for these experiments while no spatiotemporal box filter is employed. Then, the efficiency improvements provided by adopting the logical and physical plan optimizers are evaluated. Lastly, the gained performance by enabling the filtering based on 1D encoding information is measured for various spatiotemporal box sizes.

7.7.2 Experiments on Real-World Queries

In this section, the performance of the *datAcrondistributedRDFengine* is studied, by executing queries over the integrated data used in the two domains. Two queries are

---

selected, and the execution time required by the *datAcrondistributedRDFengine* in order to compute the result set for the corresponding domain field is measured.

7.7.2.1 Queries for Maritime and Aviation Domain

Two queries are selected, denoted Q_m and Q_a, to demonstrate the performance of the *datAcrondistributedRDFengine* for the maritime and aviation domain correspondingly.

The query Q_m related to the maritime domain returns all of the semantic nodes (and their related information of properties that are mentioned in the query) of vessels that sail with minimum wind direction 77.13083.

```
PREFIX  :  <http://www.datacron-project.eu/datAcron#>
SELECT *
WHERE {
    ?n  : ofMovingObject ?ves ;
    : hasGeometry ?g ;
    : hasTemporalFeature ?t ;
    : hasWeatherCondition ?w .
    ?ves  a  ?VesselType ;
    : has_vesselFixingDeviceType ?device ;
    : vesselName ?name .
    ?event  : occurs ?n .
    ?g  :hasWKT ?pos .
    ?t  : TimeStart ?time .
    ?w  : windDirectionMin "77.13083" .
}
```

The resulted data of the Q_m query include:

- The vessels' type, name, device type
- The known semantic nodes of the vessels
- The spatiotemporal position of the semantic nodes
- The events that are related to the semantic nodes
- The prevailing weather conditions

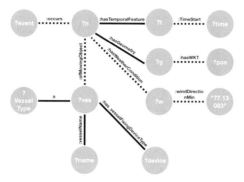

Fig. 7.6 Query graph for the Q_m query used in the experiments for the maritime domain

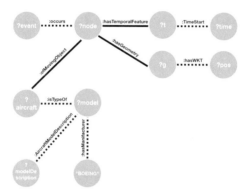

Fig. 7.7 Query graph for the Q_a query used in the experiments for the aviation domain

The query Q_a related to the aviation domain returns all of the semantic nodes (and their related properties mentioned in the query) of aircrafts manufactured by Boeing.

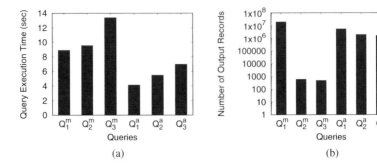

Fig. 7.8 Performance and output size for the different domain queries. (**a**) Execution time (sec).
(**b**) Output size (# Records)

```
PREFIX : <http://www.datacron-project.eu/datAcron#>
PREFIX rdfs: <http://www.w3.org/2000/01/rdf-schema#>
SELECT *
WHERE {
     ?node :ofMovingObject ?aircraft ;
     :hasGeometry ?g ;
     :hasTemporalFeature ?t .
     ?event :occurs ?node .
     ?g :hasWKT ?pos .
     ?t :TimeStart ?time .
     ?aircraft :isTypeOf ?model .
     ?model :hasManifacturer "BOEING" ;
     :AircraftModelDescription ?modelDescription .
}
```

The resulted data of the Q_a query include:

- The known semantic nodes of the aircrafts
- The events that are related to the semantic nodes
- The spatiotemporal position of the semantic nodes
- The aircraft's model and Description

7.7.2.2 Evaluation

Figures 7.6 and 7.7 depict the two RDF query graphs, for the maritime and aviation
queries, respectively. The graph edges indicated with a dashed line need a join
operation so as to be computed. The remaining edges indicated using solid lines
are stored as pre-computed joins in Property tables, thus avoiding the need for a join
operation.

Figure 7.8 presents the results for the two real-world queries. As shown in Fig. 7.8a, Q_m query, related to the maritime domain, result in higher execution times, since the input dataset contains more triples from that domain. Another factor having a strong impact on execution time is the workload of a given query; queries that require more join operations result in increased processing cost, regardless of their output sizes, which are depicted in Fig. 7.8b. Notice that Fig. 7.8b uses log scale on the y-axis.

More specifically, queries Q_1^m, Q_2^m, Q_3^m include 4, 5, and 8 join operations, respectively. The queries that need more join operations result to increased processing time regardless of the fact that their output size is decreasing. This observation justifies the need for employing a physical optimizer to pick a suitable join operator whenever needed. Similarly, the same trend is observed for the aviation queries where Q_1^a, Q_2^a, Q_3^a include 3, 6, and 10 join operations, respectively.

In the rest of this section, we arbitrarily select one query from those presented here, in order to demonstrate the gain provided by the physical optimization, the logical optimization, and the 1D encoding scheme.

7.7.3 Experiments on Physical Query Optimization

We evaluate our physical optimizer by focusing on query Q_2^a. We report the logical plan produced for this query in Fig. 7.9, along with the estimated input and output sizes for each node involved. We focus only to the 6 nodes which perform an actual join operation (i.e., they refer to relationships which are not maintained in Property Tables), namely J1, J2, J3, JS2, J4, and J5. Clearly, the estimated input size of J1, J2, J3, and J5 is larger than the estimations calculated for JS2 and J4. Our physical plan optimizer utilizes these statistical values to determine the physical join operators that should be used.

In this section, we evaluate the performance of Q_2^a, by using various physical planning strategies, as depicted in Fig. 7.10. The first strategy is to pick the Sort-Merge Join operator for all the aforementioned nodes (denoted as Sort Merge Always). The second strategy (denoted as Spark Optimizer) uses the default behavior employed by Spark optimizer which performs a Sort Merge Join on all nodes except for JS2 performing a Broadcast Hash Join. The third strategy (denoted as datAcron Optimizer) utilizes the statistics provided by our Statistics Manager to pick a physical join operator based on its estimated input sizes: If the estimated input size of one of its children is below a user-defined threshold, we opt to use a Broadcast Hash Join; otherwise a Sort Merge Join is used. Hence, the third strategy performs a Sort Merge Join on all nodes, except for JS2 and J4, which as depicted in Fig. 7.9, have low estimated input sizes. We have also experimented with a fourth strategy which performs a Broadcast Hash Join on all nodes, but its execution time is at least 4 times worse than the others. We deliberately excluded this strategy from Fig. 7.10 to enable clearer comparison between the remaining three.

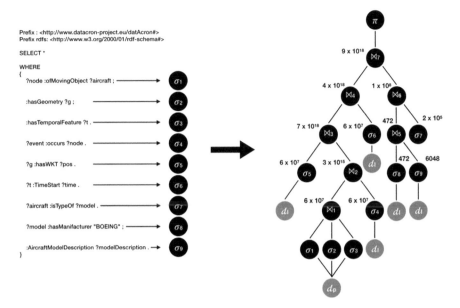

Fig. 7.9 Estimated cardinality output sizes of all operators of Q_2^a

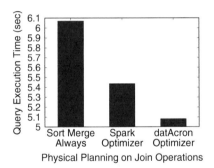

Fig. 7.10 Performance of Q_2^a when using various physical planners

Figure 7.10 demonstrates the query execution time needed for each of the aforementioned strategies. As expected, Sort Merge Always strategy is performing worse than the other two, since the processing cost for performing a Sort Merge Join on lower input sizes is larger than that of Broadcast Hash Join. Moreover, our physical optimizer outperforms the Spark optimizer, since a Broadcast Hash Join on J4 is a better choice than a Sort Merge Join. Notice that Spark optimizer also relies on statistical values obtained by Parquet files to pick a physical join operator. However, these statistics prove to be inadequate for efficiently evaluating Q_2^a in parallel.

7.7.4 Experiments on Logical Query Optimization

We select query Q_2^m to evaluate our logical plan optimizer. Figure 7.11 shows the SPARQL query, and two alternative logical plans. The plan on the right (called Basic) is the one produced when the logical optimizer is switched off, and the order of operations practically corresponds to the order of joins in the syntax of the query. Instead, the plan on the left (called Optimized) is produced by the logical optimizer. Each node in the plan is annotated with the cardinality estimation of its output, based on the maintained statistics.

As shown in Fig. 7.11, the logical plan optimizer performs join re-ordering using the cardinality estimations, in an attempt to start processing from operators estimated to produce small outputs. In turn, this results in smaller intermediate results, thereby improving performance.

Figure 7.12 demonstrates the improved efficiency provided by our logical plan optimizer. As shown in the chart, rearranging the order of join operations on Q_m^2 results in approximately 40% improved efficiency in execution time. For convenience, recall that the result size of this query is roughly 600 records. This experiment demonstrates the benefits that can be attained by using the logical plan optimizer, which is a key component towards building an efficient (distributed) RDF processing engine.

7.7.5 Experiments on Spatiotemporal Filtering

In our distributed RDF query engine, a spatiotemporal box filter can be evaluated by employing either the *1D Filtering* or the *Regular Filtering* method. The 1D Filtering method applies an approximate filter first, based on the embedded spatiotemporal information provided by our 1D encoding scheme; then, a refinement process is employed at the end of query execution to prune the false positives produced by the approximate filter. On the other hand, the Regular Filtering method prunes records by comparing the spatial and temporal columns against the query's spatiotemporal predicates.

We select query Q_2^q to compare its performance when employing 1D Filtering against Regular Filtering method. To this end, we experiment with 3 spatiotemporal boxes (SF 1, SF 2, SF 3) of increasing size, as depicted in Table 7.4. Notice that the latitude and longitude values of Table 7.4 are expressed in degrees, while the altitude is expressed in meters.

Figure 7.13a demonstrates the execution time needed for the queries to be evaluated. Clearly, our 1D filter outperforms Regular filtering, providing even better performance as the spatiotemporal box size decreases. As expected, the Regular filter consistently requires higher time to filter out non-matching records: it applies the same number of comparisons regardless of the spatiotemporal box size.

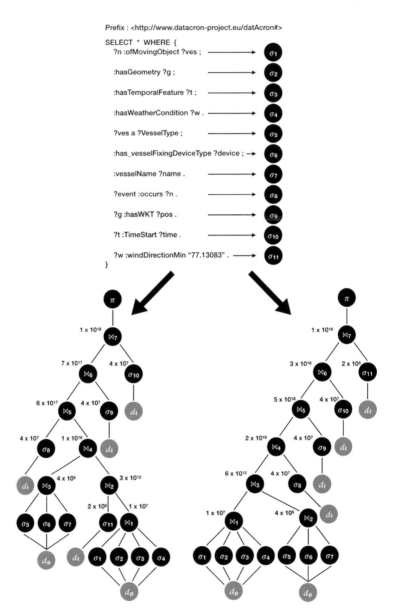

Fig. 7.11 Optimized (left) vs Basic (right) logical planning for Q_m^2

However, our 1D filter is able to quickly prune records based on the 1D encoded information, thus reducing the input size of the refinement process.

Naturally, the output size of Q_2^a is affected by the size of the spatiotemporal box filtering employed. Therefore, we report the output size for each such box in Fig. 7.13b. For convenience, we recall that the output size of Q_2^a without any

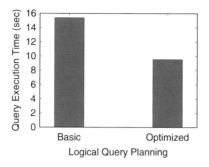

Fig. 7.12 Performance of Q_2^m for Basic and Optimized logical query planning

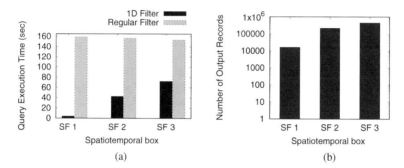

Fig. 7.13 Performance and output size for the different sizes of spatiotemporal boxes on Q_2^q. (**a**) Execution time (sec). (**b**) Output size (# Records)

Table 7.4 Spatiotemporal box boundaries for Q_2^q

	SF 1	SF 2	SF 3
Lower Spatial (lat, lon, alt)	(47, 0, 5000)	(45, -3, 4000)	(44, -4, 3000)
Upper Spatial (lat, lon, alt)	(53, 6, 9000)	(54, 7, 12000)	(55, 8, 13000)
Lower Time	2016-04-15 00:00:00	2016-04-15 00:00:00	2016-04-15 00:00:00
Upper Time	2016-04-15 23:59:59	2016-04-16 23:59:59	2016-04-17 23:59:59

filtering was roughly 2 million records, while the largest spatiotemporal box that we experimented with results to approximately 450 thousand records.

7.8   Conclusions

Scalable management of large knowledge graphs represented in RDF is a challenging topic; especially in the case of spatiotemporal RDF data, existing distributed RDF management fall short, as they are not optimized for joint processing of graph pattern queries together with spatiotemporal queries.

The design and implementation of DiStRDF, a parallel processing engine for spatiotemporal RDF data, is motivated by these limitations. To devise a scalable data processing engine many contributions were needed: a deliberate encoding scheme that captures spatiotemporal information of a moving entity in a one-dimensional value, a storage layer based on scalable technologies, and a processing layer that encompasses the basic modules of a typical database engine developed over a big data processing framework.

However, many issues remain open: First, we need to extend the capabilities of the engine so as to cover a wider class of spatiotemporal RDF queries. Then, we need to implement more efficient join algorithms and build a more mature query optimizer based on more advanced techniques for query optimization. These are some of the issues that are expected to be addressed by future work.

References

1. Abdelaziz, I., Harbi, R., Khayyat, Z., Kalnis, P.: A survey and experimental comparison of distributed SPARQL engines for very large RDF data. Proc. VLDB Endowment **10**(13), 2049–2060 (2017)
2. Bereta, K., Smeros, P., Koubarakis, M.: Representation and querying of valid time of triples in linked geospatial data. In: The Semantic Web: Semantics and Big Data, Proceedings of 10th International Conference, ESWC 2013, Montpellier, 26–30 May 2013, pp. 259–274 (2013)
3. Blanas, S., Patel, J.M., Ercegovac, V., Rao, J., Shekita, E.J., Tian, Y.: A comparison of join algorithms for log processing in MapReduce. In: Proceedings of the ACM SIGMOD International Conference on Management of Data, SIGMOD 2010, Indianapolis, IN, 6–10 June 2010, pp. 975–986 (2010). https://doi.org/10.1145/1807167.1807273
4. Dean, J., Ghemawat, S.: MapReduce: simplified data processing on large clusters. In: Proceedings of 6th Symposium on Operating Systems Design and Implementation, pp. 137–149 (2004). https://doi.org/10.1145/1327452.1327492
5. Doulkeridis, C., Nørvåg, K.: A survey of large-scale analytical query processing in MapReduce. VLDB J. **23**(3), 355–380 (2014)
6. Garbis, G., Kyzirakos, K., Koubarakis, M.: Geographica: a benchmark for geospatial RDF stores (long version). In: International Semantic Web Conference, pp. 343–359. Springer, Berlin (2013)
7. Kaoudi, Z., Manolescu, I.: RDF in the clouds: a survey. VLDB J. **24**(1), 67–91 (2015)
8. Koubarakis, M., Kyzirakos, K.: Modeling and querying metadata in the semantic sensor web: the model sTRDF and the query language stSPARQL. In: The Semantic Web: Research and Applications, Proceedings of 7th Extended Semantic Web Conference, ESWC 2010, Heraklion, Crete, 30 May–3 June 2010, Part I, pp. 425–439 (2010)
9. Koubarakis, M., Karpathiotakis, M., Kyzirakos, K., Nikolaou, C., Sioutis, M.: Data models and query languages for linked geospatial data. In: Reasoning Web. Semantic Technologies for Advanced Query Answering - Proceedings of 8th International Summer School 2012, Vienna, 3–8 Sept 2012, pp. 290–328 (2012). https://doi.org/10.1007/978-3-642-33158-9_8
10. Kyzirakos, K., Karpathiotakis, M., Bereta, K., Garbis, G., Nikolaou, C., Smeros, P., Giannakopoulou, S., Dogani, K., Koubarakis, M.: The spatiotemporal RDF store Strabon. In: Proceedings of SSTD, pp. 496–500 (2013)

11. Lim, H., Han, D., Andersen, D.G., Kaminsky, M.: MICA: a holistic approach to fast in-memory key-value storage. In: Proceedings of the 11th USENIX Symposium on Networked Systems Design and Implementation, NSDI 2014, Seattle, WA, 2–4 April 2014, pp. 429–444 (2014). https://www.usenix.org/conference/nsdi14/technical-sessions/presentation/lim
12. Liu, Q., Yuan, H.: A high performance memory key-value database based on Redis. J. Comput. **14**(3), 170–183 (2019). http://www.jcomputers.us/index.php?m=content&c=index&a=show& catid=209&id=2925
13. Naacke, H., Amann, B., Curé, O.: SPARQL graph pattern processing with apache spark. In: Proceedings of the Fifth International Workshop on Graph Data-management Experiences & Systems, GRADES@SIGMOD/PODS 2017, Chicago, IL, 14–19 May 2017, pp. 1:1–1:7 (2017)
14. Nikitopoulos, P., Vlachou, A., Doulkeridis, C., Vouros, G.A.: Parallel and scalable processing of spatio-temporal rdf queries using spark. GeoInformatica (2019). https://doi.org/10.1007/s10707-019-00371-0
15. Santipantakis, G.M., Vouros, G.A., Doulkeridis, C., Vlachou, A., Andrienko, G.L., Andrienko, N.V., Fuchs, G., Garcia, J.M.C., Martinez, M.G.: Specification of semantic trajectories supporting data transformations for analytics: the datAcron ontology. In: Proceedings of the 13th International Conference on Semantic Systems, SEMANTICS 2017, Amsterdam, 11–14 Sept 2017, pp. 17–24 (2017)
16. Schätzle, A., Przyjaciel-Zablocki, M., Berberich, T., Lausen, G.: S2X: graph-parallel querying of RDF with GraphX. In: Biomedical Data Management and Graph Online Querying - VLDB 2015 Workshops, Big-O(Q) and DMAH, Waikoloa, HI, 31 Aug–4 Sept 2015, Revised Selected Papers, pp. 155–168 (2015)
17. Schätzle, A., Przyjaciel-Zablocki, M., Skilevic, S., Lausen, G.: S2RDF: RDF querying with SPARQL on spark. Proc. VLDB Endowment **9**(10), 804–815 (2016)
18. Shi, J., Qiu, Y., Minhas, U.F., Jiao, L., Wang, C., Reinwald, B., Özcan, F.: Clash of the Titans: MapReduce vs. spark for large scale data analytics. Proc. VLDB Endowment **8**(13), 2110–2121 (2015)
19. Vlachou, A., Doulkeridis, C., Glenis, A., Santipantakis, G.M., Vouros, G.A.: Efficient spatio-temporal RDF query processing in large dynamic knowledge bases. In: Proceedings of the 34th Annual ACM Symposium on Applied Computing, SAC 2019, Limassol, 08–12 April 2019

Part IV
Analytics Towards Time Critical Mobility Forecasting

This part focuses on mobility analytics methods exploiting processed, synopsized, and enriched data streams as well as integrated mobility data that is stored in large distributed knowledge graphs: While the first two chapters present online methods for trajectory prediction and events detection, the third chapter presents offline methods focusing on trajectory analytics. Specifically, the first chapter presents online future location prediction methods and trajectory prediction methods, distinguishing between short-term predictions and challenging long-term online predictions. Online recognition of complex events for providing situation awareness and anomaly detection, are presented in the second chapter. The last chapter focuses on offline trajectory analytics, addressing trajectory clustering and detection of routes followed by mobile entities.

Chapter 8
Future Location and Trajectory Prediction

Harris Georgiou, Petros Petrou, Panagiotis Tampakis, Stylianos Sideridis, Eva Chondrodima, Nikos Pelekis, and Yannis Theodoridis

Abstract This chapter presents modern approaches and frameworks for predicting trajectories with detailed descriptions of three main research pillars. The first pillar is the problem formulation regarding two complementary tasks, namely the *Future Location Prediction* (FLP) and the *Trajectory Prediction* (TP). The second pillar tackles the issue of effectiveness, efficiency, and scalability for the corresponding predictive analytics models for big fleets of moving objects. Finally, the third pillar takes into account historical patterns and semantically rich contextual information, so as to improve the prediction accuracy, especially for long-term time windows. The overall assessment of these methods shows that the suite of FLP and TP algorithms developed addresses all the major prediction challenges regarding mobility patterns in terms of points or trajectories, respectively. It is expected that these modeling approaches can be transferred to other domains of similar challenges and with similar success.

8.1 Introduction

8.1.1 Purpose, Scope, Motivation

The increasing use of portable devices such as navigation systems and the wide range of location-aware applications have led to a huge amount of mobility data being produced on a daily basis. As a result, a plethora of research challenges have emerged in order to manage or analyze such data. One of the most challenging data analytics task is to transform data into actionable knowledge in terms of exploiting historical mobility patterns, in order to gauge what the moving entities may do in the

H. Georgiou · P. Petrou · P. Tampakis · S. Sideridis · E. Chondrodima · N. Pelekis (✉) ·
Y. Theodoridis
University of Piraeus, Piraeus, Greece
e-mail: hgeorgiou@unipi.gr; ppetrou@unipi.gr; ptampak@unipi.gr; ssider@unipi.gr;
evachon@unipi.gr; npelekis@unipi.gr; ytheod@unipi.gr

© Springer Nature Switzerland AG 2020
G. A. Vouros et al. (eds.), *Big Data Analytics for Time-Critical Mobility Forecasting*, https://doi.org/10.1007/978-3-030-45164-6_8

future and develop efficient online algorithms for short- and long-term prediction of movements [26, 43].

In real-time scenarios where new positions are traced in streaming mode, prediction algorithms evaluate whether the moving object remains on route or deviates, e.g., online outlier detection. The algorithms take advantage of offline data analytics results based on archived information to produce accurate and reliable predictions regarding future movements and events. Moreover, maritime and aviation surveillance exhibits various types of uncertainty, including spatial (e.g., due to GPS errors), temporal (e.g., different sensor refresh rates, transmission delays, asynchronous sensor clocks, and reporting frequency), or contextual (e.g., maritime or airspace regulations, adverse weather conditions that may affect the actual routes, etc.). A probabilistic treatment with error assessment and guarantees is required to address such imperfections in the raw data. In order to address these challenges, the knowledge of historical patterns of movement is used to enhance the accuracy in the prediction of moving object locations on a long-term time window.

Moreover, there may be problem-specific resources or constraints that need to be taken into account when designing predictive models for mobility and traffic flows in various domains. For example, online processing in near real time may be necessary; hence, retrieval of complex models and historic data may not be feasible in practice, and faster filtering-based models have to be employed. On the other hand, path-specific constraints may be available and, thus, these can be incorporated in the corresponding model designs (e.g., maritime routes or flight plans).

The algorithms presented here exploit various types of uncertainty, supporting probabilistic treatment of error assessment and probabilistic bounds (e.g., confidence intervals), addressing imperfections in the raw data, as well as uncertain external factors during the realization of the movement (e.g., local weather). Building upon the current state-of-the-art, novel extensions and improvements are presented here, including final decisions and optimized configurations, nominal conditions for stable and reliable operation, together with extended experiments and results.

The presentation of the material in this chapter evolves around three aspects: (a) provide a thorough but compact formalization of the main components and problems; (b) present recent developments and representative novel approaches; and (c) provide related work, comparable approaches, and discussion on the results.

Closely related to this chapter, our work regarding TP in aviation [12] introduces flight plans, localized weather, and aircraft properties as trajectory annotations that enable modeling in a space higher than the typical 4-D spatiotemporal. This results in a multi-stage hybrid approach for a new variation of the core TP task, the so-called *Future Semantic Trajectory Prediction* (FSTP). This constrained-based approach for training a predictive model per each waypoint is described in detail in Sect. 8.4.

Specifically, the structure of this chapter is as follows: After a short introduction, the background section provides all the necessary definitions for enriched points and trajectories, specifies the core problems of future location prediction (FLP) and Trajectory Prediction (TP) and associated terms, and provides a quick review of the datasets exploited. Next, the FLP problem is explored under the scope of short-term routes-agnostic (limited "memory") predictors using LSTM and long-term network-aware (increased "memory") predictors using a hybrid medoids-based approach, providing experimental results for both. For the TP problem, a semantic-aware formalization is provided for the aviation domain and the core task is explored with the exploitation of enrichment data (e.g., weather) together with the flight plans, in a hybrid medoids-based "constrained" training with historic data, providing experimental results for various linear and non-linear choices for the predictors. Finally, related work and discussion sections explore other approaches and comparable methods, with hints to future trends and prospects, and the conclusions section closes the chapter.

8.2 Background

For better understanding of the two main tasks, FLP and TP, the following definitions are provided with regard to the spatiotemporal nature of the corresponding data.

The (raw) trajectory T of an aircraft is defined as a 4-D polyline consisting of a sequence of $|T|$ pairs $< p_i, t_i >$, $i \in [0, |T| - 1]$, where p_i is a location data point (x_i, y_i, z_i) in the 2-D or 3-D space and t_i is a timestamp, assuming linear interpolation between two consecutive pairs $< p_i, t_i >$ and $< p_{i+1}, t_{i+1} >$. Having these definitions at hand, we define their semantic-aware variants.

Definition 8.1 (Enriched Point) An enriched point r_i corresponds to a (raw) pair $< p_i, t_i >$ and is defined as a triplet $< p_i, t_i, v_i >$, where v_i is a vector consisting of categorical and/or numerical variables that annotate the raw point with associated *enrichment* data.

Examples of v_i attribute values can be any user-defined label or annotation regarding the specific application, for example, generated by an event recognition module that detects the "top-of-climb" or "top-of-descent," etc. Similarly, any numerical variable can be attached to p_i, such as weather variables regarding temperature, wind speed, humidity, etc.

Definition 8.2 (Enriched Trajectory) A semantically enriched trajectory R corresponds to a (raw) trajectory T of a moving object, which is defined as the sequence of the enriched points of T.

The FLP and TP tasks build upon these definitions to formulate and describe each case separately in the following sections.

8.2.1 Future Location Prediction

As the maritime and air traffic management (ATM) domains have major impact on the global economy, a constant need is to advance the capability of systems to improve safety and effectiveness of critical operations involving a large number of moving entities in large geographical areas [9]. Towards this goal, the exploitation of heterogeneous data sources, which offer vast quantities of archival and high-rate streaming data, is crucial for increasing the accuracy of results when analyzing and predicting future states of moving entities. However, operational systems in these domains for predicting trajectories are still limited mostly to a short-term look-ahead time frame while facing increased uncertainty and lack of accuracy.

Motivated by these challenges, we present a big data solution for *online* TP by exploiting mined patterns of trajectories from historical data sources. Our approach offers predictions such as *"estimated flight of an aircraft over the next 10 minutes"* or *"predicted route of a maritime vessel in the next hour,"* based on their current movement and historical motion patterns inside a specific region of interest. The proposed framework incorporates several innovative modules, operating in streaming mode over surveillance data, to deliver accurate long-term predictions with low latency requirements. Incoming streams of moving objects' positions are cleansed, compressed, integrated, and linked with archival and contextual data by means of link discovery methods.

This part of the work includes three main contributions: (a) devise a big data methodology/algorithm that solves the FLP problem in an effective and highly scalable way; (b) design and implement this algorithm on top of state-of-the-art big data technologies, namely Spark and Kafka; and (c) conduct an extensive experimental study in large real datasets from the maritime and aviation domains. In the proposed methodology we efficiently identify the hidden mobility patterns in an offline unsupervised manner and, subsequently, the prediction algorithm exploits these patterns in order to extend the FLP temporal horizon upon these "discovered" routes. To the best of our knowledge, in contrast to related state-of-the-art systems [8] and research approaches, our approach is unique as a big data framework capable of providing long-term trajectory predictions in an online fashion.

Given definitions 8.1 and 8.2 and assuming a historic *Trajectory Database* (TD), we define the *Future Location Prediction* (FLP) problem as follows:

Definition 8.3 (Future Location Prediction) Given: (a) the incomplete trajectory $< (p_0, t_0), (p_1, t_1), \ldots, (p_{i-1}, t_{i-1}) >$ of a moving object o, consisting of its time-stamped locations recorded at past i time instances and (b) an integer value $j \geq 1$, **Predict:** $< (p_i^*, t_i), \ldots, (p_{i+j-1}^*, t_{i+j-1}) >$, i.e., the objects' anticipated locations at the following j time instances.

There are two major directions when dealing with the FLP problem: (a) *vector-based* prediction, or the spatial database management approach, and (b) *pattern-based* prediction, or the data mining and machine learning approach. Each has its own advantages and drawbacks and, most importantly, they are based on different assumptions regarding the input data and their organization.

The vector-based approaches, inspired by the spatial database management domain, aim to model current locations (and perhaps a short history) of objects as *motion functions*, in order to be able to predict future locations by some kind of extrapolation. In practice, they take into consideration space and time and predict future locations of moving objects within a given time interval using a mathematical or probabilistic model, which aims to simulate the anticipated movement. First- or second-degree physics models of movement are commonly used, employing extrapolation with velocity or velocity and acceleration components, respectively, to estimate the evolution of movement, provided that these can be assumed to be constant in a short-term look-ahead time window.

The constant-speed assumption is also very useful in the development of proper transformations of the input space, enabling time-invariant representations, e.g., via the Hough-X transformation. Essentially, the evolving position of a moving object remains a stationary point in dual space as soon as it does not change its velocity vector; thus, it can be efficiently indexed in a spatial access method.

The pattern-based approaches, inspired by the spatial data mining domain, identify and exploit motion patterns by analyzing historic data of moving objects, i.e., classification models, repetitive patterns, clusters of "similar" movements, etc., based upon historical mobility data. An important difference with respect to the vector-based approaches is that in this case the models are built upon the history of movements, not only of the object of interest, but also of the other objects moving in the same area; therefore, they are able to build better models and use them for addressing the FLP task in a more generic and data-driven way.

A more detailed description of related works to FLP is provided later in Sect. 8.5.

8.2.2   Trajectory Prediction

The problem of predictive analytics over mobility data in the aviation or maritime domains involves applications where moving objects are tracked in real time, in order to compute, e.g., short- or long-term predictions. Short-term prediction [43], which is time-critical and requires immediate response, facilitates the efficient planning, management, and control procedures while assessing traffic conditions. The latter is extremely important as safety, credibility, and cost are critical factors in immediate decision-making. On the other hand, long-term prediction [41] enhances pre-flight strategic planning to achieve cost efficiency or, when contextual information is provided (e.g., weather conditions), to ensure safety. In the future of the air traffic management (ATM), trajectories will be used as the core component of many ATM procedures.

In this part, TP is explored under the general scope of the aviation domain. Flight plans are exploited as the central element for constraint-based training of the TP models. A flight plan announced by an airline is a low-resolution trajectory that consists of the waypoints and time points that the aircraft is constrained to cross or fly by them. In order for a flown trajectory, annotated with data such as localized weather conditions, to be comparable with a flight plan, the latter has to be annotated in a similar way to encapsulate the same information as the *semantic* trajectories, e.g., weather information attached to the corresponding waypoints.

Given the previous definitions and assuming a historic *Semantic Trajectory Database* (STD), we define the *Future Semantic Trajectory Prediction* (FSTP) problem as follows:

Definition 8.4 (Future Semantic Trajectory Prediction) **Given** a STD consisting of semantic trajectories $R = < (p_l, t_l, v_l) >, l \in [0, i-1]$, a distance function $dist(.)$ that quantifies the dissimilarity between two semantic trajectories, an enriched flight plan F defined as R, and a target region G,
Predict the semantic trajectory $R_F = < (p_o^*, t_o, v_o^*) >$, where $o \geq i$ and p^* is located in G, i.e., the object's *anticipated* sequence of enriched points within G, where $R_F \in$ STD and satisfies the following property:

$$R_F = R_j = \arg \min_j \{dist(F, R_j)\}, \ \forall j \in [1, N] \tag{8.1}$$

where N is the number of points in R.

In principle, the TP problem can be approached as a generalization of the future location prediction problem (FLP) [16, 36, 40, 43], which is the task of predicting the next spatiotemporal position(s) of a moving object, based on its previous track, typically in the short-term context (up to a few minutes). On the other hand, the TP problem is to predict the entire anticipated track of the moving object given a set of constraints and/or historic data. A FLP method could be transformed to address the TP problem, given a specific granularity upon which the same method is applied iteratively. However, in that case the prediction errors are accumulated with each step (e.g., via multi-step regression), thus making the next predicted points increasingly error-saturated. In contrast, "pure" TP methods aim to forecast the trajectory itself as a whole, thus making each predicted point uniformly error-prone.

A more detailed description of related approaches to TP is provided later in Sect. 8.5.

8.2.3 Related Datasets Used

The experimental setup for validating the FLP proposed framework is based on aviation and maritime data. We conducted experiments against real datasets, namely IFS message and AIS messages, correspondingly. Table 8.1 summarizes some basic statistics about these datasets.

Table 8.1 Dataset Description

	Aviation (IFS)	Maritime (AIS)
Number of points	455,000	16,000,000
Number of objects	680 flights	5055 MMSI
Spatial coverage	Spain (Madrid–Barcelona flights)	Brest Area
Time span	April 2016 (1 week)	6 months

The experimental setup for validating the FSTP proposed framework is based on a set of flights between Madrid (LEMD) and Barcelona (LEBL). More specifically, the flight plans (the latest submitted before departure), the IFS radar tracks,[1] NOAA weather data and additional aircraft properties (aircraft type, wake category/size), and calendar (weekday) were included in the enriched trajectories dataset from April 2016. The specific pair of airports was selected as the one with the heaviest traffic on a monthly basis compared to any other airport pair in Spain[2] and because it involves different flight plans (reference waypoints) and multiple takeoff and landing approaches.

Table 8.2 summarizes the dataset used in the experimental study for FSTP; Fig. 8.1 illustrates an example of a flight with matched flight plan (F) and actual trajectory (R) reference waypoints; Fig. 8.2 shows the IFS (red) and flight plan (blue) points of the entire dataset; and Fig. 8.3 presents the medoids (colored) of all the clusters generated (stage-1).

Table 8.2 Datasets used

Element	Description	Comments
Airport pair	Madrid/Barcelona (LEMD/LEBL), 1–31 April 2016 → 696 flights	
Flight plans F	Pre-takeoff latest submitted for each flight.	Each F consists of 11–18 waypoints.
Actual route R	Reference waypoints from the full-resolution IFS radar track actual route R matched against F (closest waypoints)	Waypoint matching was conducted only on the spatiotemporal basis
Weather W	Latest NOAA weather parameters estimated via interpolation upon each F waypoint	Wind speed, wind direction, temperature, humidity
Other information S	Additional parameters used in the enrichment process	Aircraft type, wake category (size), weekday

8.3   Distributed Online Future Location Prediction (FLP)

There are various aspects and trade-offs in the design of FLP methods, primarily with regard to "memory" versus look-ahead time and, subsequently, simpler/faster versus more complex/demanding algorithms, lower or higher volume/rate of available input data, etc.

---

[1]For the area of the utilized dataset, i.e., flights between Barcelona and Madrid in Spain, the true geodesic resolution (mean) is 111.133 km/deg Lat and 83.921 km/deg Lon.

[2]LEMD/LEBL geodesic ranges: $Lat = [40, \ldots, 43]^o$, $Lon = [-3, \ldots, +3]^o$, $Alt = [0, \ldots, 40K]$ ft.

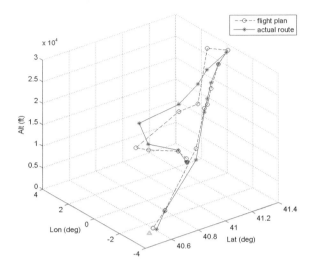

Fig. 8.1 Example of per-waypoint spatial comparison between flight plan (blue) and actual route (red)

Fig. 8.2 LEMD/LEBL dataset, April 2016, IFS tracks (red) and flight plans (blue)

Fig. 8.3 LEMD/LEBL dataset, April 2016, cluster medoids of the enriched trajectories

In this section, two major viewpoints of FLP are presented, namely one for short-term routes-agnostic (limited "memory") FLP and one for long-term network-aware (increased "memory") FLP.

8.3.1 Short-Term FLP: Routes-Agnostic Approach

8.3.1.1 Long Short-Term Memory (LSTM)

Over the last decade, neural networks (NNs) have attracted a renewed research interest, in order to reveal their true power on forecasting aircraft locations. Recent literature review has highlighted that a large number of papers employ the power of the special recurrent neural network (RNN) architecture. RNNs have become the state-of-the-art for sequence modeling, due to the fact that they naturally handle temporal and sequential data.

More specifically, an RNN is a natural generalization of feed-forward neural networks (FFNN) applied to sequences. A standard FFNN is comprised of an input layer, one or more hidden layers, and one output layer. The structure of an RNN is similar to that of an FFNN with the distinction of employing *weighted feedback connections* between the hidden units or/and the output units, in order to maintain a kind of internal *memory*. However, RNNs suffer from the well-known "vanishing" and "exploding" gradients [17]. In order to remedy the RNNs' drawbacks, i.e., hinder the processing of long sequences, the long short-term memory (LSTM) network [17] architecture, a special kind of RNNs was proposed.

Motivated by LSTM networks, Shi et al. [30] proposed an LSTM-based flight prediction method, which can accurately predict 3D and 4D flight trajectories without using the physical model of the aircraft. In [20], in order to predict aircraft 4D trajectories, high-dimension meteorological features and information from the last filed flight plans before departure (latitude, longitude, altitude, latitude speed, and longitude speed) were fed into an NN generative model, which consists of a multi-layer encoder LSTM, a multi-layer decoder LSTM, and a set of convolutional layers. Another LSTM encoder–decoder mechanism has been employed to model flight routes by analyzing sequences of legitimate ADS-B messages [14]. Considering the scene movement state of an aircraft as three-type (acceleration, deceleration, and uniform movement), Ma et al. [22] introduced the attenuation memory window, in order to enhance the prediction accuracy of an LSTM model. Zhang et al. [42] achieved more accurate flight prediction by employing the Ant Lion optimization algorithm to optimize the initial weights and thresholds of the LSTM network.

Motivated by the aforementioned works, we propose a "vanilla" LSTM-based framework for the aircraft FLP task. This framework provides satisfactory predictions of aircraft future locations by using only historical IFS data, i.e., without requiring other information such as flight plans.

Vanilla LSTM

LSTM constitutes a special kind of RNNs, which are explicitly designed to avoid the long-term dependency problem. Unlike RNNs, LSTM contains special units, called memory blocks, and each memory block is composed of gates. A lot of variants of LSTM architectures have been introduced, which use slightly different versions. The initial version of the LSTM block [17, 30] included input and output gates. However, the most commonly used LSTM block includes three gates: forget **f**, input **i**, output **o**.

For the sake of consistency, we briefly state the update rules for the employed vanilla LSTM layer below. For more details, the interested reader is referred to the original publication [17]. An LSTM cell, for each time-step t, is fed with the input vector $\mathbf{u}(t)$ and the updating process can be described by the following equations as also shown in Fig. 8.4:

$$\mathbf{i}(t) = \sigma(\mathbf{W}_i \cdot \mathbf{u}(t) + \mathbf{R}_i \cdot \mathbf{h}(t-1) + \mathbf{b}_i)$$
$$\mathbf{f}(t) = \sigma(\mathbf{W}_f \cdot \mathbf{u}(t) + \mathbf{R}_f \cdot \mathbf{h}(t-1) + \mathbf{b}_f)$$
$$\mathbf{o}(t) = \sigma(\mathbf{W}_o \cdot \mathbf{u}(t) + \mathbf{R}_o \cdot \mathbf{h}(t-1) + \mathbf{b}_o)$$
$$\tilde{\mathbf{c}}(t) = \tanh(\mathbf{W}_g \cdot \mathbf{u}(t) + \mathbf{R}_g \cdot \mathbf{h}(t-1) + \mathbf{b}_g) \qquad (8.2)$$
$$\mathbf{c}(t) = \mathbf{f}(t) \odot \mathbf{c}(t-1) + \mathbf{i}(t) \odot \tilde{\mathbf{c}}(t)$$
$$\mathbf{h}(t) = \mathbf{o}(t) \odot \tanh(\mathbf{c}(t))$$

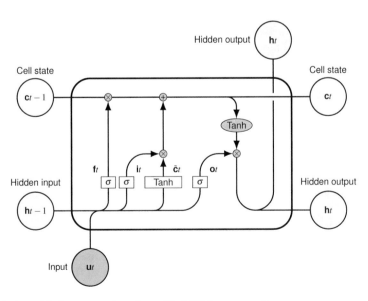

Fig. 8.4 A graphical representation of a vanilla LSTM memory cell

where:

- $\mathbf{c}$ is the cell state vector,
- $\tilde{\mathbf{c}}$ is the candidate value for the states of the memory cells,
- $\mathbf{W}_f$, $\mathbf{W}_i$, $\mathbf{W}_o$, $\mathbf{W}_c$ are the input-to-hidden weight matrices,
- $\mathbf{R}_f$, $\mathbf{R}_i$, $\mathbf{R}_o$, $\mathbf{R}_c$ are the state-to-state recurrent weight matrices,
- $\mathbf{b}_f$, $\mathbf{b}_i$, $\mathbf{b}_o$, $\mathbf{b}_c$ are the bias terms,
- $\odot$ is the Hadamard product (element-wise product) of the vectors, and
- $\sigma()$, tanh() are the sigmoid and Hyperbolic tangent functions, respectively.

Problem Formulation: LSTM-based Aircraft FLP

The FLP problem as stated in Definition 8.3 can be refined here for LSTM, timestamp-focused and oriented to the aviation domain. Specifically, given the features of timestamp $t_j^s(p)$, longitude $\tilde{x}_j^s(p)$, latitude $\tilde{y}_j^s(p)$, and altitude $\tilde{z}_j^s(p)$, for each aircraft s, for each trajectory j, for each record-point p, we seek to learn a model that at the next record-point $p + 1$, i.e., at timestamp $t_j^s(p + 1)$, predicts the 3-D aircraft trajectories composed of longitude $\tilde{x}_j^s(p + 1)$, latitude $\tilde{y}_j^s(p + 1)$, and altitude $\tilde{y}_j^s(p + 1)$ coordinates.

In order to allow Euclidean geometry computations, for each record-point p, the geodetic coordinates longitude and latitude are converted according to the Universal Transverse Mercator (UTM) system, into Cartesian coordinates. Also, to enhance the performance of the LSTM network, the time information together with the coordinates are incorporated to the network by employing the first-order differential processing between two consecutive points described in [22]. For instance, the time difference between p and $p - 1$ entries in the trajectory j of aircraft s can be calculated as: $\Delta t_j^s(p) = t_j^s(p) - t_j^s(p - 1)$. Similarly, the intervals of UTM longitude, UTM latitude, and UTM altitude are denoted by $\Delta x_j^s(p)$, $\Delta y_j^s(p)$, and $\Delta z_j^s(p)$, respectively. Moreover, in order to better predict an aircraft next location, the Euclidean distance $dist_j^s(p)$ between two consecutive points p and $p - 1$ is fed to the network.

To this end, the proposed model, for each aircraft s and each trajectory j, predicts the 3D vector $[\Delta x_j^s(p + 1), \Delta y_j^s(p + 1), \Delta z_j^s(p + 1)]$, by using the input vector $[\Delta t_j^s(p + 1), \Delta t_j^s(p), \Delta x_j^s(p), \Delta y_j^s(p), \Delta z_j^s(p), dist_j^s(p)]$. Subsequently, the predicted coordinates intervals are transformed to the actual coordinates. Note that the input feature $\Delta t_j^s(p + 1)$ indicates the desired time interval that the prediction must take place.

Furthermore, to demonstrate the inherent effectiveness of the LSTM networks we do not use sliding windows, or a constant length window for all instances, but instead, we feed the network with the whole trajectory. Due to the fact that flight trajectories are of variable length, a zero pre-padding procedure is employed to extend all trajectories to match the length of the longest trajectory.

Experimental results and overall assessment of the proposed Vanilla LSTM approach are presented in Sect. 8.4.3.

8.3.2   Long-Term FLP: Routes-Based Approach

In this section we describe the architecture of our proposed framework [28, 29], which follows a typical lambda architecture that combines streaming and batch layers to implement an end-to-end big data prediction solution. The proposed framework, as depicted in Fig. 8.5, consists of two main modules, namely pattern extraction (PE) and FLP. All modules are built on top of big data engines, so that they can be scalable and offer low latency. Kafka is used as an integration network for online toolboxes and a shared storage (i.e., Apache Hadoop HDFS) is used in order to update existing patterns or add new ones. Subsequently, the FLP module can "read" these patterns and execute the prediction pipeline.

At first, each moving object sends its location via traditional network protocols and then a Kafka producer collects all positions and pushes them to a Kafka topic. The PE module identifies "common routes," in an offline manner. Finally, these "typical routes" are broadcast among all workers and the FLP module combines them with the live incoming stream of data in order to predict the future location for each object.

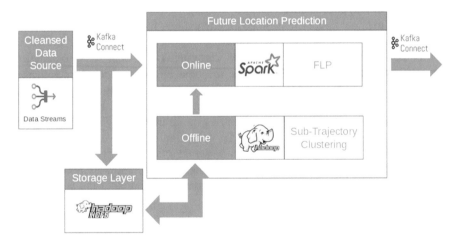

Fig. 8.5 Architecture of the proposed framework

8.3.2.1   Offline Step: Routes Network Discovery (Medoids)

The goal of this module is to identify frequent patterns of movement that will assist the FLP module to increase the accuracy of the predictions. The research so far has focused mainly in methods that aim to identify specific collective behavior patterns among moving objects, such as flocks, convoys, and swarms [43], or methods that try to identify patterns that are valid for the entire lifespan of the moving objects [23]. However, discovering clusters of entire trajectories can overlook significant patterns that might exist only for small portions of their lifespan.

Furthermore, most of the approaches either operate at specific predefined temporal "snapshots" of the dataset and ignore the movement between these "snapshots" and/or ignore the temporal dimension and perform spatial-only clustering and/or assume that the length (number of samples) of the trajectories and the sampling rate are fixed, which is unrealistic. Another thing that should be taken into account when designing a prediction-oriented trajectory clustering algorithm is that the resulting clusters should have a small extent in order for the predictions to be more accurate. Obviously, this rules out a large number of approaches that perform density-based clustering which might lead to spatially extended clusters through expansion.

For the previous reasons, the desired specifications that such a trajectory clustering algorithm should hold, in order to be able to predict the movement of future trajectories, are the following:

- Discovering of clusters of *subtrajectories*, instead of whole trajectories.
- Spatiotemporal clustering, instead of spatial only.
- Support of trajectories with variable sampling rate, length, and with temporal displacement.
- Distance-based clustering of trajectories.

There have been some approaches to deal with the problem of subtrajectory clustering in a centralized way [1, 19]; however, all these do not scale with the size of today's trajectory data, thus calling for parallel and distributed algorithms. For this reason, we utilize the work presented in [35], coined *DSC*, which introduces an efficient and highly scalable approach to deal with the problem of *Distributed Subtrajectory Clustering*, by means of MapReduce. More specifically, the authors of [35] split the original problem to three sub-problems, namely *Subtrajectory Join*, *Trajectory Segmentation*, and *Clustering and Outlier Detection*, and deal with each one in a distributed fashion by utilizing the MapReduce programming model (cf. Chap. 10).

To elaborate more, the *Subtrajectory Join* step aims at retrieving for each trajectory $r \in D$ all the moving objects, with their respective portion of movement, that moved close enough in space and time with r, for at least some time duration. Subsequently, the *Trajectory Segmentation* step takes as input the result of the *Subtrajectory Join* step, which is actually a trajectory and its neighboring trajectories and targets at segmenting each trajectory $r \in D$ into a set of subtrajectories in a neighborhood-aware fashion, meaning that a trajectory will be segmented whenever its neighborhood changes significantly. Finally, the third step takes as input the output of the first two steps and the goal is to create clusters of similar subtrajectories and at the same time identify subtrajectories that are significantly dissimilar from the others (outliers).

For more details about the algorithms involved in *DSC* and an extensive experimental study, please refer to Chap. 10 of this book, as well as to [34] and [35].

8.3.2.2 Online Step: Network-Based Prediction

In this section, based on the observation that moving objects often follow the same routes, we describe how the FLP module takes advantage of an individual's typical movement, referred to as cluster *medoids*, in order to predict the future location of a moving object in an online and streaming fashion. This observation is well-fitted to the maritime and aviation domains where sea vessels or aircraft have more or less strict routes between ports or airports, either implied due to route optimization (e.g., minimizing ship's fuel consumption) or explicitly required as official regulation (flight plans). The FLP module, as described earlier, aims to make an accurate estimation of the next position of a moving object within a specific look-ahead time frame.

Most approaches do not take advantage of any other historic data available, either from the object itself or other "similar" objects moving within the same area and context, making them susceptible to errors associated with noise, artifacts, or outliers in the input. This results in inaccurate predictions and only with a short horizon (seconds or few minutes). A very different approach for the FLP problem is making the associated predictive models less adaptive but more reliable, by introducing specific "memory" based on historic data of an entire fleet of objects relevant to the context at hand.

On the other hand, this requires a combination of historical and streaming data which is not a trivial task. A big challenge of our proposed framework is how to handle thousands of records efficiently in the context of online streaming data, join each object with the appropriate medoids, and finally do all the necessary model calculations to produce predictions for the future locations of an object. In practice, several such medoids are pre-computed and stored in an efficient way (partitioned by object identifier), so that they can be retrieved on demand or even kept in-memory for several thousands of objects, making long-term FLP feasible in a large scale. This task is addressed by employing a big data engine that is designed to conduct fast joins between streaming data and historical data. Spark module (SQL or Streaming) can efficiently join historical and streaming data, either with map-side-join (a.k.a broadcast join) or using dataset (Spark structure) metadata to achieve extra optimizations. For example, if the medoids can be sent to all workers (broadcast) at the initial phase, it is recommended to replicate medoids in each worker and for each object in MapReduce phase we select its medoids to perform prediction. On the other hand, if the medoids' size cannot stored in each workers' memory, we partition the medoids by the objects' identifier in order to have quick access for a specific object and create distributed structures that can be easily joined with streaming data via Spark SQL API.

In the first step (medoid matching) we try to match the object's recent history with the medoids. More specifically, for all the medoids, we find the closest to the object's current trajectory. In the second step (prediction) the algorithm has already identified the last point from the best-matched medoid, according to the previous stage. Then, it follows the medoid's points one by one until it reaches the prediction horizon.

In summary, the FLP approach described here is inherently intuitive and self-explanatory. It relies on past routes of the same or similar objects in order to forecast how a specific object will move while it is already residing on a specific frequently traversed route. In order to assign a moving object to the most similar medoid we adopt the similarity function proposed in [24].

The algorithm itself can be implemented in Spark MapReduce API as follows:

1. Receiving and parsing messages from input Kafka topic (map),
2. Reduce by object identifier over a window period,
3. Join objects streaming data with the proper medoids,
4. Map partition (process each object for the current window) in order to perform prediction.

Step 3 is required only for the dataset join, otherwise (broadcast join) step 3 is performed inside step 4. Figure 8.6 illustrates an example of this FLP approach over a flight between Madrid and Barcelona, where the red points are the actual data and the blue points are the predictions.

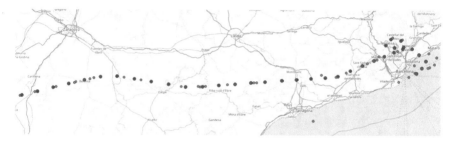

Fig. 8.6 Madrid–Barcelona flight example of the FLP-L approach. Red points/locations are real data and blue points are the predictions

8.3.3 Experimental Study

8.3.3.1 Short-Term FLP: Routes-Agnostic Approach

LSTM: Experimental Setup and Results

In this section, we present the results of our experimental study for the routes-agnostic FLP approach, employed in the aviation domain. In this work, a set of flights between Madrid (LEMD) and Barcelona (LEBL) reported in Sect. 8.2.3 was employed for experimental purposes. The available trajectories were allocated to three subsets (training, validation, and testing) randomly, using 50% of the data for training, 25% for validation, and 25% for testing the proposed network [2]. Moreover, the employed NN architecture consists of an input layer, a vanilla LSTM

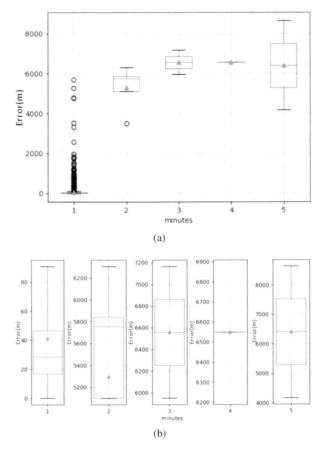

Fig. 8.7 Performance over the testing set for each prediction horizon by using boxplots (**a**) including outliers and (**b**) excluding outliers

layer of 50 neurons, a fully connected layer of 20 neurons and an output fully connected layer of 3 neurons.

The results were validated by calculating the RMSE and MAE of the Euclidean distance between the original locations and the predicted ones [30]. Figure 8.7a reports graphically the performance over the testing set by using boxplots, one for each prediction horizon. The error, which is the Euclidean distance between the actual and the predicted points in meters is shown in the vertical axis. The circles above and below the whiskers represent the outliers. Also, the same results, excluding the outliers, are depicted in Fig. 8.7b. Obviously, as the time horizontally increases, the error increases too. Furthermore, the errors (RMSE and MAE) and per individual spatial dimension are represented in Table 8.3.

Table 8.3 Performance over the testing set for each dimension in Root Mean Squared Error (RMSE) and Mean Absolute Error (MAE)

Dimension	RMSE(m)	MAE(m)
All	77.30	41.39
x (longitude)	59.55	26.84
y (latitude)	46.51	20.53
z (altitude)	16.31	10.19

8.3.3.2 Long-Term FLP: Routes-Based Approach

In this section, we present the results of our experimental study for the network-aware FLP approach, employed in the aviation and maritime domains. Our cluster consists of 10 nodes (1 master, 9 workers) with 5 executor cores per worker and 4 GB memory per worker. Input streams are provided by a Kafka topic and FLP is implemented on top of Spark SQL Streaming engine and Apache Yarn used as a resource manager. Spark SQL streaming tasks are processed using a micro-batch processing engine, which processes data streams as a series of small batch jobs, thereby achieving low latency and exactly once guarantees.

Based on the optimal Spark/Kafka configuration described in Fig. 8.8, the total delay originates almost entirely from the processing time, which asymptotically stabilizes at around 5 s. This essentially translates to 60,000 Kafka messages (points) per 10 s or 6000 points/s, which corresponds to 8-min look-ahead window. In other words, with an average sampling rate of 5 s for each moving object, this system configuration of the FLP module can accommodate up to 30,000 moving objects with 5-s update and 8-min look-ahead predictions.

As described above, in this option an FLP approach is employed for exploiting the cluster medoids as "guidelines" for providing online predictions, e.g., as the actual flight evolves in real time. The general clustering method in this case is the same as described in Sect. 8.3.2.1. We use up to 14 clusters in order to perform future location prediction. The FLP module uses sliding windows of 2 min of past positions in order to optimally match the most recent segment of the current trajectory to one of the available medoids, using a custom spatiotemporal similarity function. Then, the best-matched medoid is used as the maximum-likelihood trajectory evolution and the predicted positions are taken along its path for a specific (user-defined) look-ahead step.

Figure 8.9 illustrates the histogram of the horizontal error, i.e., the distribution of errors, for all the trajectories in the aviation (Madrid/Barcelona) and maritime (Brest Area) datasets and with spatial-only comparison (point-wise Euclidean distance). Specifically, they illustrate the boxplots of the per-complete-trajectory mean error for multiple look-ahead steps (1, 2, 4, 8, 16, 32 min). Additionally, the notation of the boxplot provides hints of the underlying error distributions, i.e., means, medians, upper/lower quartiles, non-outlier ranges, etc. These verify that the prediction errors are indeed in accordance with the expected shape of the distribution, i.e., a typical extreme value (EV) with medium/low skewness (Gaussian-like) towards the lower limit and an asymptotically decreasing right tail, i.e., accumulate and expand exponentially as the look-ahead span doubles.

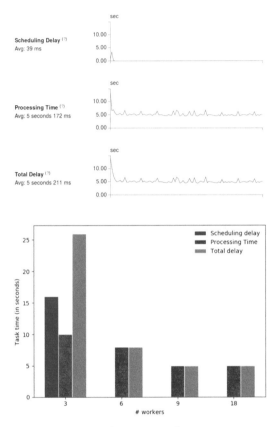

Fig. 8.8 Performance metrics for $16 \cdot 10^6$ points, $6 \cdot 10^3$ points/s, batch interval 10 s, 9 workers and 60 partitions

8.4 Semantics-Driven Trajectory Prediction (FSTP)

In this part, TP is presented under the scope of the aviation domain, specifically for aircraft TP using enriched flight plans and past history, in accordance with Definition 8.4 presented in Sect. 8.2.2. This is a challenging and inherently data-driven time-series modeling problem. Adding annotation or enrichment parameters further increases the search space complexity, especially when "blind" optimization algorithms are employed.

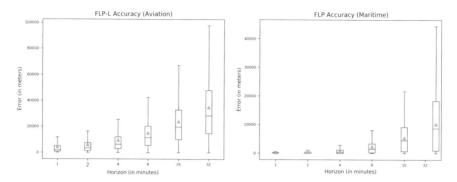

Fig. 8.9 Mean error for multiple look-ahead steps (1, 2, 4, 8, 16, 32 min), with 90%-threshold outliers removed

Despite the fact that a flight plan is generally considered as a strict guideline for the actual path of the aircraft in civil aviation, surveillance data show that deviations from the reference waypoints are in the range of up to 12–15 km or more, compared to each individual reference waypoint of the submitted flight plan. This is verified by the surveillance (IFS) dataset used in our work, as well as other state-of-the-art works using similar datasets and setups [4, 20]. This is due to adverse weather conditions and local airspace restrictions w.r.t. traffic, which lead to in-flight alterations to the initially submitted flight plan in pre-flight, e.g., a last-minute change in the landing approach due to local weather conditions. As a result, the flight plan data are a useful guideline but cannot be used as-is for trajectory prediction, as the actual route of the aircraft is severely altered by stochastic factors. The proposed approach builds upon the enrichment parameters, e.g., local weather, and introduces flight plans as an important input for the predictive models. More specifically, given the flight plan submitted by an airline for a specific flight, we make a prediction of the entire flight trajectory from departure to destination airport including all phases, e.g., top-of-climb, top-of-descent, etc.

Instead of following the typical approach of operating on the raw surveillance data collected from various sensors, in this work we make use of *semantically enhanced* data, i.e., exploiting trajectories enriched/annotated with associated information, for example, weather. This way, the raw trajectory data is transformed into multi-dimensional sequences (*semantic* trajectory data) that form a more realistic representation model of the complex real-world flight tracks; mobility of aircrafts belongs to this broad class of (BDA).

Our methodology exploits trajectory clustering and prediction tasks in relation to the FSTP problem. In particular, the merits and contributions of our work are summarized as follows:

- We define the FSTP problem and propose a multi-stage "pipelined" processing, with clustering in the first stage dealing with dimensionality reduction from the

N-dim enriched trajectory space to a set of "similar" (compact) clusters upon which the predictors are trained;

- We devise a novel predictive modeling representation for semantic trajectories, exploiting flight plans as input;
- We introduce an inherently parallelizable design with separate lightweight predictors per reference waypoint at the subsequent stages, working with a non-uniform graph-based grid of much lower complexity compared to the current state-of-the-art practices;
- We provide an extensive experimental study with real aviation surveillance data (IFS radar tracks, flight plans, weather, etc.), assessing the overall prediction accuracy of the proposed approach in contrast to current state-of-the-art alternatives, illustrating at least a sixfold improvement.

This work presents a novel multi-stage approach to tackle this challenge and verifies its effectiveness with experimental results, illustrating a full-trajectory expected error (RMSE) in the order of <2 km. It should be noted that any other proposed work with similar setup, e.g., [4] but with no use of flight plans and [20] using flight plans as input, achieves comparative errors of 10–12 km and 92 km, respectively.

8.4.1 Clustering Enriched Trajectories for FSTP

Clustering enriched trajectories imply the partitioning of a Semantic Trajectory Database (STD) into clusters (groups), so that each cluster contains "similar" enriched trajectories according to a specific similarity measure. Two semantic trajectories of aircrafts may be considered similar in many ways; they may have common departure and/or destination airports, they may fully or partly be close to each other throughout the flight, they may be fully or partly synchronous, or they may be disjoint in time but with similar behavior (e.g., the same control operations as these are represented by their aircraft intent, etc.).

This clustering phase of "discovering" medoids is based on the method described earlier in the context of long-term FLP (see Sect. 8.3.2.1). However, here there is a somewhat different way to define trajectory similarity exploiting several of the enrichment dimensions (e.g., weather localized conditions), instead of only spatiotemporal as described in long-term FLP.

For our task, we adopt the SemT-OPTICS approach, which is driven by the popular OPTICS clustering method [26] tuned by the parameters, $minPts$, describing the number of elements required to form a cluster, and eps, describing the maximum distance (radius) to consider for a sufficiently dense cluster. The outcome is also a reachability plot, upon which we automatically extract clusters and outliers using the ξ-clustering method, originally proposed in [3].

For our purposes, the (dis)similarity between two enriched points is composed of two parts, one regarding their spatiotemporal distance and another regarding their dissimilarity on the enrichment components. In particular, we adopt an appropriate modification of the function proposed in [25], which in its turn is a variant of $Edit distance with Real Penalty$ (ERP):

Definition 8.5 (Distance D_r Between Enriched Points) Given two enriched points r_i and r_j, their distance $D_r(r_i, r_j)$ is defined as:

$$D_r(r_i, r_j) = \lambda \cdot dist_e(r_i, r_j) + (1 - \lambda) \cdot dist_v(r_i, r_j) \qquad (8.3)$$

$$dist_e(r_i, r_j) = \frac{\sqrt{w_1 \cdot ||p_i - p_j||^2 + \frac{w_2}{w_1} \cdot (t_i - t_j)^2}}{max\,Euclidean\,Distance(STD)} \qquad (8.4)$$

$$dist_v(r_i, r_j) = 1 - \frac{v_i \bullet v_j}{||v_i||^2 + ||v_j||^2 - v_i \bullet v_j} \qquad (8.5)$$

where $dist_e(.)$ is the Euclidean distance in the 4-D spatiotemporal domain (x, y, z, t); user-defined weights w_1 and $w_2 = 1 - w_1$ are regularization factors in [0,1] for the spatial versus the temporal dimension; $dist_v(.)$ is the Jaccard distance of the enrichment components; in Eq. (8.4) the $max\,Euclidean\,Distance(STD)$ is the coverage of the database in the 4-D space acting as a normalization factor for output within [0,1]; and $\lambda \in [0,1]$ is a user-defined parameter that tunes the relative importance between the two components.

Specifically for the semantic distance $dist_v(.)$, the part of vector v_i consisting of the numerical variables is normalized per dimension to exclude scaling effects, whereas each categorical variable described by a set of keywords (e.g., aircraft type) is transformed to a set of numerical values, the cardinality of which corresponds to the vocabulary of all distinct keywords in STD. Thus, each keyword corresponds to a numeric value that is calculated by TF-IDF. Overall, v_i is the concatenation of these two, numerical and categorical subvectors. As the Jaccard distance maps the semantic similarity to the range [0,1], it follows that $D_r(r_i, r_j)$ always results into [0,1].

Having defined the distance D_r between two enriched points, distance D_R between two enriched trajectories is defined as follows:

Definition 8.6 (Distance D_R Between Enriched Trajectories) Given two semantic trajectories R_i and R_j of arbitrary length (i.e., arbitrary number of enriched points), their distance $D_R(R_i, R_j)$ is defined as :

$$D_R(R_i, R_j) = min \left\{ \begin{array}{l} D_R\left(\tau(R_i), \tau(R_j)\right) + D_r(r_{i,1}, r_{j,1}) \\ D_R\left(\tau(R_i), \tau(R_j)\right) + D_r(r_{i,1}, h) \\ D_R\left(\tau(R_i), \tau(R_j)\right) + D_r(h, r_{j,1}) \end{array} \right\} \qquad (8.6)$$

where $\tau(R_i)$ denotes the *tail* of R_i, namely the enriched points of R_i after removing the first enriched point of the i-th semantic trajectory $(r_i, 1)$, and h or gap is a virtual enriched point whose spatiotemporal value is the origin of the 4-D space of the entire dataset, while its semantic component corresponds to the zero vector.

The value of the *gap* element is given in a way similar to the Edit distance, where it is determined as the first value of the timescale for the time series (i.e., typically $gap = 0$). Following a similar approach as in [25], it is trivial to prove that D_R is a metric.

Given the distance D_R and a corresponding clustering result C consisting of K clusters (noise and/or outliers may also be considered as a separate cluster), we define the average distance $\overline{D_{R_l}^{C_k}}$ of a member R_l, $(1 \leq l \leq m = |C_k|)$ of a cluster $C_k \in C, (k = 1, \ldots, K)$ from all other $m - 1$ members of the same cluster as:

$$\overline{D_{R_l}^{C_k}} = \frac{1}{m-1} \sum_{i \in [m], i \neq l} D_R(R_i, R_j) \tag{8.7}$$

The member R_l which has the minimum average distance from all other members in cluster C_k is considered the *medoid* $R_\mu^{C_k}$ of cluster C_k, formally defined as:

$$R_\mu^k = R_\mu^{C_k} = \arg\min_l \{\overline{D_{R_l}^{C_k}}\}, l \in [1, m] \tag{8.8}$$

Thus, each cluster C_k of enriched trajectories R can be represented by its corresponding medoid R_μ^k.

8.4.2 Predictive Models

In this work, flight plans, localized weather, and aircraft properties are introduced as trajectory annotations that enable modeling in a space higher than the typical 4-D spatiotemporal. A multi-stage hybrid approach is employed for a new variation of the core FSTP task, including clustering the enriched trajectory data using a semantic-aware similarity function as distance metric. Subsequently, a separate predictive model is trained for each cluster, using a non-uniform graph-based grid that is formed by the waypoints of each flight plan. In practice, flight plans constitute a constrained-based training of each predictive model, one for each waypoint, independently.

The proposed method is formulated and experimentally validated with real aviation dataset (flight plans and IFS radar tracks) and localized weather data for a 1-month time frame of flights in the Spanish airspace. Various types of predictive models are tested, including Hidden Markov Model (HMM), linear regressors (LR), regression trees (CART), and feed-forward neural networks (NN).

The following sections describe briefly the predictive models employed in the core FSTP task, trained and evaluated per-cluster, using the enriched flight plans as input.

1st Step: Cluster semantically
annotated trajectories

2nd Step: For each
cluster, train a HMM

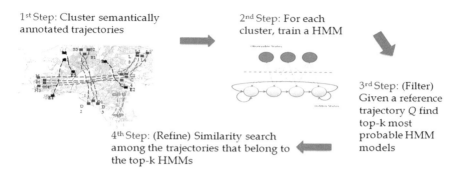

3rd Step: (Filter)
Given a reference
trajectory Q find
top-k most
probable HMM
models

4th Step: (Refine) Similarity search
among the trajectories that belong to
the top-k HMMs

Fig. 8.10   Overview of the proposed FSTP framework, HMM case

In summary, our multi-stage approach for addressing the FSTP problem, according to Definition 8.4, is as follows:

1. **Clustering (stage-1):** The historic STD is clustered into a set of clusters **C** by using $dist(.)$, separating the set of enriched trajectories into K distinct groups.
2. **Predictive models (stage-2):** For each cluster created, we build a predictive model, e.g., HMM or other.
3. **Evaluate/Query (stage-3):** Given F an enriched flight plan as input/query, we probabilistically assign it to the correct cluster C_k based on $dist(.)$ and R_F is estimated from the corresponding predictive model.
4. **Refinement (stage-4):** Within cluster C_k, perform a top-j $(1 \leq j \leq |C_k|)$ nearest neighbor query w.r.t. its members using $dist(.)$.

Figure 8.10 illustrates schematically the approach when HMM is used as predictive model in the second stage.

The first two stages constitute the training phase. The clustering stage is described in detail in Sect. 8.4.1, while the various choices over the predictive modeling are described in detail in Sect. 8.4.2. The last two stages refer to the use of the already trained system for predicting the actual trajectory of an aircraft given its flight plan. Stage-3 is a classification problem, where the set of classes corresponds to the set of cluster identifiers **C**: $< C_1, \ldots, C_K >$. More specifically, given an unclassified flight plan F and the set θ^k, $1 \leq k \leq K$, of the representative models for the clusters (i.e., classes), we search for the model that maximizes the likelihood of being associated with F:

$$C_F = C_k : \arg \max_k \{\mathbf{P}(F|\theta^k)\}, \quad k \in [1, K] \qquad (8.9)$$

In the case of using HMM as the representative model in stage-2, this problem can be straightforwardly solved by applying the Forward algorithm. Similarly, if regressors are used instead, the model that is selected is the one maximizing the likelihood of generating R from F, where R is the actual trajectory flown (IFS radar track). Finally, in stage-4 we focus on the cluster that is the most probable to find the solution and retrieve the top-j most similar semantic trajectories from the stage-3 output by applying $dist(.)$.

In this study, stage-4 is not included in the experimental work, as the main focus here is to assess the performance of clustering and the various TP models, i.e., stages 1–3, without any positive or negative effects from the top-j nearest neighbor query in the STD. In practice, this means that lookup in the historical database of flights is disabled, since this would artificially increase the overall accuracy and somewhat hide the actual performance of the proposed algorithms in stages 1–3. Hence, the rest of this study is focused on the first three stages, without the solution refinement stage. In the following subsections, we provide details for each step of our methodology.

Four different designs have been investigated in this study regarding the predictive modeling in stage-2 (and, subsequently, stage-3) of the proposed approach. Each of them is described separately in the following sections.

8.4.2.1 Hidden Markov Models (HMM)

In time-series modeling, an HMM formulates the evolution of a system by a set of states and transitions between them, each one accompanied by a probability that is typically extracted by analyzing historic data. In the context of FSTP, the flight path and all the associated information (weather, semantic data, etc.) are usually transformed into discrete values that constitute the HMM states. Then, the trajectory itself is treated as an evolution of transitions between these states, using the trajectory data of a large set of flights for training, plus spatiotemporal constraints (locality) to reduce the dimensionality of the problem and, thus, the total number of HMM states. This approach has already been tested for trajectory prediction from raw surveillance data and recent studies show that its results on real aviation data are very promising (e.g., see [4, 5]).

Formally, a HMM is defined by $M = |S|$ distinct states S, S_t being the state at time t (stationary, first-order Markov chain), $U = |O|$ distinct or continuous values O that can be observed at each state transition, O_t being the observed value produced by the HMM at time t, the $M \times M$ matrix $A = \{a_{i,j}\}$ of transition probabilities where $a_{i,j} = \mathbf{P}\{S_t = j | S_{t-1} = i\}$ with $1 \leq \{i, j\} \leq M$, the $M \times U$ matrix $B = \{b_i(o)\}$ of the probabilities of emissions o at state i where $b_i(o) = \mathbf{P}\{O_t = o | S_t = i\}$, and the set $\Pi = \{\pi_i\}$ of prior state probabilities, where $\pi_i = \mathbf{P}\{S_1 = i\}, i \in [1, M]$. The observations O can be a continuous-valued set, in which case matrix B becomes a continuous probability distribution function rather than a discrete matrix. Therefore, a HMM θ is specified as a triple $\theta = \{A, B, \Pi\}$. According to this formulation, a HMM is the functional form:

$$\mathbf{P}\{S, O\} = \mathbf{P}\{S_1\} \prod_{t=2}^{t_{\max}} \mathbf{P}\{S_t | S_{t-1}\} \prod_{t=1}^{t_{\max}} \mathbf{P}\{O_t | S_t\} \qquad (8.10)$$

In order to apply a HMM-based modeling to already clustered enriched trajectories, the following steps are performed:

- *Training (stage-2):* For each enriched trajectory cluster C_k, $k \in [1, K]$, a HMM θ^k is trained, i.e., the model parameters $\theta^k = \{A^k, B^k, \pi^k\}$ are estimated for maximum likelihood upon observation sequences for the training subset corresponding to cluster C_k. This is the HMM training phase and it is realized by applying the EM algorithm.
- *Evaluation (stage-3):* For a flight plan F_Q that is used as a query, the model likelihood of F_Q is computed for all possible HMM models θ^k, i.e., $\mathbf{P}_{Q,k} = \mathbf{P}\{F_Q|\theta^k\}, k \in [1, K]$. Then F_Q is assigned to the cluster with the highest likelihood $\mathbf{P}_{Q,k}$.

The final result is the predicted (future) trajectory R_Q corresponding to F_Q and it can be either the medoid R_μ^k of the selected cluster or the synthetic trajectory generated by the maximum-likelihood emissions sequence O_t. The physical meaning of the HMM emissions is typically the observed output of the system. In this approach, given the availability of both the flight plans F and the corresponding actual trajectories R, the emissions are designed as the deviation between these two, since this is the actual output observed from the evolution of a flight. More specifically, each pair of F and R defines a sequence of waypoints $wp_t = <r_t^F, r_t^R>$ where t is now a sequence number instead of actual timestamp. This wp_t sequence of the medoid R_μ^k of each cluster defines the states and transitions of the corresponding θ^k for this cluster. In other words, each wp_t is translated to a state in S and each transit between $< wp_t, wp_{t+1} >$ is translated to an observation in O, thus completely defining $\theta^k = \{A^k, B^k, \pi^k\}$ as described above.

In this work, the actual O_t is the 3-D deviation between F and R at wp_t where $t \in [1, L_k]$, $k \in [1, K]$, and $L_k = |F| = |R|$ is the number of waypoints in members of C_k. The 3-D deviation between F and R is calculated with the typical 2-D Haversine formula instead of simple Euclidean, extended with trapezoid approximation for any arbitrary pair of points within the Earth's atmosphere, for accurate spherical geometry even at long range. An emissions model was designed for each spatial dimension (Lat/Lon/Alt), so that each dimension can be predicted separately by the HMM.

Since multiple flights are included in each cluster, the values of the O_t emissions produce an empirical probability distribution function per state (i.e., wp_t), which under moderate statistical assumptions ($m = |C_k| \geq 30$) can be approximated by a parametric Gaussian. Using such a parametric probability distribution as the (continuous) emissions model upon each wp_t of R_μ^k, the maximum-likelihood $\mathbf{P}_{Q,k}$ can be estimated for each cluster k given a query F_Q. Then, the corresponding R_Q is the trajectory generated by the associated HMM θ^k.

Using the formulation above, proper confidence intervals can be estimated for every $\mu_{k,t}$.

8.4.2.2 Linear Regressors (LR)

As described earlier, the HMM approach for stage-2 in the proposed FSTP framework is based on an emissions model that translates deviations between flight plans and actual routes into probabilistic predictions in the maximum-likelihood sense. In other words, the continuous emissions model generates the "most likely" per-waypoint deviation from the intended flight path F and, hence, the future route R.

A more generic approach is to incorporate multiple dimensions in the input and also to approximate R itself, instead of the deviation from F. The simplest and straightforward way to do this is by designing proper linear regressors (LR) [38] to replace the HMM counterparts for each waypoint. LR is defined as the first-order approximation of an over-determined linear system of equations in a way that minimizes a global error criterion, typically the least squares error (LSE).

In this study, LR is employed as the mapping function between a composite input vector which is based on F and an output that is associated with R, designed and trained per waypoint, i.e., independently for each target value. Figure 8.11 illustrates the two main differences of LR compared to the HMM approach for stage-2, namely (a) using multiple waypoints of F in the input vector and (b) targeting R itself instead of the deviations from F.

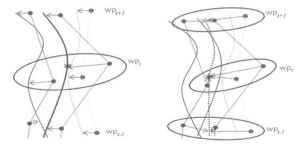

Fig. 8.11 HMM versus LR designs. HMM (left) uses only the current wp_t of F and R is estimated as deviation; LR (right) uses multiple wp_* and R is estimated directly. Vectors illustrate the pairing in calculating the distances of nearest points to actual trajectories w.r.t. group "representatives" (cluster medoids)

Formally, the LR approach for stage-2 can be defined by Eqs. (8.11) and (8.12) for training (stage-2) and evaluation (stage-3), respectively:

$$\phi_{j,t}^{R} = \mathbf{W}_{j,t}^{F} \cdot \mathbf{b}_{k,t} + \mathbf{b}_{k,0} , \quad \forall j \in C_k, \ t \in L_k \tag{8.11}$$

$$\phi_{Q,t}^{R} = \mathbf{W}_{Q,t}^{F} \cdot \mathbf{b}_{k,t} + \mathbf{b}_{k,0} , \quad t \in L_k \tag{8.12}$$

where C_k, L_k, $\phi_{j,t}^{R}$ are defined as for HMM, $\mathbf{W}_{j,t}^{F}$ is the (composite) input vector based on pre-flight data, and $\mathbf{b}_{k,*}$ are the LR coefficients that minimize the LSE.

Based on the way the input vector $\mathbf{b}_{k,*}$ is defined, three LR variants were designed for use in stage-2 of the proposed FSTP framework:

- **LR(1):** $\mathbf{W}_{j,t}^F = \phi_{j,T}^F$
- **LR(3):** $\mathbf{W}_{j,t}^F = \mathbf{p}_{j,T}^F = \langle x_{j,T}^F, y_{j,T}^F, z_{j,T}^F \rangle$
- **LR(4):** $\mathbf{W}_{j,t}^F = \langle \mathbf{p}_{j,T}^F, \mathbf{v}_{j,*}^F \rangle$

where $T \subseteq \{1, \ldots, t, \ldots, L_k\}$ (one or more waypoints used), $p_{j,t}^F$ refers to the 3-D waypoint (x_t, y_t, z_t) of flight plan j and $\mathbf{v}_{j,*}^F$ refers to the associated semantic information available in pre-flight. The three LR(.) variants were used in order to test the effects on the final accuracy of each regressor w.r.t. the input space, i.e., when using only same dimension ϕ as the target or the full 3-D $\mathbf{p}$ or the 3-D $\mathbf{p}$ plus the semantic part $\mathbf{v}$. In this study, the main waypoint-related semantics are incorporated and exploited by the similarity function of Definition 8.5 Alternatively, in order to reduce the complexity of the clustering task, all waypoint-invariant inputs can be included here instead, namely aircraft properties (type and wake/size category) and calendar variables (weekday)—hence the use of asterisk instead of sequencing t in the $\mathbf{v}_{j,*}^F$ notation.

It is worth noting that the HMM emissions model described in Sect. 8.4.2.1 can be considered functionally as a special case of LR(1), with $\mathbf{b}_{k,t} \equiv \mathbf{1}$ and $\mathbf{b}_{k,0} \equiv \widehat{\mu_{k,t}}$.

8.4.2.3 Regression Trees (CART)

In this study, standard *Classification and Regression Trees* (CART) [21] in regression mode have been employed as an alternative to LR. More specifically, the LR(4) formulation for input/output specifications was also employed with CART regressors, in order to investigate the local non-linearities of the input (composite) space and to address this problem with hierarchical subspace partitioning. In practice, this means that CART is inherently implementing feature selection (dimensionality reduction) in each node and optimally trained thresholds define the proper subspace partition for the final prediction of $\phi_{j,t}^R$. Figure 8.12 illustrates the root portion of such a CART model, comparing the altitude ("x44") of a specific waypoint (wp_{14}) of flight plan F against various threshold values (x100 ft) and in various levels of the tree, in order to end up with the optimal estimation leaf for the altitude in the same waypoint of the predicted (future) route R.

8.4.2.4 Feed-Forward Neural Network Regressors (NN-MLP)

Additionally to the HMM and LR predictor models, feed-forward NN regressors, specifically *Multi-Layer Perceptrons* (NN-MLP), were employed as replacements. Again, the LR(4) formulation for input/output specifications was employed with NN-MLP as with LR regressors, using a typical topology of one hidden layer of 10 neurons with softmax-designed activation function, as illustrated in Fig. 8.13.

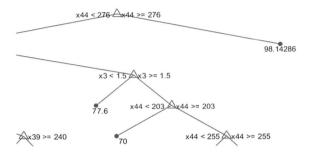

Fig. 8.12 Example of trained CART (partial) regressor predicting altitude; leaf values indicate goodness-of-fit (0–100)

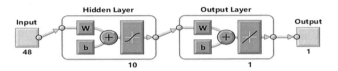

Fig. 8.13 NN-MLP topology used for non-linear regression, using NN(4) input specification with 15 waypoints (3-D) plus three global enrichments

8.4.3 Experimental Study

The general experimental setup and datasets used for evaluating the FSTP approach is described in Sect. 8.2.3. The experiments were conducted using a variety of software tools and programming platforms.[3] The core software for each stage of the proposed FSTP framework, including clustering, HMM, and LR models, are currently ported to R for cross-platform prototyping, as well as to Spark (Scala) platform.

8.4.3.1 Clustering Stage

As described earlier, the experimental work was focused on evaluating the core stages 1 through 3 of the proposed FSTP framework, i.e., without the (optional) top-k retrieval in stage 4, for reasons explained in Sect. 8.4.2.

In clustering (stage-1), the parameters of the composite distance metric described according to Definitions 8.5 and 8.6 were established after extensive experimentation and evaluation of the quality (size versus compactness) of the resulting clusters. More specifically, the spatiotemporal part in Eq. (8.3) was preferred over the semantic part ($\lambda = \frac{3}{4}$), equally weighted spatial dimensions ($w_1 = \frac{1}{3}$) and time-

---

[3]Mathworks MATLAB v9.2/R2017a (x64); Octave v4.2.1; R v3.4.3; WEKA v3.8.2; custom Java & C/C++ tools.

invariant trajectory matching ($w_2 = 0$) were employed. These design choices for the distance function were specifically selected as a compromise between clustering compactness versus ease of visualization, in order for the standard prediction error metrics *Mean Absolute Prediction Error* (MAPE) and *Root Mean Squared Error* (RMSE) to be easily interpreted in the 3-D spatial-only sense. The best clustering result w.r.t. Silhouette [38] includes a partitioning of $\mathbf{C} = \{255, 228, 138, 75\}$, $K = 4$ and was used as baseline throughout this experimental work.

8.4.3.2   Predictive Modeling Stage

The main reason for using HMM in stage-2 of the proposed FSTP framework was, as described in Sect. 8.4.2.1, to investigate the nominal confidence intervals for error estimations by proper statistical methods. Since HMM is the simplest of all the other options (LR, CART, NN-MLP), these estimations can be considered relevant to these other models too, especially LR, as HMM can be formulated as a special case of LR(1) (see Sect. 8.4.2.2). Figure 8.14 and Table 8.4 present the *Half-Width Confidence Interval* (HWCI) [33] estimations for the best clustering result (stage-1), including 4 main clusters of 696 flights and one of 7 outliers (excluded).

Table 8.4   HMM Half-width Confidence Interval

C_k	$|C_k|$	L_k	Mean [HWCI]
1	255	13	208.5
2	228	14	285.3
3	138	15	460.9
4	75	11	665.9

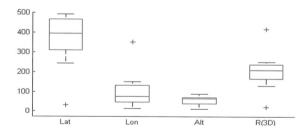

Fig. 8.14   Half-width confidence intervals (m) for HMM accuracy estimations per spatial dimension and in 3-D

It should be noted that, according to the current state-of-the-art in similar setups [4, 20], splitting the 3-D space into separate dimensions is the standard approach. In our work, the other reason we do this is precisely to look at how each individual dimension affects the overall prediction error. As a result, we verify again what other works in trajectory prediction in aviation mention [11]: altitude errors are of much less importance compared to Lat/Lon dimensions (see Tables 8.5, 8.6, Fig. 8.14). On the other hand, using predictors with full 3-D input for each reference waypoint as in LR(4) is also investigated (see Sect. 8.4.2.2) and illustrated comparatively to the 1-D alternative of LR(1).

As an example of prediction error tracking along the sequence of waypoints, Fig. 8.15 presents the MAPE and RMSE for the LR(4) model (stage-2), trained on the same 4-cluster partitioning of the data (stage-1).

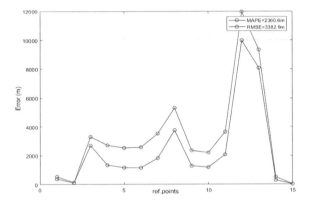

Fig. 8.15 Example MAPE and RMSE (m) plots of LR predictor (stage-2) along the waypoints

For CART regressors, described in detail in Sect. 8.4.2.3, the training was implemented with both node merging and tree post-pruning enabled (parent size 10), using *Mean Squared Error* (MSE) as the node splitting criterion [38].

For NN-MLP regressors, described in detail in Sect. 8.4.2.4, the training was implemented using Bayesian regularization back-propagation for better convergence and generalization capabilities, while *tansig* activation was used in the hidden layer neurons as the softmax-like function. The training itself included k-fold cross-validation [38] with $k = 10$ folds, i.e., 90% training and 10% testing randomized subsets in each run, along with some additional training configurations of fixed splits down to 50% training subsets, in order to explore the true generalization of the NN-MLP regressors in this problem.

Tables 8.5 and 8.6 present the best performances for all stage-2 predictor models using the same set of 696 flights (excluding outliers), non-clustered and clustered (K=4), respectively. The NN-MLP model is presented comparatively but separately from the others, since its performance was asserted by a slightly different experimental protocol with a k-fold cross-validation scheme ($k = 10$).

Table 8.5 Prediction accuracies in wp-averaged RMSE (m), *non-clustered* dataset

Model	$R_k : Lat$	$R_k : Lon$	$R_k : Alt$	$R_k : 3D$
HMM	3986.0	1072.3	587.3	4169.3
LR(1)	3660.1	999.3	528.3	3830.7
LR(3)	3090.7	741.8	391.0	3202.4
LR(4)	3074.3	736.7	380.8	3184.2
CART	2830.2	1396.9	316.9	3172.0
NN-MLP	1555.7	960.1	203.9	1877.4

Table 8.6 Prediction accuracies in wp-averaged RMSE (m), *clustered* dataset ($K=4$)

Model	$R_k : Lat$	$R_k : Lon$	$R_k : Alt$	$R_k : 3D$
HMM	3154.6	847.3	418.9	3294.6
LR(1)	3047.3	806.7	403.9	3179.9
LR(3)	2736.7	662.4	330.8	2837.4
LR(4)	2697.8	652.6	321.5	2796.4
CART	2661.4	1673.0	289.3	3377.1
NN-MLP	1527.6	1204.7	178.3	1953.6

Specifically for the NN-MLP regressors, which are the best-performing model for stage-2, Table 8.7 presents a summary of all the per-waypoint prediction errors for one cluster, while Fig. 8.16 shows the exact distribution of prediction errors (signed MAPE) for one such waypoint. In the unsigned form, the histogram of MAPE is clearly associated with a probability distribution similar to the *Generalized Extreme Value* (GEV) family, i.e., with mode close to zero and heavy right tail, as expected.

It is worth noting that the per-waypoint prediction error remains fairly close to the mean (RMSE) value, not only in 3-D but also for each individual spatial dimension. This is particularly important, since these results prove the robustness of the NN-MLP predictions along the entire flight path and the validity of using the flight plan as the main element in constraint-based training.

Finally, Fig. 8.17 presents the summary of the performance of all stage-2 predictor models for non-clustered and clustered dataset. The corresponding per-model run times[4] for combined training and testing, as illustrated in Table 8.8, vary between only few msec for the non-clustered HMM models to 10–11 ms for clustered LR or CART equivalents with $K = 4$ (sum over all clusters). The corresponding run times for NN-MLP in stage-2 are significantly higher, ranging up to almost a second, i.e., two orders of magnitude larger, due to the computationally

---

[4]Platform used: Intel i7 quad-core @ 2.0 GHz / 8 GB RAM / MS-Windows 8.1 (x64).

Table 8.7 NN-MLP accuracies in per-wp RMSE (m) for cluster 1 (example)

wp	$R_k : Lat$	$R_k : Lon$	$R_k : Alt$	$R_k : 3D$
1	279.7	70.0	37.2	290.7
2	511.3	149.2	113.1	544.5
3	1780.0	422.5	246.1	1845.9
4	1810.6	608.4	256.7	1927.3
5	1031.7	1518.9	334.7	1866.4
6	1072.7	2346.8	214.6	2589.2
7	1354.8	3709.2	64.4	3949.4
8	2076.0	1148.3	85.1	2373.9
9	1610.6	164.7	205.3	1632.0
10	2163.1	250.3	189.9	2185.8
11	1868.3	331.2	184.8	1906.4
12	319.2	3187.1	64.4	3203.7
13	46.7	34.8	8.4	58.8
Mean	1225.0	1072.4	154.2	1874.9

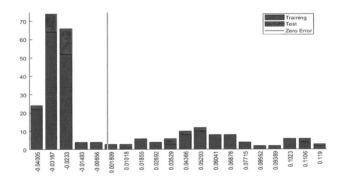

Fig. 8.16 Example NN-MLP distribution of prediction errors (signed MAPE)(deg) for Lat for one waypoint (wp_{10})

intensive NN training process. Faster training algorithms may be employed instead of the Bayesian regularization back-propagation used here, but at the expense of training accuracy and robustness, while the order of magnitude remains the same for the training times. It should be noted that these numbers do not include stage-1 processing (clustering), which is in the order of several seconds, as this runs only once, regardless of the model selected for stage-2.

It is worth noting that, as Table 8.7 shows, the per-waypoint prediction errors of the NN-MLP regressors remain fairly close to the mean (RMSE) value, not only in 3-D but also for each individual spatial dimension. This is particularly important, since these results prove the robustness of the NN-MLP predictions along the entire flight path and the validity of using the flight plan as the main element of constraint-based training in the proposed FSTP framework.

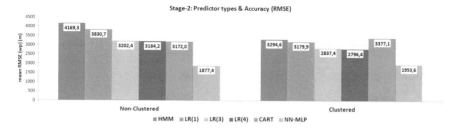

Fig. 8.17 Overview of all the stage-2 regressors accuracy in wp-averaged RMSE (m) (see Tables 8.5 and 8.6)

Table 8.8 Training and testing times (msec per wp per dimension) for all stage-2 models, averaged over 100 runs; $C(0)$ is for non-clustered dataset and $C(4)$ is for clustered dataset with $K=4$

	HMM	LR(1)	LR(3)	LR(4)	CART	NN
$C(0)$	2.1	4.8	5.2	10.3	5.2	870.6
$C(4)$	4.6	6.1	5.7	8.1	11.3	680.6

The use of flight plans for constrained-based training, specifically their waypoints as reference points for designing independent predictors for each one, essentially downscales the original FSTP problem to a much smaller non-uniform graph-based grid. As presented above, a roughly 1-h flight between Madrid and Barcelona translates to a set of 680–730 data points of the raw IFS radar track for each flight, which is down-scaled to only 11–18 waypoints of a typical flight plan for this route. Additionally, the clustering stage partitions the input space into smaller, more compact groups of trajectories and at the same time incorporates the enrichment part into this process, so that the predictive models that are to be trained subsequently can be designed in much smaller dimensionality, even the 3-D spatial-only if necessary. These three design aspects, i.e., independent per-waypoint model training, dimensionality reduction, and input space partitioning via clustering, constitute this proposed approach inherently parallelizable and highly scalable to very large volumes and rates of data. The estimated run times presented in Table 8.8 provide some experimental evidence for this, as the best LR and CART regressors can be trained on-the-fly per-waypoint in the order of few msec.

8.5   Related Work

Recently, there has been plenty of work on FLP and TP in the mobility [11], especially in the aviation domain. The proposed approaches include systems-engineering view [31], splitting the flight phases [13], collaborative TP via conflict avoidance and resolution (CA&R) [6, 41], anomaly detection [7], etc. Not surprisingly, the vast majority of methods are domain-specific in order to take advantage of the properties of the moving objects. The issue of exploiting additional data or *enrichments* in TP has created the notion of *semantic-aware* TP or FSTP as described earlier, which enables better estimations for departure and arrival times and, hence, more robust scheduling and logistics, especially in the congestion points (airports, major waypoints, etc.).

A typical example of a FLP method is presented in [36], where the authors propose TPR*-tree (index-based), which derives from TPR-tree, and exploits the characteristics of dynamic moving objects in order to retrieve only those which will meet specific spatial criteria within the given time interval, i.e., query window, in the future.

During the last few years, there is a mainstream trend of using stochastic models for TP, with Hidden Markov Models (HMM) being the most popular, as it has proved its efficiency in modeling a wide range of trajectory types. In general terms, a system is assumed to have the Markovian property if its future situations depend only upon its current state. Exhibiting high accuracy in modeling sequential data, the HMM approach has given rise to a wide range of applications, such as speech recognition, music retrieval, human activity recognition, consumer pattern recognition, etc.

Ayhan and Samet [4] introduce a novel stochastic approach to aircraft trajectory prediction problem, which exploits aircraft trajectories modeled in space and time by using a spatiotemporal data grid. They represent airspace in 4-D joint data cubes consisting of aircraft's motion parameters (i.e., latitude, longitude, altitude, and time) enriched by weather conditions. In their experimental study, they demonstrate that their methodology predicts aircraft trajectories efficiently by comparing the prediction results with the ground truth aligned trajectories, with the error being reasonably low for 1-h flights, i.e., within the grid's single-cell boundaries (8–13 km).

Two of the most widely explored approaches in TP are regression and clustering, separately or in combination, with some also exploring the use of weather or other data. These include methods based on generalized linear model (GLM) [9], typical regression-based short-/mid-term TP [18, 37], combination of clustering and Kalman filters [32], etc. Neural networks have also been used successfully for the climb/vertical TP [10] or in relation to the air traffic flows for estimated time of arrival (ETA).

Regarding en route climb TP, one of the major aspects of decision support for ATM, Coppenbarger [8] discusses the exploitation of real-time aircraft data, such as aircraft state variables, aircraft performance, pilot intent, and atmospheric data, for improving ground-based TP. The problem of climb TP is also discussed in Thipphavong et al. [39], as it also constitutes a very important challenge in ATM. In that work, an algorithm dynamically adjusts modeled aircraft weights, exploiting the observed track data to improve the accuracy of TP for climbing flights. Real-time evaluation with actual air traffic data shows a significant improvement in the altitude-only prediction of the trajectory, as well as the time to reach the top-of-climb. In another work by Ayhan and Samet [5], the authors investigate the applicability of HMM for TP on only one phase of a flight, specifically the climb after takeoff. A stochastic approach such as the HMM can address the TP problem by taking environmental uncertainties into account and training a model, using historical trajectory data along with weather observations. The results show robust performance and high TP accuracy, proving that HMM can be applied equally well for single-phase prediction, as well as complete-flight prediction. There are also numerical approaches to the problem of climb-phase TP, e.g., Hadjaz et al. [15].

Despite the plethora of various TP approaches in the aviation domain, most of them are either based on discovered mobility patterns from historical data or "blind" predictive modeling along entire flight paths in the sense that no constraint is imposed to the search space. In this work, the FSTP problem is addressed by exploiting not only trajectory enrichment (e.g., weather and semantic information) but, most importantly, flight plans. According to our research, only one very recent work [20] exploits flight plans and enrichment data (e.g., weather) as input for TP in the aviation domain. In particular, the authors propose a tree-based matching algorithm to construct image-like feature maps from meteorological datasets, then model the track points on trajectories as conditional Gaussian mixtures with parameters to be learned from a deep generative model, which is an end-to-end convolutional recurrent neural network (RNN) that consists of a long short-term memory (LSTM) encoder network and a mixture density LSTM decoder network. During the inference process, beam search, adaptive Kalman filter, and Rauch–Tung–Striebel smoother algorithms are used to prune the variance of generated trajectories. Although promising, this approach requires significant computing resources and employs an image-like uniform spatiotemporal grid of high dimensionality for every input component. In terms of accuracy, this deep learning approach is demonstrated with a very dataset similar to ours (direct continental flights between two heavy-traffic airports) and the resulting prediction error is in the order of 92 km, i.e., about 7–9 times worse than the best HMM-based state-of-the-art approaches [4, 5] that do not use flight plans.

Discovering clusters of complete trajectories can overlook significant patterns that might exist only for portions of their lifespan. To deal with this, the authors of [19] propose TraClus, a partition-and-group framework for clustering 2-D moving objects which segments the trajectories based on their geometric features and then clusters them by ignoring the temporal dimension. A more recent approach to the problem of subtrajectory clustering is S^2T-Clustering [27]. A similar method is adopted in [1], where the authors aim at identifying common portions between trajectories, with respect to some constraints and/or objectives, by taking into account the "neighborhood" of each trajectory.

The approach presented for FSTP combines several aspects and ideas from the methods cited above, in order to develop a highly adaptive, long-term, big data framework for FLP which is experimentally evaluated with datasets from both the maritime and the aviation domain. More specifically, this two-stage approach includes: (a) mobility pattern discovery from the historical movement of the moving objects and (b) employ optimal estimations of FLP in the sense of maximum likelihood [38], as they are dictated by the identified patterns. Furthermore, some promising experimental results are presented for real datasets from both domains, as well as performance indicators for deployment in a big data platform.

8.6 Discussion

The results presented for the FLP task address two different use cases: (a) when there is no historic data or not adequate time available for processing them and constructing network-aware predictive models and (b) when such data and time are indeed available and common routes are discovered and included in the predictive models. For (a) the LSTM-based approach is presented here as routes-agnostic/short-term FLP, while for (b) the two-stage clustering/prediction approach is presented as network-aware/long-term FLP. Both options are extremely valuable in real-world applications, especially when treating widely different domains such as aviation and maritime.

According to these results, the LSTM-based FLP shows very promising performance in terms of prediction accuracy. The true power of LSTM is the automated and robust discovery of mobility patterns from the raw data, with little to no assumptions about their statistical properties (e.g., noise, trends, etc.), hence to need to impose artificial constraints that are commonly implied in short-term FLP, such as the constant-speed assumption for "adequately small" dt. It should also be noted that the method proposed here is an inherently variable-rate approach, using the position data (input) as they arrive, with no need for reconstructing the raw trajectory via fixed-rate resampling, as it is usually the case with NN and other regression models. This means that the discovered mobility patterns are not "imposed" by any resampling algorithm, e.g., linear, quadratic, etc.

For the long-term FLP, it is important to emphasize that the proposed framework relies end-to-end on big data technologies. Nevertheless, this framework is directly applicable and valid in the aviation domain too, especially since the discovery of medoids is based upon some form of spatiotemporal clustering to form groups and common motion patterns, either with or without considering flight plans as input in the predictive models. The accuracy in both domains, as well as the performance results, proves that it is a very efficient and scalable big data solution for real-world applications, easily adaptable to various other domains.

Similarly, the results presented for the FSTP approach verify the applicability and performance of the proposed FSTP framework in the aviation domain, specifically in the context of pre-flight trajectory prediction, exploiting all the available information from flight plans, localized weather data and other semantics, e.g., aircraft type and category, weekday, etc. This approach was designed from the start as lightweight, fully parallelizable and compatible with distributed computing platforms for big data real-world applications. HMM and LR regressors are all valid within these design specifications, as they are all low-complexity models in terms of both training and size. Combined with the partitioning of the input data via properly designed semantic-aware clustering, this modular approach is highly scalable and adaptable to any type of transit route and takeoff/landing patterns, provided that the associated flight plan is available.

It should be noted that the current data-driven methods for long-term FSTP, e.g., as in [4] with "blind" HMM, produce cross-section 3-D prediction errors in the order of 8–13 km. Although this proposed approach is not directly comparable, using flight plans for constrained-based FSTP as described here produces per-waypoint 3-D prediction errors consistently in the order of 2–3 km and even lower when NN-MLP regressors are used (see Fig. 8.17, Table 8.7).

8.7 Conclusion

In this work, a novel multi-stage hybrid approach was presented for the FLP and TP tasks. More specifically, FLP is considered under the short-term/routes-agnostic and the long-term/network-aware variants, while FSTP is incorporating extensive exploitation of data enrichments and flight plans as constraints in the aviation domain.

In FLP, the LSTM-based short-term variant demonstrates how streaming trajectory data can be exploited to train predictive models based on variable-rate "common" mobility patterns that are discovered from the data itself. In contrast, the two-stage clustering/prediction long-term variant demonstrates how a rich set of historic mobility data can be used to model extensive "common" routes in the form of connected nodes, i.e., a routes network, which is subsequently exploited as the base for long-term trajectory predictions for any given moving object within the same region of interest. These two approaches are complementary and can be adapted to various domain-specific tasks in maritime, aviation, etc.

In FSTP, clustering is introduced for grouping together "similar" enriched trajectories, using a properly designed semantic-aware similarity function. This enrichment includes localized weather data (e.g., wind speed & direction), aircraft properties (e.g., type), and other external factors (e.g., weekday). Subsequently, a set of independent predictive models are trained for each cluster, addressing the task of TP in the context of each reference waypoint of the flight plans. HMM, linear, and non-linear regressors were employed as base for the predictive models, exploring the trade-off between having very simple predictors and moderate accuracies versus more complex predictors and higher accuracies.

Acknowledgments This work was partially supported by the projects *datACRON*[5] (No:687591), *DART*[6] (No:699299), *Track & Know*[7] (No:780754), which have received funding from the European Union's Horizon 2020 (H2020) programme, and *MASTER*[8] (No:777695), which has received funding from the Marie Sklowdoska-Curie programme. This research has also been co-

---

[5]http://datacron-project.eu/.

[6]http://dart-research.eu/.

[7]http://trackandknowproject.eu/.

[8]http://www.master-project-h2020.eu/.

financed by the European Regional Development Fund of the European Union and Greek National Funds through the Operational Program Competitiveness, Entrepreneurship and Innovation, under the call RESEARCH-CREATE-INNOVATE (id:T1EDK-3268).

References

1. Agarwal, P.K., Fox, K., Munagala, K., Nath, A., Pan, J., Taylor, E.: Subtrajectory clustering: models and algorithms. In: PODS, pp. 75–87 (2018)
2. Alexandridis, A., Chondrodima, E., Giannopoulos, N., Sarimveis, H.: A fast and efficient method for training categorical radial basis function networks. IEEE Trans. Neural Netw. Learn. Syst. **28**(11), 2831–2836 (2017). http://doi.org/10.1109/TNNLS.2016.2598722
3. Ankerst, M., Breunig, M., et al.: Optics: ordering points to identify the clustering structure. In: Proc. of SIGMOD 1999 (1999)
4. Ayhan, S., Samet, H.: Aircraft trajectory prediction made easy with predictive analytics. In: Proc. of ACM SIGKDD 2016 (2016)
5. Ayhan, S., Samet, H.: Time series clustering of weather observations in predicting climb phase of aircraft trajectories. In: Proc. of the IWCTS 2016 (2016)
6. Chen, X., Landry, S., Nof, S.: A framework of enroute air traffic conflict detection and resolution through complex network analysis. Comput. Ind. **62**(8), 787–794 (2011)
7. Ciccio, C.D., var der Aa, H., Cabanillas, C., et al.: Detecting flight trajectory anomalies and predicting diversions in freight transportation. Decis. Support Syst. **88**, 1–17 (2016)
8. Coppenbarger, R.: En route climb trajectory prediction enhancement using airplane flight-planning information. In: American Institute of Aeronautics and Astronautics (AIAA-99-4147) (1999)
9. de Leege, A., Paassen, M.V., Mulder, M.: A machine learning approach to trajectory prediction. In: Proc. of AIAA GNC 2013 (2013)
10. Fablec, Y.L., Alliot, J.: Using neural networks to predict aircraft trajectories. In: Proc. of the ICIS 1999 (1999)
11. Georgiou, H., Karagiorgou, S., Kontoulis, Y., Pelekis, N., Petrou, P., Scarlatti, D., Theodoridis, Y.: Moving objects analytics - survey on future location and trajectory prediction methods. Tech. rep., Data Science Lab, University of Piraeus. arXiv:1807.04639 (2018)
12. Georgiou, H., Pelekis, N., Sideridis, S., Scarlatti, D., Theodoridis, Y.: Semantic-aware aircraft trajectory prediction using flight plans. Int. J. Data Sci. Anal. 1–14 (2019). http://doi.org/10.1007/s41060-019-00182-4
13. Gong, C., McNally, D.: A methodology for automated trajectory prediction analysis. In: 2004 AIAA Guidance, Navigation, and Control Conference and Exhibit (AIAA 2004), 16–19 August 2004, Providence. https://aviationsystemsdivision.arc.nasa.gov/publications/more/analysis/gong_08_04.pdf
14. Habler, E., Shabtai, A.: Using LSTM encoder-decoder algorithm for detecting anomalous ADS-B messages. Comput. Secur. **78**, 155–173 (2018). https://doi.org/10.1016/j.cose.2018.07.004
15. Hadjaz, A., Marceau, G., Saveant, P., et al.: Online learning for ground trajectory prediction. CoRR abs/1212.3998 (2012)
16. Hamed, M., Gianazza, D., Serrurier, M., Durand, N.: Statistical prediction of aircraft trajectory: regression methods vs point-mass model. In: Proc. of the ATM 2013 (2013)
17. Hochreiter, S., Schmidhuber, J.: Long short-term memory. Neural Comput. **9**(8), 1735–1780 (1997). http://dx.doi.org/10.1162/neco.1997.9.8.1735
18. Krumm, J., Horvitz, E.: Predestination: inferring destinations from partial trajectories. In: Proc. of the UbiComp 2003 (2003)
19. Lee, J., Han, J., Whang, K.: Trajectory clustering: a partition-and-group framework. In: SIGMOD, pp. 593–604 (2007)

20. Liu, Y., Hansen, M.: Predicting aircraft trajectories: a deep generative convolutional recurrent neural networks approach. Tech. rep., Institute of Transportation Studies, University of California, arXiv:1812.11670 (2018)
21. Loh, W.: Regression trees with unbiased variable selection and interaction detection. Stat. Sin. **12**, 361–386 (2002)
22. Ma, Z., Yao, M., Hong, T., Li, B.: Aircraft surface trajectory prediction method based on LSTM with attenuated memory window. J. Phys. Conf. Ser. **1215**, 012003 (2019). http://doi.org/10.1088/1742-6596/1215/1/012003
23. Nanni, M., Pedreschi, D.: Time-focused clustering of trajectories of moving objects. J. Intell. Inf. Syst. **27**(3), 267–289 (2006)
24. Panagiotakis, C., Pelekis, N., Kopanakis, I.: Trajectory voting and classification based on spatiotemporal similarity in moving object databases. In: IDA, pp. 131–142. Springer, Berlin (2009)
25. Patroumpas, K., Alevizos, E., Artikis, A., Vodas, M., Pelekis, N., Theodoridis, Y.: Online event recognition from moving vessel trajectories. GeoInformatica **21**(2), 389–427 (2017)
26. Pelekis, N., Theodoridis, Y.: Mobility Data Management and Exploration. Springer, Berlin (2014)
27. Pelekis, N., Tampakis, P., Vodas, M., Panagiotakis, C., Theodoridis, Y.: In-DBMS sampling-based sub-trajectory clustering. In: EDBT, pp. 632–643 (2017)
28. Petrou, P., Nikitopoulos, P., Tampakis, P., Glenis, A., Koutroumanis, N., Santipantakis, G., Patroumpas, K., Vlachou, A., Georgiou, H., et al: Argo: a big data framework for online trajectory prediction. In: 16th International Symposium on Spatial and Temporal Databases (SSDT19) (2019)
29. Petrou, P., Tampakis, P., Georgiou, H., Pelekis, N., Theodoridis, Y.: Online long-term trajectory prediction based on mined route patterns. In: 2019 European Conference on Machine Learning and Principles and Practice of Knowledge Discovery in Database (ECML/PKDD19) (2019)
30. Shi, Z., Xu, M., Pan, Q., Yan, B., Zhang, H.: LSTM-based flight trajectory prediction. In: 2018 International Joint Conference on Neural Networks (IJCNN), pp. 1–8 (2018). http://doi.org/10.1109/IJCNN.2018.8489734
31. Sip, S., Green, S.: Common trajectory prediction capability for decision support tools. In: ATM 5th USA/Europa R&D Seminar (2003)
32. Song, Y., Cheng, P., Mu, C.: An improved trajectory prediction algorithm based on trajectory data mining for air traffic management. In: Proc. of the IEEE ICIA 2012 (2012)
33. Spiegel, M., Schiller, J., Srinivasan, R.: Probability and Statistics, 3rd edn. McGraw-Hill, New York (2009)
34. Tampakis, P., Doulkeridis, C., Pelekis, N., Theodoridis, Y.: Distributed subtrajectory join on massive datasets (2019). http://arxiv.org/abs/1903.07748
35. Tampakis, P., Pelekis, N., Doulkeridis, C., Theodoridis, Y.: Scalable distributed subtrajectory clustering (2019). http://arxiv.org/abs/1906.06956
36. Tao, Y., Faloutsos, C., Papadias, D., Liu, B.: Prediction and indexing of moving objects with unknown motion patterns. In: Proc. of the ACM SIGMOD 2004 (2004)
37. Tastambekov, K., Puechmorel, S., Delahaye, D., et al.: Aircraft trajectory forecasting using local functional regression in Sobolev space. Transp. Res. Part C Emerg. Technol. **39**, 1–22 (2014)
38. Theodoridis, S., Koutroumbas, K.: Pattern Recognition, 4th edn. Academic Press, London (2008)
39. Thipphavong, D., Schultz, C., et al.: Adaptive algorithm to improve trajectory prediction accuracy of climbing aircraft. J. Guid. Control Dyn. **36**(1), 15–24 (2013)
40. Trasarti, R., Guidotti, R., Monreale, A., Giannotti, F.: MyWay: location prediction via mobility profiling. Inf. Syst. **64**, 350–367 (2017)

41. Vouros, G., Vlachou, A., Santipantakis, G., et al.: Big data analytics for time critical mobility forecasting: recent progress and research challenges. In: Proc. of the EDBT 2018 (2018)
42. Zhang, Z., Yang, R., Fang, Y.: LSTM network based on Antlion optimization and its application in flight trajectory prediction. In: 2018 2nd IEEE Advanced Information Management, Communicates, Electronic and Automation Control Conference (IMCEC), pp. 1658–1662 (2018). http://10.1109/IMCEC.2018.8469476
43. Zheng, Y.: Trajectory data mining: an overview. Trans. Intell. Syst. Technol. **6**(3), 1–41 (2015)

Chapter 9
Event Processing for Maritime Situational Awareness

Manolis Pitsikalis, Konstantina Bereta, Marios Vodas, Dimitris Zissis, and Alexander Artikis

Abstract Numerous illegal and dangerous activities take place at sea, including violations of ship emission rules, illegal fishing, illegal discharges of oil and garbage, smuggling, piracy and more. We present our efforts to combine two stream reasoning technologies for detecting such activities in real time: a formal, computational framework for composite maritime event recognition, based on the Event Calculus, and an industry-strong maritime anomaly detection service, capable of processing daily real-world data volumes.

9.1 Introduction

Numerous illegal and dangerous activities take place at sea, such as pollution (illegal discharges of oil and garbage, violations of ship emission rules, etc.), illegal fishing, smuggling (drugs, arms, oil, etc.), piracy and many more [8]. Often the vessels involved in such activities attempt to behave as common commercial ships, concealing their true intentions. Unlike in the past, though, when there was no way of detecting these activities as they happened, today numerous monitoring systems,

M. Pitsikalis
Institute of Informatics & Telecommunications, NCSR Demokritos, Athens, Greece
e-mail: manospits@iit.demokritos.gr

K. Bereta · M. Vodas
MarineTraffic, Athens, Greece
e-mail: konstantina.bereta@marinetraffic.com; marios.vodas@marinetraffic.com

D. Zissis
Department of Product & Systems Design Engineering, University of the Aegean, Syros, Greece
MarineTraffic, Athens, Greece
e-mail: dzissis@marinetraffic.com

A. Artikis (✉)
Department of Maritime Studies, University of Piraeus, Piraeus, Greece
Institute of Informatics & Telecommunications, NCSR Demokritos, Athens, Greece
e-mail: a.artikis@unipi.gr

© Springer Nature Switzerland AG 2020
G. A. Vouros et al. (eds.), *Big Data Analytics for Time-Critical Mobility Forecasting*, https://doi.org/10.1007/978-3-030-45164-6_9

such as the automatic identification system (AIS), produce constant streams of surveillance data, often revealing the true intentions of suspicious vessels.

"Collaborative" monitoring systems use equipment that is mostly installed aboard the vessels that should be monitored and rely on the collaboration of the vessels' crew. This equipment is used for reporting the position of the vessel, along with information about its navigational status (e.g., destination, speed, heading, course over ground, etc.). The automatic identification system (AIS) [6], for example, is such a system and is based on the transmission of messages through VHF transponders installed aboard the vessels themselves and received using VHF receivers installed on other vessels sailing close by, coastal stations and/or satellites. Other collaborative systems include the long-range identification and tracking (LRIT) system [7], and the vessel monitoring system (VMS) [4], which is a system designed for managing fishing activity. 'Non-collaborative' monitoring systems, on the other hand, do not rely on the collaboration of the crew and include systems such as coastal and high-frequency radar, active and passive sonar, ground- and vessel-based (e.g., thermal) cameras, satellite and airborne Earth Observation systems, such as optical and synthetic aperture radar systems.

The data streams produced by maritime monitoring systems may be consumed by stream reasoning systems, in order to support *Maritime Situational Awareness* [20, 21], i.e. the effective understanding of activities, events and threats in the maritime environment that could impact the global safety, security, economic activity and the environment (Chap. 1 of this book provides further details on maritime data sources and operational needs). Terosso-Saenz et al. [18], for example, presented a system detecting abnormally high or low vessel speed, as well as when two vessels are in danger of colliding. SUMO [5] is an open-source system combining AIS streams with synthetic aperture radar images for detecting illegal oil dumping, piracy and unsustainable fishing. van Laere et al. [19] evaluated a workshop aiming at the identification of potential vessel anomalies, such as tampering, rendez-vous between vessels and unusual routing. Mills et al. [14] described a method for identifying trawling using speed and directionality rules, thus helping in studies of how trawling impacts on species, habitats and the ecosystem. Millefiori et al. [13] proposed a distributed framework that uses AIS data to identify port operational regions.

In this chapter, we present our effort to combine two stream reasoning technologies for maritime situational awareness. First, a formal, computational framework for composite maritime event recognition, based on the Event Calculus [16].[1] Second, an industry-strong maritime anomaly detection service, processing daily real-world data volumes [8].[2] Our integrated system aims to pave the way for the real-time recognition of a wide variety of maritime events of high significance, including forms of illegal and suspicious vessel activity. We build on our previous work in this domain by presenting the architecture and implementation details of our approach, as used in real-world operational conditions to analyse actual

---

[1] http://cer.iit.demokritos.gr/cermm.

[2] www.marinetraffic.com.

sensor data, received from the MarineTraffic network. Our novel approach combines automated reasoning (based on complex event processing) with massive amounts of historical observational data used for the extraction of contextual information, so as to improve maritime situational awareness and assist an operational end user in the decision-making process by providing real-time notifications. The novel data processing architecture presented here supports large-scale multistage analysis of data generated by a distributed network of sensors and in situ data (from computations). In this context we address the modalities of processing combinations of data stored in massive observational archives and streaming data. The multistage workflow demonstrates how several state-of-the-art technologies can be combined (synopsis engine, distributed processing, etc.) to achieve real-time response times.

The remainder of this chapter is structured as follows: Section 9.2 presents the proposed architecture of the integrated stream reasoning technology. Then, Sect. 9.3 presents a set of maritime events that may be identified by our technology. Subsequently, Sect. 9.4 presents a service making available the detected events to users. Finally, in Sect. 9.5 we summarise our work and present further work directions.

9.2   System Architecture

9.2.1   Setting the Scene

We support maritime situational awareness following two online tasks/steps: (a) computing a set of spatial relations among vessels, such as proximity, and among vessels and areas of interest (e.g., fishing areas), as described in Chap. 6 of this book and (b) labelling position signals of interest as 'critical'—such as when a vessel changes its speed, turns, stops, moves slowly or stops transmitting its position, as Chap. 4 of this book describes. Figure 9.1 illustrates these steps. Streaming-in AIS position signals go through a spatial preprocessing step, for the computation of the spatial relations required by maritime situational awareness [17]. These relations are displayed at the top of Table 9.1. Then, the relevant position signals are annotated as critical—see the middle part of Table 9.1. Subsequently, the position signals may be consumed by our stream reasoning technology either directly (see 'enriched AIS stream' in Fig. 9.1) or after being compressed, that is, after removing all signals that have not been labelled as critical (see 'critical point stream' in Fig. 9.1).

Critical point labelling is performed as part of trajectory synopsis generation, whereby major changes along each vessel's movement are tracked (cf. Chap. 4). This process instantly identifies critical points along each trajectory, such as a stop, a turn, or slow motion. Using the retained critical points, we may reconstruct a vessel trajectory with small acceptable deviations from the original one. Empirical results have indicated that 70–80% of the input data may be discarded as redundant, while

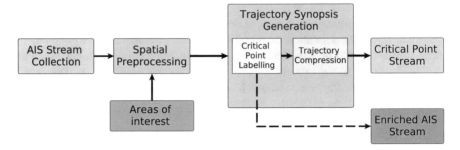

Fig. 9.1 Steps required for maritime situational awareness

compression ratio can be up to 99% when the frequency of position updates is high [15].

9.2.2 Maritime Stream Reasoning

We have been developing two stream reasoning technologies for maritime situational awareness. More precisely, we have been constructing a formal, computational framework for composite maritime event recognition, based on RTEC [1], a logic programming implementation of the Event Calculus [16]. RTEC is designed to compute continuous narrative assimilation queries for pattern matching on data streams and includes several optimisation techniques allowing for real-time event recognition. To facilitate the interaction of RTEC with state-of-the-art distributed systems, we have been re-implementing RTEC in the Scala programming language. This way, we may integrate RTEC with the industry-strong maritime anomaly detection service of MarineTraffic [8]. This service is based on a hybrid architecture, comprised of stream and batch processing components. The stream processing component is based on the actor model—specifically, the Akka framework—for concurrency and event-driven processing. In what follows, we briefly present RTEC and illustrate its use for maritime stream reasoning. Then, in Sect. 9.4 we present MarineTraffic's service, and the way these two stream reasoning technologies are being integrated.

The 'Event Calculus for Run-Time reasoning' (RTEC) is an Event Calculus dialect optimised for composite event recognition over high-velocity data streams [1]. For example, RTEC may detect the composite events displayed at the bottom of Table 9.1. The time model in RTEC is linear and includes integer time-points. An *event description* includes rules that define the event instances with the use of the happensAt(predicate, the effects of events on *fluents*—time-varying properties— with the use of the initiatedAt(and terminatedAt(predicates and the values of the fluents with the use of the holdsAt(and holdsFor(predicates. Table 9.2 summarises the main predicates of RTEC.

Table 9.1 Events for maritime situational awareness: Input events are presented above the double horizontal line, while the output stream is presented below this line. The input events above the single horizontal line are detected at the spatial preprocessing step, while the remaining ones are detected by the trajectory synopsis generator (critical events). With the exception of *proximity*, all items of the input stream are instantaneous, while all output activities are durative

		Event/Activity	Description
Input	Spatial	$entersArea(V, A)$	Vessel V enters area A
		$leavesArea(V, A)$	Vessel V leaves area A
		$proximity(V_1, V_2)$	Vessels V_1 and V_2 are close
	Critical	$gap\_start(V)$	Vessel V stopped sending position signals
		$gap\_end(V)$	Vessel V resumed sending position signals
		$slow\_motion\_start(V)$	Vessel V started moving at a low speed
		$slow\_motion\_end(V)$	Vessel V stopped moving at a low speed
		$stop\_start(V)$	Vessel V started being idle
		$stop\_end(V)$	Vessel V stopped being idle
		$change\_in\_speed\_start(V)$	Vessel V started changing its speed
		$change\_in\_speed\_end(V)$	Vessel V stopped changing its speed
		$change\_in\_heading(V)$	Vessel V changed its heading
Output	Composite	$highSpeedNC(V)$	Vessel V has high speed near coast
		$anchoredOrMoored(V)$	Vessel V is anchored or moored
		$drifting(V)$	Vessel V is drifting
		$trawling(V)$	Vessel V is trawling
		$tugging(V_1, V_2)$	Vessels V_1 and V_2 are engaged in tugging
		$pilotBoarding(V_1, V_2)$	Vessels V_1 and V_2 are engaged in pilot boarding
		$rendez\text{-}Vous(V_1, V_2)$	Vessels V_1 and V_2 are having a rendez-vous
		$loitering(V)$	Vessel V is loitering
		$sar(V)$	Vessel V is engaged in a search and rescue (SAR) operation

Fluents are 'simple' or 'statically determined'. In brief, simple fluents are defined by means of initiatedAt(and terminatedAt(rules, while statically determined fluents are defined by means of application-dependent holdsFor(rules, along with the interval manipulation constructs of RTEC: union_all(, intersect_all(and relative_complement_all(. See Table 9.2 for a brief explanation of these constructs and Fig. 9.2 for an example visualisation. Composite events/activities are typically durative; thus, the task generally is to compute the maximal intervals for which a fluent expressing a composite activity has a particular value continuously. Below, we discuss the representation of fluents/composite maritime activities and briefly present the way in which we compute their maximal intervals.

Table 9.2 Main predicates of RTEC. '$F = V$' denotes that fluent F has value V

Predicate	Meaning
happensAt(E, T)	Event E occurs at time T
holdsAt($F = V, T$)	The value of fluent F is V at time T
holdsFor($F = V, I$)	I is the list of the maximal intervals for which $F = V$ holds continuously
initiatedAt($F = V, T$)	At time T a period of time for which $F = V$ is initiated
terminatedAt($F = V$, T)	At time T a period of time for which $F = V$ is terminated
union_all(L, I)	I is the list of maximal intervals produced by the union of the lists of maximal intervals of list L
intersect_all(L, I)	I is the list of maximal intervals produced by the intersection of the lists of maximal intervals of list L
relative_complement_all (I', L, I)	I is the list of maximal intervals produced by the relative complement of the list of maximal intervals I' with respect to every list of maximal intervals of list L

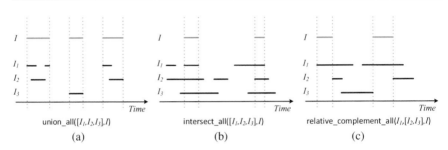

Fig. 9.2 A visual illustration of the interval manipulation constructs of RTEC. In these examples, there are three input streams, I_1, I_2 and I_3, coloured black. The output of each interval manipulation construct I is coloured light blue. (**a**) Union. (**b**) Intersection. (**c**) Relative complement

9.3 Maritime Events

To analyse the behaviour at sea, we need to spatially define the implicated sites, such as ports, sea structures, fishing areas and sea zones. In [13], for example, we presented a data-driven implementation of a distributed method for calculating port operational areas and other activity areas. Then, we need to define the spatiotemporal patterns of vessel activity of interest. Such patterns have been documented in various papers, including [8, 13, 16, 18, 19]. In this section, we present a set of indicative patterns in the language of RTEC following [16]. The hierarchy supported by our integrated stream reasoning technology is displayed in Fig. 9.3. In this figure, an arrow from pattern A to pattern B denotes that A is used in the specification of B.

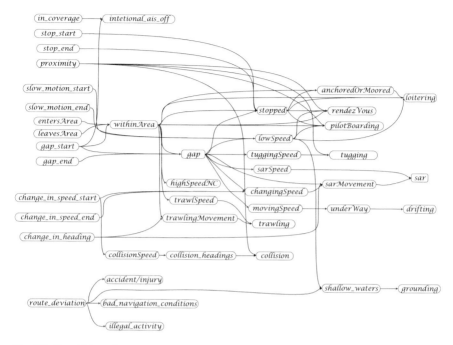

Fig. 9.3 Event hierarchy

9.3.1   Building Blocks

We begin by presenting a set of building blocks that will be later used for the construction of more involved patterns.

9.3.1.1   Vessel Within Area of Interest

Calculating the intervals during which a vessel is in an area of some type, such as a (protected) Natura 2000, fishing or anchorage area, is particularly useful in maritime (e.g., fishing) patterns. Consider the formalisation below:

$$
\begin{aligned}
&\mathsf{initiatedAt}(withinArea(Vessel, AreaType) = \mathsf{true},\, T) \leftarrow \\
&\quad \mathsf{happensAt}(entersArea(Vessel, AreaID),\, T), \\
&\quad areaType(AreaID, AreaType). \\[4pt]
&\mathsf{terminatedAt}(withinArea(Vessel, AreaType) = \mathsf{true},\, T) \leftarrow \\
&\quad \mathsf{happensAt}(leavesArea(Vessel, AreaID),\, T), \\
&\quad areaType(AreaID, AreaType). \\[4pt]
&\mathsf{terminatedAt}(withinArea(Vessel, AreaType) = \mathsf{true},\, T) \leftarrow \\
&\quad \mathsf{happensAt}(gap\_start(Vessel),\, T).
\end{aligned}
\tag{9.1}
$$

Variables start with an uppercase letter, while predicates and constants start with a lower-case letter. $withinArea(Vessel, AreaType)$ is a simple fluent indicating that a *Vessel* is within an area of some type. $entersArea(Vessel, AreaID)$ and $leavesArea(Vessel, AreaID)$ are input events computed at the spatial preprocessing step (see the top part of Table 9.1), indicating that a *Vessel* entered (respectively, left) an area with *AreaID*. $areaType(AreaID, AreaType)$ is an atemporal predicate storing the areas of interest of a given dataset. $withinArea(Vessel, AreaType) =$ true is initiated when a *Vessel* enters an area of *AreaType* and terminated when the *Vessel* leaves the area of *AreaType*. $withinArea(Vessel, AreaType) =$ true is also terminated when the trajectory synopsis generator produces a *gap_start* event (see the middle part of Table 9.1), indicating the beginning of a communication gap (in the subsection that follows we discuss further communication gaps). In this case we chose to make no assumptions about the location of the vessel. With the use of rule-set (9.1), RTEC computes the *maximal intervals* during which a vessel is said to be within an area of some type.

9.3.1.2 Communication Gap

According to the trajectory synopsis generator, a communication gap takes place when no message has been received from a vessel for at least 30 min. All numerical thresholds, however, may be tuned—for example, by machine learning algorithms—to meet the requirements of the application under consideration. A communication gap may occur when a vessel sails in an area with no AIS receiving station nearby, or because the transmission power of its transceiver allows broadcasting in a shorter range, or when the transceiver is deliberately turned off. The rules below present a formalisation of communication gap:

$$
\begin{aligned}
&\mathsf{initiatedAt}(gap(Vessel) = nearPorts, T) \leftarrow \\
&\quad \mathsf{happensAt}(gap\_start(Vessel), T), \\
&\quad \mathsf{holdsAt}(withinArea(Vessel, nearPorts) = true, T). \\
&\mathsf{initiatedAt}(gap(Vessel) = farFromPorts, T) \leftarrow \\
&\quad \mathsf{happensAt}(gap\_start(Vessel), T), \\
&\quad \mathsf{not\ holdsAt}(withinArea(Vessel, nearPorts) = true, T). \\
&\mathsf{terminatedAt}(gap(Vessel) = \_Value, T) \leftarrow \\
&\quad \mathsf{happensAt}(gap\_end(Vessel), T).
\end{aligned}
\tag{9.2}
$$

gap is a simple, multi-valued fluent, gap_start and gap_end are input critical events (see Table 9.1), 'not' expresses Prolog's negation-by-failure [3], while variables starting with '_', such as *_Value*, are free. We chose to distinguish between communication gaps occurring near ports from those occurring in the open sea, as the first ones usually do not have a significant role in maritime monitoring. According to rule-set (9.2), a communication gap is said to be initiated when the synopsis generator emits a 'gap start' event and terminated when a 'gap end' is detected. Given this rule-set, RTEC computes the maximal intervals for which a vessel is not sending position signals.

9.3.2 Maritime Situational Indicators

Our aim is to detect in real-time maritime situational indicators [9], that is, composite maritime activities of special significance, building upon the blocks presented above. As mentioned earlier, Fig. 9.3 displays the hierarchy of our formalisation, that is, the relations between the indicators' specifications.

9.3.2.1 Vessel with High Speed Near Coast

Several countries have regulated maritime zones. In French territorial waters, for example, there is a 5 knots speed limit for vessels or watercrafts within 300 m from the coast. One of the causes of marine accidents near the coast is vessels sailing with high speed; thus, the early detection of violators ensures safety by improving the efficiency of law enforcement. Figure 9.4 displays the case of a vessel not conforming to the above regulations. Consider the following formalisation:

$$
\begin{aligned}
&\mathsf{initiatedAt}(highSpeedNC(Vessel\, text) = \mathsf{true},\, T) \leftarrow \\
&\quad \mathsf{happensAt}(velocity(Vessel, Speed, \_CoG, \_TrueHeading), T), \\
&\quad \mathsf{holdsAt}(withinArea(Vessel, nearCoast) = \mathsf{true},\, T), \\
&\quad threshold(v_{hs}, V_{hs}),\quad Speed > V_{hs}. \\
&\mathsf{terminatedAt}(highSpeedNC(Vessel) = \mathsf{true},\, T) \leftarrow \hspace{2cm} (9.3)\\
&\quad \mathsf{happensAt}(velocity(Vessel, Speed, \_CoG, \_TrueHeading), T), \\
&\quad threshold(v_{hs}, V_{hs}),\quad Speed \leq V_{hs}. \\
&\mathsf{terminatedAt}(highSpeedNC(Vessel) = \mathsf{true},\, T) \leftarrow \\
&\quad \mathsf{happensAt}(end(withinArea(Vessel, nearCoast) = \mathsf{true}), T).
\end{aligned}
$$

highSpeedNC(Vessel) is a Boolean simple fluent indicating that a *Vessel* is exceeding the speed limit imposed near the coast. *velocity* is input contextual information expressing the speed, course over ground (CoG) and true heading of a vessel. This information is attached to each incoming AIS message. Recall that variables starting with '_' are free. *withinArea(Vessel, nearCoast)* = true expresses the time periods during which a *Vessel* is within 300 m from the French coastline (see rule-set (9.1) for the specification of *withinArea*). *threshold* is an auxiliary atemporal predicate recording the numerical thresholds of the maritime patterns. The use of this predicate supports code transferability, since the use of different thresholds for different applications requires only the modification of the *threshold* predicate, and not the modification of the patterns. end$((F = V))$ (respectively, start$((F = V))$) is an RTEC built-in event indicating the ending (resp. starting) points for which $F = V$ holds

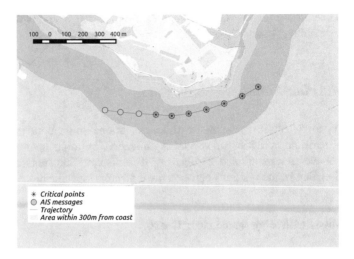

Fig. 9.4 A vessel near the port of Brest, France, with speed above the 5 knots limit. The marked circles denote the AIS position signals that are labelled as 'critical' by the synopsis generator

continuously. According to rule-set (9.3), therefore, $highSpeedNC(Vessel) = $ true is initiated when the *Vessel* sails within 300 m from the French coastline with speed above 5 knots and terminated when its speed goes below 5 knots, sails away (further than 300 m) from the coastline or stops sending position signals (recall that *withinArea* is terminated/ended by *gap_start*).

9.3.2.2 Anchored or Moored Vessel

A vessel lowers its anchor in specific areas; for example, when waiting to enter into a port or taking on cargo or passengers where insufficient port facilities exist. Figure 9.5 displays an example of a vessel stopped in an anchorage area. Furthermore, a vessel may be moored, that is, when a vessel is secured with ropes in any kind of permanent fixture such as a quay or a dock. Consider the specification below:

$$
\begin{aligned}
&\text{holdsFor}(anchoredOrMoored(Vessel(=\text{true}, I) \leftarrow \\
&\quad \text{holdsFor}(stopped(Vessel(=farFromPorts, I_{sffp}), \\
&\quad \text{holdsFor}(withinArea(Vessel, anchorage(=\text{true}, I_{wa}), \\
&\quad \text{intersect\_all}([I_{sffp}, I_{wa}], I_{sa}), \\
&\quad \text{holdsFor}(stopped(Vessel) = nearPorts, I_{sn}), \\
&\quad \text{union\_all}([I_{sa}, I_{sn}], I_i), \\
&\quad threshold(v_{aorm}, V_{aorm}), \quad intDurGreater(I_i, V_{aorm}, I).
\end{aligned}
\tag{9.4}
$$

anchoredOrMoored(Vessel) is a statically determined fluent, that is, it is specified by means of a domain-dependent holdsFor(predicate and interval manipulation

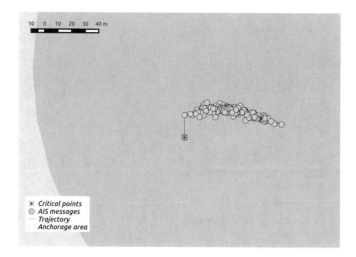

Fig. 9.5 Anchored vessel

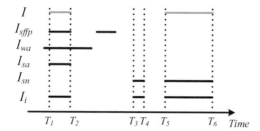

Fig. 9.6 Interval computation example of *anchoredOrMoored*

constructs—intersect_all(and union_all(in this case, that compute, respectively, the intersection and union of lists of maximal intervals (see Table 9.2 and Fig. 9.2). *stopped* is a fluent recording the intervals in which a vessel is stopped—this may be far from all ports or near some port(s). *intDurGreater(I', V_t, I)* is an auxiliary predicate keeping only the intervals I of list I' with length greater than V_t. *anchoredOrMoored(Vessel)*= true, therefore, holds when the *Vessel* is stopped in an anchorage area or near some port, for a time period greater than some threshold (see V_{aorm} in rule (9.4)). The default value for this threshold is set to 30 min, as suggested by domain experts.

Figure 9.6 illustrates with the use of a simple example the computation of the *anchoredOrMoored* intervals. The displayed intervals I, I_{sffp}, etc., correspond to the intervals of rule (9.4). In the example of Fig. 9.6, the second interval of I_i, $[T_3, T_4]$, is discarded since it is not long enough according to the V_{aorm} threshold.

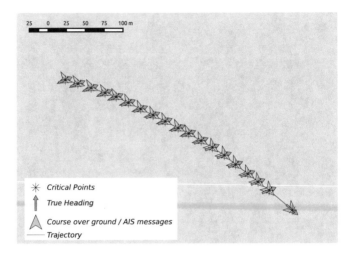

Fig. 9.7 A drifting vessel. In this example, all AIS position signals have been labelled as 'critical' (*change_in_heading*(

9.3.2.3 Drifting Vessel

A vessel is drifting when its course over ground, that is, the direction calculated by the GPS signal, is heavily affected by sea currents or harsh weather conditions, or when the vessel is not under control (say due to engine failure). Typically, as illustrated in Fig. 9.7, when the course over ground deviates from the true heading of a sailing vessel, that is, the direction of the ship's bow, then the vessel is considered drifting. Consider the formalisation below:

$$
\begin{aligned}
&\text{initiatedAt}(drifting(Vessel) = \text{true}, T) \leftarrow \\
&\quad \text{happensAt}(velocity(Vessel, \_Speed, CoG, TrueHeading), T), \\
&\quad angleDiff(CoG, TrueHeading, Ad), \\
&\quad threshold(v_{ad}, V_{ad}), \quad Ad > V_{ad}, \\
&\quad \text{holdsAt}(underWay(Vessel) = \text{true}, T). \\
&\text{terminatedAt}(drifting(Vessel) = \text{true}, T) \leftarrow \\
&\quad \text{happensAt}(velocity(Vessel, \_Speed, CoG, TrueHeading), T), \\
&\quad angleDiff(CoG, TrueHeading, Ad), \\
&\quad threshold(v_{ad}, V_{ad}), \quad Ad \leq V_{ad}. \\
&\text{terminatedAt}(drifting(Vessel) = \text{true}, T) \leftarrow \\
&\quad \text{happensAt}(end((underWay(Vessel) = \text{true}), T).
\end{aligned}
$$

(9.5)

drifting is a Boolean simple fluent, while, as mentioned earlier, *velocity* is input contextual information, attached to each AIS message, expressing the speed, course over ground (CoG) and true heading of a vessel. *angleDiff(A, B, C)* is an auxiliary predicate calculating the absolute minimum difference C between two angles A and B. The use of *underWay* in the initiation and termination conditions of *drifting* (see rule-set (9.5)) expresses the constraint that only moving vessels can be considered to be drifting.

9.3.2.4 Tugging

A vessel that should not move by itself, such as a ship in a crowded harbour or a narrow canal, or a vessel that cannot move by itself is typically pulled or towed by a tug boat. Figure 9.8 shows an example. During tugging, the two vessels are typically close and their speed is lower than normal, for safety and manoeuvrability reasons. We have formalised tugging as follows:

$$
\begin{aligned}
&\textsf{holdsFor}(tugging(Vessel_1, Vessel_2) = \textsf{true}, I) \leftarrow \\
&\quad oneIsTug(Vessel_1, Vessel_2), \\
&\quad \textsf{holdsFor}(proximity(Vessel_1, Vessel_2) = \textsf{true}, I_p), \\
&\quad \textsf{holdsFor}(tuggingSpeed(Vessel_1) = \textsf{true}, I_{ts1}), \\
&\quad \textsf{holdsFor}(tuggingSpeed(Vessel_2) = \textsf{true}, I_{ts2}), \\
&\quad \textsf{intersect\_all}([I_p, I_{ts1}, I_{ts2}], I_i), \\
&\quad threshold(v_{tug}, V_{Tug}), \quad intDurGreater(I_i, V_{Tug}, I).
\end{aligned}
\tag{9.6}
$$

tugging is a relational fluent referring to a pair of vessels, as opposed to the fluents presented so far that concern a single vessel. *oneIsTug(V_1, V_2)* is an auxiliary predicate stating whether one of the vessels V_1, V_2 is a tug boat. *proximity* is a durative input fluent computed at the spatial preprocessing step (see Table 9.1), expressing the time periods during which two vessels are 'close' (that is, their distance is less than a 100 m). *tuggingSpeed* is a simple fluent expressing the intervals during which a vessel is said to be sailing at tugging speed. According to rule (9.6), two vessels are said to be engaged in tugging if one of them is a tug boat, and, for at least V_{Tug} time-points, they are close to each other and sail at tugging speed.

9.3.2.5 Vessel rendez-vous

A scenario that may indicate illegal activities, such as illegal cargo transfer, is when two vessels are nearby in the open sea, stopped or sailing at a low speed. See Fig. 9.9 for an illustration. A specification of a potential 'rendez-vous', or

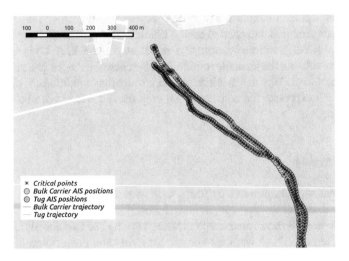

Fig. 9.8 Example of bulk carrier tugging. In this example, all position signals are labelled as 'critical'

'ship-to-ship transfer', may be found below:

$$
\begin{aligned}
&\text{holdsFor}(rendez\text{-}Vous(Vessel_1, Vessel_2) = \text{true}, I) \leftarrow \\
&\quad \text{not}oneIsTug(Vessel_1, Vessel_2), \\
&\quad \text{holdsFor}(proximity(Vessel_1, Vessel_2) = \text{true}, I_p), \\
&\quad \text{holdsFor}(lowSpeed(Vessel_1) = true, I_{l1}), \\
&\quad \text{holdsFor}(stopped(Vessel_1) = farFromPorts, I_{s1}), \\
&\quad \text{union\_all}([I_{l1}, I_{s1}], I_1), \\
&\quad \text{holdsFor}(lowSpeed(Vessel_2) = true, I_{l2}), \\
&\quad \text{holdsFor}(stopped(Vessel_2) = farFromPorts, I_{s2}), \\
&\quad \text{union\_all}([I_{l2}, I_{s2}], I_2), \\
&\quad \text{intersect\_all}([I_1, I_2, I_p], I_f), \\
&\quad \text{holdsFor}(withinArea(Vessel_1, nearPorts) = true, I_{np1}), \\
&\quad \text{holdsFor}(withinArea(Vessel_2, nearPorts) = true, I_{np2}), \\
&\quad \text{holdsFor}(withinArea(Vessel_2, nearCoast) = true, I_{nc1}), \\
&\quad \text{holdsFor}(withinArea(Vessel_2, nearCoast) = true, I_{nc2}), \\
&\quad \text{relative\_complement\_all}(I_f, [I_{np1}, I_{np2}, I_{nc1}, I_{nc2}], I_i), \\
&\quad threshold(v_{rv}, V_{rv}), \quad intDurGreater(I_i, V_{rv}, I).
\end{aligned}
\tag{9.7}
$$

rendez-Vous is a relational fluent, while *lowSpeed* is a fluent recording the intervals during which a vessel sails with a speed between 0.5 and 5 knots. relative_complement_all(is an interval manipulation construct of RTEC (see Table 9.2 and Fig. 9.2). According to rule (9.7), $rendez - Vous(V_1, V_2)$ holds when neither of the two vessels V_1, V_2 is a tug boat, V_1, V_2 are close to each other, and they are

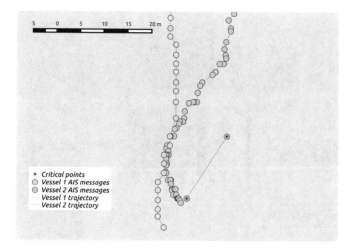

Fig. 9.9 Fishing vessels in close proximity. In this example, the vessels started sailing at a low speed before they came close to each other. Hence, these critical events (*slow_motion_start*) are not displayed in the figure

stopped or sail at low speed far from the coast and ports. Depending on the chosen distance thresholds for *nearCoast* and *nearPorts*, a vessel may be 'far' from the coastline and at the same time 'near' some port. Moreover, a vessel may be 'far' from all ports and 'near' the coastline.

We require that the two vessels are not near the coastline since illegal ship-to-ship transfer typically takes place far from the coast. We also require that both vessels are far from ports, as two slow moving or stopped vessels near some port would probably be moored or departing from the port.

9.4   Anomaly Detection Service

As mentioned earlier, we have been re-implementing RTEC in the Scala programming language, in order to integrate it with the industry-strong maritime anomaly detection service of MarineTraffic [8]. MarineTraffic is currently the world's leading platform offering vessel tracking services and actionable maritime intelligence. It offers an end-to-end service that tracks vessel positions across the globe based on AIS and disseminates this information to the general public though its interactive website, www.MarineTraffic.com. With an open community network of more than 3200 coastal AIS stations, MarineTraffic is capable of tracking vessels on their journeys across the coastlines of more than 140 countries. While the MarineTraffic terrestrial-based AIS network provides excellent coverage of several thousands of ports, the limited range of AIS results in restricted ocean coverage. To address this,

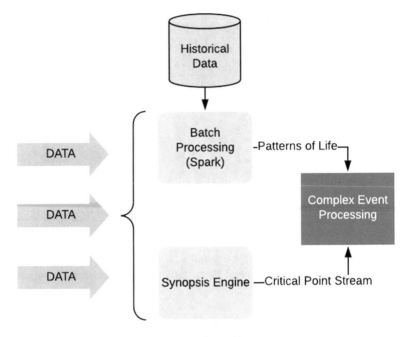

Fig. 9.10 MarineTraffic anomaly detection service architecture

Terrestrial AIS is combined with Satellite AIS in order to support almost global vessel monitoring.

On top of the vessel tracking services, MarineTraffic has deployed an anomaly detection service, which is based on a modified Lambda architecture [11]. This scheme allows the decoupling of batch processing, performed upon historical data, and online streaming analysis, which typically exploits the knowledge extracted from the batch processing (see Fig. 9.10).

The speed layer involves the trajectory synopsis generation engine as described in the previous section.

The batch layer performs the analysis of historical positional data of vessels and extracts the so-called Patterns of Life, that is, 'normal' maritime activity [2]. Batch processing is a long-running process which takes several hours to complete. Once completed, the Patterns of Life are fed into the online layer in order to accommodate detection of vessel anomalies in real time. Upon detection, those incidents are displayed to the end user. Furthermore, historical data are sent from this layer back to the batch layer at specific time intervals, defined from the seasonality of the data, thus replacing previously constructed patterns with new ones.

The online component is based on the actor model, specifically the Akka framework, for concurrency and event-driven processing. Actors are versatile, light-weight objects that have a state, communicate with each other and process the messages sent to them sequentially. By designing different types of actors and flows of information between them, we can create quite complex topologies, such

Fig. 9.11 Area's incidents page: the incidents/events occurring in an area are displayed on the map. Red dots, for example, display potential 'rendez-vous' incidents

as that displayed in Fig. 9.3, where each actor is responsible for recognising a maritime event and may be implemented in (the Scala version of) RTEC. Our current implementation includes 'ship actors' and 'cell actors'. The former type of actor receives the stream of AIS messages of a particular ship and detects propositional events concerning that ship, such as 'high speed near coast' (see Sect. 9.3.2.1) and 'drifting' (see Sect. 9.3.2.3). The latter type of actor detects relational events, where data from different ships within a grid cell need to be compared, such as tugging (Sect. 9.3.2.4) and potential 'rendez-vous' (Sect. 9.3.2.5). In the occasion of a recognition of a maritime event, users registered in the anomaly detection service are automatically alerted via email, or through the graphical user interface (see Figs. 9.11 and 9.12). The main goal of the design of the user interface was to reduce fatigue and the cognitive overload of the operators when having to search through numerous surveillance datasets and alerts. Towards this, the entire structure and format of the produced maritime situational awareness picture follows an interactive goal-driven approach.

An Example of a Real Case *SBI Jaguar* is a vessel sailing under the flag of Marshall Islands. On the 28th of March 2018, as the vessel was heading out of the Western Scheldt, a cylinder of the steering gear sheared off and the vessel was not under command. Then, *SBI Jaguar* ran aground on a sandbar in line with Perkpolder, and seven tugs were engaged in the vessel's re-floating operation. Subsequently, the vessel was towed to the Everingen's anchorage for an underwater inspection.

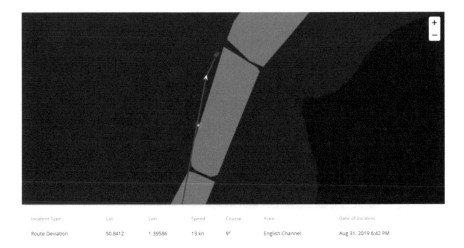

Incident Type	Lat	Lon	Speed	Course	Area	Date of incident
Route Deviation	50.8412	1.39586	13 kn	9°	English Channel	Aug 31, 2019 6:42 PM

Fig. 9.12 Incident's details page. This screenshot shows a 'route deviation' incident/event (see also Fig. 9.3). Such an event is recognised when a ship is found travelling outside its expected route or speed patterns in a given area and time. The past track (in blue colour) and the normal route (in blue-green colour) are displayed on the map together with the vessel's speed, course over ground and the time the incident occurred

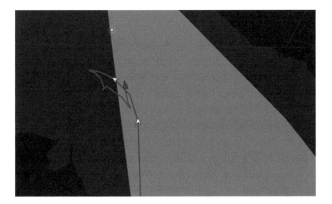

Fig. 9.13 The *SBI Jaguar* grounding

Figure 9.13 showcases that the vessel was sailing within its 'safe path', displayed by the polygon with turquoise colour, when the cylinder sheared off. At that point, the vessel started drifting having multiple changes of course over ground, which was detected by our service.

9.5   Summary & Further Work

We presented our on-going efforts towards integrating two state-of-the-art stream reasoning technologies for maritime situational awareness. A formal, computational framework has been re-implemented in the Scala programming language to allow for the interaction with the distributed implementation of MarineTraffic, thus paving the way for the real-time recognition of a wide variety of maritime events.

There are several directions for current and further work. First, we are currently evaluating the performance of the integrated system on real-world and synthetic datasets. (Results on the individual technologies have been already documented and indicate their capability for real-time performance [8, 16].) Second, we are developing online machine learning techniques for continuously refining maritime patterns given new streaming data [10, 12]. Finally, we are integrating satellite images with position signals and geographical information for a more complete account of maritime situational awareness.

Acknowledgments This work was supported by the datACRON and the INFORE projects, which have received funding from the European Union's Horizon 2020 research and innovation programme, under grant agreements No 687591 and No 825070.

References

1. Artikis, A., Sergot, M.J., Paliouras, G.: An event calculus for event recognition. IEEE Trans. Knowl. Data Eng. **27**(4), 895–908 (2015)
2. Chatzikokolakis, K., Zissis, D., Vodas, M., Spiliopoulos, G., Kontopoulos, I.: A distributed lightning fast maritime anomaly detection service. In: OCEANS 2019 - Marseille, June 2019, pp. 1–8 (2019)
3. Clark, K.L.: Negation as failure. In: Gallaire, H., Minker, J. (eds.) Logic and Databases, pp. 293–322. Plenum Press, New York (1978)
4. FAO. VMS for fishery vessels. http://www.fao.org/fishery/topic/18103/en. Accessed 15 May 2019
5. Greidanus, H., Alvarez, M., Santamaria, C., Thoorens, F.-X., Kourti, N., Argentieri, P.: The sumo ship detector algorithm for satellite radar images. Remote Sens. **9**(3), 246 (2017)
6. IMO. Technical characteristics for an automatic identification system using time division multiple access in the vhf maritime mobile frequency band. Tech. rep., ITU (2017)
7. IMO. Long-range identification and tracking system. Tech. rep., IMO (2018)
8. Improving maritime situational awareness through big data analytics, machine learning and artificial intelligence. Anomaly detection white paper, MarineTraffic Research (2019). https://www.marinetraffic.com/research/publication/anomaly-detection-white-paper/
9. Jousselme, A.-L., Ray, C., Camossi, E., Hadzagic, M., Claramunt, C., Bryan, K., Reardon, E., Ilteris, M.: Maritime use case description, h2020 datACRON project deliverable d5.1. http://datacron-project.eu/ (2016)
10. Katzouris, N., Artikis, A., Paliouras, G.: Online learning of event definitions. Theory Pract. Log. Program. **16**(5–6), 817–833 (2016)
11. Marz, N., Warren, J.: Big Data: Principles and Best Practices of Scalable Realtime Data Systems, 1st edn. Manning Publications Co., Shelter Island (2015)

12. Michelioudakis, E., Artikis, A., Paliouras, G.: Semi-supervised online structure learning for composite event recognition. Mach. Learn. **108**(7) , 1085–1110 (2019)
13. Millefiori, L.M., Zissis, D., Cazzanti, L., Arcieri, G.: A distributed approach to estimating sea port operational regions from lots of AIS data. In: 2016 IEEE International Conference on Big Data, BigData 2016, Washington DC, 5–8 December 2016, pp. 1627–1632 (2016)
14. Mills, C.M., Townsend, S.E., Jennings, S., Eastwood, P.D., Houghton, C.A.: Estimating high resolution trawl fishing effort from satellite-based vessel monitoring system data. ICES J. Mar. Sci. **64**(2), 248–255 (2007)
15. Patroumpas, K., Alevizos, E., Artikis, A., Vodas, M., Pelekis, N., Theodoridis, Y.: Online event recognition from moving vessel trajectories. GeoInformatica **21**(2), 389–427 (2017)
16. Pitsikalis, M., Artikis, A., Dreo, R., Ray, C., Camossi, E., Jousselme, A.: Composite event recognition for maritime monitoring. In: Proceedings of the 13th ACM International Conference on Distributed and Event-based Systems, DEBS 2019, Darmstadt, 24–28 June 2019, pp. 163–174 (2019)
17. Santipantakis, G.M., Vlachou, A., Doulkeridis, C., Artikis, A., Kontopoulos, I., Vouros, G.A.: A stream reasoning system for maritime monitoring. In: 25th International Symposium on Temporal Representation and Reasoning, TIME 2018, Warsaw, 15–17 October 2018, pp. 20:1–20:17 (2018)
18. Terroso-Saenz, F., Valdes-Vela, M., Skarmeta-Gomez, A.F.: A complex event processing approach to detect abnormal behaviours in the marine environment. Inf. Syst. Front. **18**(4), 765–780 (2016)
19. van Laere, J., Nilsson, M.: Evaluation of a workshop to capture knowledge from subject matter experts in maritime surveillance. In: Proceedings of FUSION, pp. 171–178 (2009)
20. Vouros, G.A., Vlachou, A., Santipantakis, G.M., Doulkeridis, C., Pelekis, N., Georgiou, H.V., Theodoridis, Y., Patroumpas, K., Alevizos, E., Artikis, A., Claramunt, C., Ray, C., Scarlatti, D., Fuchs, G., Andrienko, G.L., Andrienko, N.V., Mock, M., Camossi, E., Jousselme, A., Garcia, J.M.C.: Big data analytics for time critical mobility forecasting: recent progress and research challenges. In: Proceedings of the 21th International Conference on Extending Database Technology, EDBT 2018, Vienna, 26–29 March 2018, pp. 612–623 (2018)
21. Vouros, G.A., Vlachou, A., Santipantakis, G.M., Doulkeridis, C., Pelekis, N., Georgiou, H.V., Theodoridis, Y., Patroumpas, K., Alevizos, E., Artikis, A., Fuchs, G., Mock, M., Andrienko, G.L., Andrienko, N.V., Claramunt, C., Ray, C., Camossi, E., Jousselme, A.: Increasing maritime situation awareness via trajectory detection, enrichment and recognition of events. In: Web and Wireless Geographical Information Systems - Proceedings of 16th International Symposium, W2GIS 2018, A Coruña, 21–22 May 2018, pp. 130–140 (2018)

Chapter 10
Offline Trajectory Analytics

Panagiotis Tampakis, Stylianos Sideridis, Panagiotis Nikitopoulos,
Nikos Pelekis, Christos Doulkeridis, and Yannis Theodoridis

Abstract In recent years, there has been observed an "explosion" of trajectory data production due to the proliferation of GPS-enabled devices, such as mobile phones and tablets. This massive-scale data generation has posed new challenges in the data management community in terms of storing, querying, analyzing, and extracting knowledge out of such data. Knowledge discovery out of mobility data is essentially the goal of every mobility data analytics task. Especially in the maritime and aviation domains, this relates to challenging use-case scenarios, such as discovering valuable behavioral patterns of moving objects, identifying different types of activities in a region of interest, environmental fingerprint, etc. In order to be able to support such scenarios, an analyst should be able to apply, at massive scale, several knowledge discovery techniques, such as trajectory clustering, hotspot analysis, and frequent route/network discovery methods.

10.1 Introduction

Location aware devices such as mobile phones, tablets, and automobiles carry numerous networked sensors, which create huge amounts of data that represent some kind of mobility. In addition, the massive participation of individuals on location-based social networks will continue to fuel exponential growth in the production of this kind of data. This enormous volume of data has posed new challenges in the world of mobility data management in terms of storing, querying, analyzing, and extracting knowledge out of them in an efficient way.

P. Tampakis · S. Sideridis · P. Nikitopoulos · N. Pelekis (✉) · C. Doulkeridis · Y. Theodoridis
University of Piraeus, Piraeus, Greece
e-mail: ptampak@unipi.gr; ssider@unipi.gr; nikp@unipi.gr; npelekis@unipi.gr; cdoulk@unipi.gr; ytheod@unipi.gr

© Springer Nature Switzerland AG 2020
G. A. Vouros et al. (eds.), *Big Data Analytics for Time-Critical Mobility Forecasting*, https://doi.org/10.1007/978-3-030-45164-6_10

10.1.1   About Mobility Data Analytics

Concerning the analysis of mobility data, mobility data analytics aim to describe the mobility of objects, to extract valuable knowledge by revealing motion behaviors or patterns, to predict future mobility behaviors or trends, and in general, to generate various perspectives out of data, useful for many other scientific fields. To serve its purpose, mobility data analytics follows a series of steps. Having assured the collection and efficient storage of mobility data, the next step for an analyst is to familiarize with the mobility data by employing a number of techniques (e.g., statistics, data visualization, and visual analytics) to form a compact and complete picture of the available mobility data. Afterwards, the analyst, depending on the application requirements, proceeds to the appropriate preprocessing steps. The goal is to bring mobility data in a form that serves its later usage by various processes and algorithms that respond to given questions. Data preparation is essential for successful mobility data analytics, since low-quality data typically result in incorrect and unreliable conclusions, as mentioned in chapters in Part I. Finally, mobility data are ready for the application of knowledge extraction methods that will satisfy the given application requirements. There are already several analytical methods and algorithms available from the scientific community and an analyst has the capability either to employ some of the existing techniques or implement some ad hoc solutions that better serve the problems' needs.

The overall objective is to develop advanced, beyond current state-of-the-art data analytics methods and tools over the repository of trajectories of moving objects. The challenge to be addressed here is that information is not purely spatiotemporal; it is contextually enhanced by exploiting integrated data. The big data solutions proposed in this chapter focus onto the problems of cluster analysis and motion pattern detection, hotspot analysis, and semantic-aware mobility network discovery.

In more detail, we designed and implemented a scalable distributed trajectory join method, which utilizes the popular MapReduce distributed programming model. This approach plays a key role as it is the building block upon which our clustering analytics methods are based, as it tackles the scalability bottleneck problem present in mobility data. In addition, we devised and implemented a novel, scalable distributed (sub)trajectory clustering method, which utilizes the aforementioned distributed trajectory join method in order to cluster massive-scale datasets of trajectories. The goal of this approach, that is the upshot of our clustering methods, is to provide a hybrid solution for the whole-trajectory as well as for the subtrajectory clustering problems in an efficient and scalable way. Furthermore, we designed and implemented a scalable distributed trajectory-based hotspot analysis method. In this line of research, we followed a different clustering approach that provides statistical guarantees for the identified clusters. More interestingly, with this approach we solve a different clustering problem that is also inherent in the maritime and aviation domains. Specifically, as a proof of concept, with this approach, we were able to discover hotspots that in the maritime domain can be used to measure the fishing pressure at sea, while in the aviation domain it identifies air-blocks that present demand-capacity problems. Finally, we designed and implemented a method that discovers mobility networks, which consist of synthetic, pattern-based, compact representations of data. This method employs a semantic-aware methodology, applicable for big contextually enhanced trajectory data (actually the synopses generated the respective component), which results in a network representation of mobility data (actually, spatial graphs enhanced with thematic annotations) that can be utilized to support prediction/forecasting problems.

The rest of the chapter is organized as follows: Section 10.2 familiarizes the reader with some background knowledge. In more detail, Sect. 10.2.1 presents sufficient background knowledge about (sub)trajectory clustering, Sect. 10.2.2 about hotspot analysis, and Sect. 10.2.3 about enriched mobility networks. Subsequently, in Sect. 10.3 we present two big data solutions to the problem of (sub)trajectory clustering and more specifically Sect. 10.3.1.1 employs off-the-shelf clustering algorithms provided by Spark MLib in order to identify clusters of entire trajectories, while Sect. 10.3.2 presents a highly scalable solution to the problem of *Distributed Subtrajectory Clustering*. Moreover, Sect. 10.4 introduces the reader with a big data solution to the problem *Distributed Hotspot Analysis* and Sect. 10.5 with the discovery of *Semantic-aware Mobility Networks* in a distributed way. Section 10.6 presents the related state-of-the-art approaches to the problems identified in this chapter, and finally, Sect. 10.7 concludes the chapter.

10.2   Background

10.2.1   (Sub)trajectory Clustering

Given a set D of moving object trajectories, a trajectory $r \in D$ is a sequence of timestamped locations $\{r_1, \ldots, r_N\}$. Each $r_i = (loc_i, t_i)$ represents the i-th sampled point, $i \in 1, \ldots, N$ of trajectory r, where N denotes the length of r (i.e., the number of points it consists of). Moreover, loc_i denotes the spatial location (2D or 3D) and t_i the time coordinate of point r_i, respectively.

A subtrajectory $r_{i,j}$ is a subsequence $\{r_i, \ldots, r_j\}$ of r which represents the movement of the object between t_i and t_j, where $i < j$ and $i, j \in 1, \ldots, N$. Let $d_s(r_i, s_j)$ denote the spatial distance between two points $r_i \in r$, $s_j \in s$. In our case we adopted the Euclidean distance; however, other distance functions might be applied. Also, let $d_t(r_i, s_j)$ denote the temporal distance, defined as $|r_i.t - s_j.t|$. Furthermore, let Δt_r symbolize the duration of trajectory r (similarly for subtrajectories).

10.2.2 Hotspot Analysis

In geospatial analysis, a hotspot is a geographic area that contains unusually high concentration of activities (e.g., moving objects). The difference between a hotspot and a cluster is that the former aims to discover areas that are statistically significant, whereas the latter focuses on grouping similar objects.

Statistical significance determines whether the relationship between two or more variables is caused by chance. For example, the number of vehicles moving in a specific geospatial area is statistically significant, if it can be proved that it is not the result of chance. In statistical hypothesis testing, the statistical significance can be determined by testing the null hypothesis. To this end, a p-value (probability value) needs to be calculated which represents the probability that the relationship between two or more variables rejects the null hypothesis. In geospatial analysis a p-value can be determined by calculating a z-score value for a given geospatial area. Such z-scores can be calculated by using several geospatial statistics defined in literature, namely the Getis–Ord statistic [23] or the Moran's I [18].

Motivated by the need for big data analytics over trajectories of moving objects, we focus on discovering *trajectory hotspots* in the maritime domain, as this relates to various challenging use-case scenarios [5], as, for instance, detecting fishing pressure, as discussed in Part I of this book. Trajectory hotspot analysis is related to geospatial hotspot analysis, since both discover hotspots on geographical areas. However, the former analysis has two main differences: (a) it considers an additional variable, namely the temporal dimension in the z-score calculation, and (b) it discovers hotspots based on trajectories of objects rather than individual traced object locations. In the following we formally define the problem of trajectory hotspot analysis.

10.2.3 Data-Enriched Mobility Networks

In both the maritime and the aviation domain, we notice a plethora of moving objects. At the same time with the advances in tracking technology there is an abundance of information, not always valuable, concerning the movement of such objects. This has enabled a wide spectrum of novel applications and services. Among them is the process of using the traces of moving entities to produce maps of transportation networks.

This quantity of information leads to the need of discovering new efficient and effective ways to infer the underlying transportation network driven by the data of moving objects itself. Although both ships and aircraft should follow predefined movement plans, there exist many cases where, for various reasons (weather, protected areas, congestion, etc.), objects do not follow these routes plans. Such deviations are crucial to be instantly identified so that preventative measures can be taken to avoid the occurrence of safety-compromising events, such as natural disasters or accidents.

10.3   Distributed Trajectory Clustering

As mentioned in Sect. 10.1, one of the challenges when trying to extract knowledge
out of mobility data is cluster analysis, which aims at identifying clusters of moving
objects, as well as detecting moving objects that demonstrate abnormal behavior and
can be considered as outliers. By discovering these clusters, the underlying hidden
patterns of collective behavior can be revealed.

10.3.1   Distributed Whole-Trajectory Clustering

The research so far has focused mainly on dealing with the trajectory clustering
problem in a centralized way. However, the problem that we are trying to deal with in
this section is that of whole-trajectory clustering in a distributed way. The intuition
here is to use the off-the-shelf clustering algorithms provided by Spark MLlib, such
as k-means, Gaussian mixture model, power iteration clustering, and bisecting k-
means, in order to identify clusters of entire trajectories. To achieve this, we need to
transform each trajectory into a vector that will be given as input to the respective
clustering algorithm of Spark MLlib.

10.3.1.1   Clustering Trajectories in a Distributed Way with Spark MLlib

The solution to the distributed whole-trajectory clustering problem proposed here
is pretty simple, since it utilizes off-the-shelf algorithms of Spark MLlib. In order
to achieve this, we just need to transform each trajectory into a N-dimensional
vector, where N is the number of samples that constitute a trajectory. Subsequently,
these vectors will be given as input to the respective clustering algorithm of Spark
MLlib. The actual challenge here is how to transform a dataset of trajectories into
vectors, which will be the input for a series of clustering algorithms included in
Spark MLlib. In order to vectorize each trajectory, some kind of resampling needs
to be performed, due to the fact that different trajectories might have different
number of samples. Selecting the resampling method is a crucial decision because
it determines what kind of clustering (spatial-only or spatiotemporal) we will
end up in performing. This applies because, implicitly, the position j inside a
vectorized trajectory determines the time of the observation of v_j. In this analysis,
two alternative resampling methods are adopted: the interpolation method and the
normalized interpolation method.

    More specifically, a trajectory vector v is a vector (2D or 3D) representing a
trajectory. The size of the vector for each trajectory $T \in D$ has a fixed length s
which is defined a priori. Each element v_j of v, where $j \in 1, \ldots, s$, represents the
spatial location (2D or 3D) of the respective trajectory. The interpolation method,
as depicted in Algorithm 10.1, takes as input a set of trajectories D and the vector
length s and outputs a set D' of vectorized trajectories. In more detail, for each
trajectory T (line 3) it creates a vector of size s (line 4) by selecting s timestamps
st_j, with $j \in 1, \ldots, s$, of duration $(TN.t - T1.t)/s$, where $(TN.t - T1.t)$ is the

duration of trajectory T, starting from $st_1 = T1.t$ and ending at $st_s = TN.t$. For each st_j, the algorithm finds the corresponding position of T by performing cubic interpolation and stores this position in v_j (lines 5–6). Finally, v is added to D' (line 6) and when all trajectories of D get vectorized, D' is returned (line 9).

Algorithm 10.1 Vectorization by interpolation

1: **Input:** D, s
2: **Output:** D'
3: **for each** $T \in D$ **do**
4:     **create** v;
5:     **for** $j = 1 \ldots s$ **do**
6:         $st_j = T1.t + (j - 1)(TN.t - T1.t)/s$;
7:         $v_j \leftarrow cubic\_interpolation(T, st_j)$;
8:     $v \rightarrow D'$;
9: **return** D';

Now, let us consider two trajectories q and r and their vectorized versions $v(q)$ and $v(r)$, respectively, with $q_N.t - q_1.t \neq r_N.t - r_1.t$. This means that for any given $j \in 1, \ldots, s$, $v(q)_j$ will depict the position of trajectory q at a different timestamp than the position depicted for trajectory r by $v(r)_j$. However, in order for the algorithms that will be employed to perform spatiotemporal clustering, for any given pair of trajectories q and r and for any given $j \in 1, \ldots, s$, it must hold that $v(q)_j$ and $v(r)_j$ represent the position of trajectory q and r at the same timestamp. For this reason, the vectorization by interpolation is used to perform spatial-only clustering. However, if the goal is to perform spatiotemporal trajectory clustering the interpolation vectorization method has the aforementioned shortcoming and in order to overcome this, another vectorization method needs to be employed. For this reason, we propose the normalized interpolation vectorization method which takes as input a set of trajectories D, the vector length s, and the time of the temporally first and last observed sample $D.t_i$ and $D.t_e$, respectively, of D and output a set D' of vectorized trajectories.

In more detail, as depicted in Algorithm 10.2, we first create the universal resampling vector rsv (line 3) by selecting s timestamps st_j, with $j \in 1, \ldots, s$, of duration $(D.t_e - D.t_i)/s$, where $(D.t_e - D.t_i)$ is the duration of the dataset, starting from $st_1 = D.t_i$ and ending at $st_s = D.t_e$ (lines 4–6). The utility of rsv is to help resample each $T \in D$ in such a way so that for any given pair of trajectories q and r and for any given $j \in 1, \ldots, s$, it holds that $v(q)_j$ and $v(r)_j$ represent the position of trajectory q and r at the same timestamp. Subsequently, for each trajectory T we create a vector v of size s (lines 7–8). Then, for each sample j, we examine whether the corresponding time in rsv_j is contained by the lifespan of T (lines 9–10). If it is contained, then we find the corresponding position of T by performing cubic interpolation and we store this position in v_j (line 11). If rsv_j is not contained by the lifespan of T and rsv_j is less or equal to the first timestamp of the trajectory $T_1.t$, then the first recorded position of T is stored in v_j (lines 13). Otherwise, if rsv_j is not contained by the lifespan of T and rsv_j is greater or equal to the last

---

Algorithm 10.2 Vectorization by normalized interpolation

1: **Input:** $D, s, D.t_i$ and $D.t_e$
2: **Output:** D'
3: **create** rsv;
4: **for** $j = 1 \ldots s$ **do**
5:      $st_j = D.t_i + (j-1)(D.t_e - D.t_i)/s$;
6:      $rsv_j \leftarrow st_j$;
7: **for each** $T \in D$ **do**
8:      **create** v;
9:      **for** $j = 1 \ldots s$ **do**
10:        **if** $rsv_j \in (T_1.t, T_N.t)$ **then**
11:          $v_j \leftarrow cubic\_interpolation(T, rsv_j)$;
12:        **else if** $rsv_j \leq T_1.t$ **then**
13:          $v_j \leftarrow T_1.p$
14:        **else**
15:          $v_j \leftarrow T_N.p$
16:      $v \rightarrow D'$;
17: **return** D';

---

timestamp of the trajectory $T_N.t$, then the last recorded position of T is stored in v_j (line 15). Finally, v is added to D' (line 16) and when all trajectories of D get vectorized, D' is returned (line 17).

10.3.2 Distributed Subtrajectory Clustering

However, identifying clusters of complete trajectories can result in disregarding significant patterns that might exist only for some portions of their lifespan. The following motivating example shows the merits of subtrajectory clustering.

Example 10.1 (Subtrajectory Clustering) Figure 10.1a illustrates six trajectories moving in the xy-plane, where each one of them has a different origin–destination pair. More specifically, these pairs are $A \rightarrow B, A \rightarrow C, A \rightarrow D, B \rightarrow A, B \rightarrow C$, and $B \rightarrow D$. These six trajectories have the same starting time and similar speed. A typical trajectory clustering technique would fail to identify any clusters. However, the goal of a subtrajectory clustering method is to identify 4 clusters ($A \rightarrow O$ (red), $B \rightarrow O$ (blue), $O \rightarrow C$ (purple), $O \rightarrow D$ (orange)) and 2 outliers ($O \rightarrow A$ and $O \rightarrow B$ (black)), as depicted in Fig. 10.1b.

The problem of subtrajectory clustering is shown to be NP-Hard (cf. [1]). In addition, the objects to be clustered are not known beforehand (as in entire-trajectory—from now on—clustering algorithms), but have to be identified through a trajectory segmentation procedure. Efforts that try to deal with this problem in a centralized way do exist [1, 15, 28]; however, applying these centralized algorithms over massive data in a scalable way is far from straightforward. This calls for parallel and distributed algorithms that address the scalability requirements.

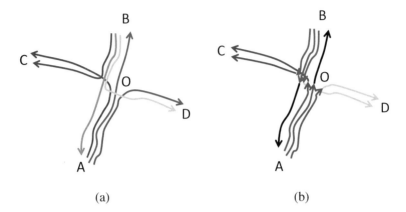

Fig. 10.1 (**a**) Six trajectories moving in the xy-plane and (**b**) 4 clusters (red, blue, orange, and purple) and 2 outliers (black)

Algorithm 10.3 $DSC(D)$

1: **Input:** D
2: **Output:** set C of clusters, set O of outliers
3: **Preprocessing:** *Repartition D*;
4: **for each** partition $D_i \in \cup_{i=1}^{P} D_i$ **do**
5:     **perform** *Point-level Join*;
6: **group by** *Trajectory*;
7: **for each** Trajectory $r \in D$ **do**
8:     **perform** *Subtrajectory Join*
9:     **perform** *Trajectory Segmentation*;
10: **group by** D_i;
11: **for each** subtrajectory $r' \in D_i$ **do**
12:     **calculate** $Sim(r', s') \, \forall s' \in D_i$;
13: **perform** *Clustering*;
14: **perform** *Refine Results*;
15: **return** C and O;

10.3.2.1 Definitions

Subtrajectory clustering relies on the use of a similarity function between subtrajectories. Although various similarity measures have been defined in the literature, our choice of similarity function is motivated by the following (desired) requirements: **variable sampling rate and lack of alignment**, **variable trajectory length**, **temporal displacement**, **symmetry**, and **efficiency**.

In order to meet with the aforementioned specifications we utilize the longest common subsequence (LCSS) for trajectories, as defined in [36].

$$Sim(r, s) = \frac{LCSS_{\epsilon_t, \epsilon_{sp}}(r, s)}{min(|r|, |s|)} \tag{10.1}$$

where $|r|$ $(|s|)$ is the length of r (s, respectively). Moreover, it holds that $Sim(r, s)$ $= Sim(s, r)$.

However, LCSS considers as equally similar all the points that exist within an ϵ_{sp} spatial range from r, which is a fact that might compromise the quality of the clustering results. Ideally, given two matching points $r_i \in r$ and $s_j \in s$, s_j (r_i, respectively) should contribute to $LCSS_{\epsilon_t,\epsilon_{sp}}(r, s)$, proportionally to the distance $d_s(r_i, s_j)$. For this reason, we propose a "weighted" LCSS similarity between trajectories, that incorporates the aforementioned distance proportionality. In more detail:

$$Sim(r, s) = \frac{\sum_{k=1}^{min(|r|,|s|)} \left(1 - \frac{d_s(r_k,s_k)}{\epsilon_{sp}}\right)}{min(|r|, |s|)} \quad (10.2)$$

where (r_k, s_k) is a pair of matched points.

Our approach to subtrajectory clustering splits the problem into three steps. The first step is to retrieve for each trajectory $r \in D$ all the moving objects, with their respective portion of movement, that moved close enough in space and time with r, for at least some time duration. This is a well-defined problem in the literature of mobility data management, known as *subtrajectory join* (the case of self-join). The subtrajectory join will return for each pair of (sub)trajectories, all the common subsequences that have at least some time duration, which are actually candidates for the longest common subsequence. Formally:

Problem 10.1 (Subtrajectory Join) Given a temporal tolerance ϵ_t, a spatial threshold ϵ_{sp}, and a time duration δt, retrieve all pairs of subtrajectories (r', s') $\in D$ such that: (a) for each pair $\Delta t_{r'}, \Delta t_{s'} \geq \delta t$, (b) $\forall r_i \in r'$ there exists at least one $s_j \in s'$ so that $d_s(r_i, s_j) \leq \epsilon_{sp}$ and $d_t(r_i, s_j) \leq \epsilon_t$, and (c) $\forall s_j \in s'$ there exist at least one $r_i \in r'$ so that $d_s(s_j, r_i) \leq \epsilon_{sp}$ and $d_t(s_j, r_i) \leq \epsilon_t$.

The second step takes as input the result of the first step and aims at segmenting each $r \in D$ into a set of subtrajectories. In our case, the way that a trajectory is segmented into subtrajectories is neighborhood-aware, meaning that a trajectory will be segmented every time its neighborhood changes significantly.

Problem 10.2 (Trajectory Segmentation) Given a trajectory r, discover the set of timestamps CP (cutting points), where the neighborhood (the density or the composition) of r changes significantly. Then according to CP, r is partitioned to a set of subtrajectories $\{r'_1, \ldots, r'_M\}$, where $M = |CP| + 1$ is the number of subtrajectories for a given trajectory r, such that $r = \cup_{k=1}^{M} r'_k$ and $k \in [1, M]$.

Given the output of Problem 10.1, applying a trajectory segmentation algorithm for the trajectories D will result in a new set of subtrajectories D'. The third step takes as input D' and the goal is to create clusters (whose cardinality is unknown) of similar subtrajectories and at the same time identify subtrajectories that are significantly dissimilar from the others (outliers). More specifically, let $C = \{C_1, \ldots, C_K\}$ denote the clustering, where K is the number of clusters, and for every pair of clusters C_i and C_j, with $i, j \in [1, K]$, it holds that $C_i \cap C_j = \emptyset$. Now, let us assume that each cluster $C_i \in C$ is represented by one subtrajectory $R_i \in C_i$, called *Representative*. Furthermore, let R denote the set of all representatives. Actually, the problem of clustering is to discover clusters of objects such that the intra-cluster similarity is maximized and the inter-cluster similarity is minimized. Therefore, if we ensure that the similarity between the representatives is zero, then the problem of subtrajectory clustering can be formulated as an optimization problem as follows.

Problem 10.3 (Subtrajectory Clustering and Outlier Detection) Given a set of subtrajectories D', partition D' into a set of clusters C, and a set of outliers O, where $D' = C \cup O$, in such a way so that the sum of similarity between cluster members and cluster representatives (*SSCR*) is maximized:

$$SSCR = \Sigma_{\forall R_i \in R} \Sigma_{\forall r'_j \in C_i} Sim(R_i, r'_j) \tag{10.3}$$

However, trying to solve Problem 10.3 by maximizing Eq. (10.3) is not trivial, since the problem to segment trajectories to subtrajectories, select the set of representatives R and its cardinality $|R|$ that maximizes Eq. (10.3), has combinatorial complexity.

Here, we address the challenging problem of subtrajectory clustering in a distributed setting, where the dataset D is distributed across different nodes, and centralized processing is prohibitively expensive.

Problem 10.4 (Distributed Subtrajectory Clustering) Given a distributed set of trajectories, $D = \cup_{i=1}^{P} D_i$, where P is the number of partitions of D, perform the subtrajectory clustering task in a parallel manner.

Actually, Problem 10.4 can be broken down to solving Problems 10.1–10.3 (in that order) in a parallel/distributed way. In what follows, we adopt this approach and outline a solution that is based on MapReduce.

The overall subtrajectory clustering algorithms are presented in Algorithm 10.3. Each of the major steps in this algorithm is presented in subsequent paragraphs. Initially, we *Repartition* the data into P equi-sized, temporally sorted temporal partitions (files), which are going to be used as input to the distributed subtrajectory join algorithm (line 3). Note that this is actually a preprocessing step that only needs to take place once for each dataset D. However, it is essential as it enables load balancing by addressing the issue of temporal skewness in the input data. Subsequently, for each partition $D_i \in \cup_{i=1}^{P} D_i$ and for each trajectory (the reference trajectory) we discover parts of other trajectories that move close enough in space and time (line 5). Successively, we group by reference trajectory in order to perform the subtrajectory join (line 8). At this phase, since our data is already grouped by trajectory, we also perform trajectory segmentation in order to split each trajectory to subtrajectories (line 9). In turn, we utilize the temporal partitions created during the *Repartition* phase and re-group the data by temporal partition. For each $D_i \in \cup_{i=1}^{P} D_i$ we calculate the similarity between subtrajectories and perform the clustering procedure (line 12). At this point we should mention that if a subtrajectory intersects the borders of multiple partitions, then it is replicated in all of them. This will result in having duplicate and possibly contradicting results. For this reason, as a final step, we specify the *Refine Results* procedure (line 14). Finally, a set C of clusters and a set O of outliers are produced.

10.3.2.2 Distributed Subtrajectory Join

As already mentioned, the first step is to perform the subtrajectory join in a distributed way. For this reason, we exploit the work presented in [34], coined *DTJ* (Distributed subTrajectory Join), which introduces an efficient and highly scalable approach to deal with the subtrajectory join problem (Problem 10.1) by means of MapReduce. More specifically, *DTJ* is comprised of a *Repartitioning* phase and a *Query* phase. The *Repartitioning* phase is a preprocessing step that takes place only once and it is independent of the actual parameters of the problem, namely ϵ_{sp}, ϵ_t, and δt. The goal for this step is to produce load balanced temporal partitions. The idea is to construct an equi-depth histogram based on the temporal dimension, where each of the M bins contains the same number of points and the borders of each bin correspond to a temporal interval $[t_i, t_j)$. The histogram is constructed by taking a sample of the input data. Then, the input data is partitioned to processing tasks based on the temporal intervals of the histogram bins. This guarantees temporal locality in each partition, as well as equi-sized partitions, thus balancing the load fairly.

In the *Query* phase, the actual join processing takes place. It consists of two steps, the *Join* and the *Refine* step, which are implemented as a *Map* and a *Reduce* function, respectively. The output of this MapReduce job is, for each reference trajectory $r \in D$, all the moving objects, with their respective portion of movement, that moved close enough with r in space and time for at least some time duration. In Fig. 10.2, the *DTJ* query corresponds to Job 1 until the Refine() procedure.

For more technical details about the algorithms involved in *DTJ*, their complexity and an extensive experimental study, we refer to [34].

10.3.2.3 Distributed Trajectory Segmentation

The *Trajectory Segmentation* Algorithm (TSA) takes as input a single trajectory, along with information about its neighborhood, and partitions it to a set of subtrajectories. We propose two alternative segmentation algorithms. The first algorithm, coined TSA_1, identifies the beginning of a new subtrajectory whenever the density of its neighborhood changes significantly. For this purpose, we use the concept of *voting* as a measure of density of the surrounding area of a trajectory. For a given point r_i and any trajectory s, the voting $V(r_i)$ is defined as:

$$V(r_i) = \Sigma_{\forall s \in D} \frac{d_s(r_i, s_k)}{\epsilon_{sp}} \tag{10.4}$$

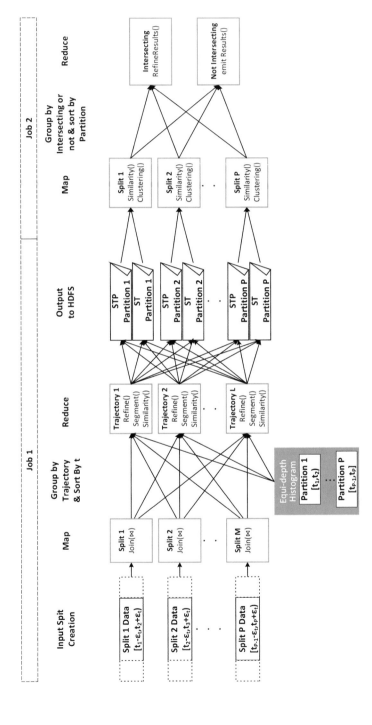

Fig. 10.2 The *DSC* algorithm. (Job 1) *DTJ* and *Trajectory Segmentation* and (Job 2) *Clustering* and *Refine Results*

where s_k is the matching point of s with r_i, as emitted by the subtrajectory join procedure.

Finally, the voting of a trajectory (or subtrajectory) is defined as:

$$V(r) = \frac{1}{N}\Sigma_{i=1}^{N} V(r_i) \tag{10.5}$$

The second segmentation algorithm, coined TSA_2, identifies the beginning of a new subtrajectory whenever the composition of its neighborhood changes substantially. This algorithm takes as input a list $L(r_i)[]$ of the trajectory ids that have been produced as output by the DTJ procedure. The following example explains intuitively the difference between the two segmentation algorithms.

Example 10.2 Consider the example of Fig. 10.3a that illustrates five trajectories: $A \rightarrow B$, $A \rightarrow C$, $A \rightarrow D$, $C \rightarrow B$, and $D \rightarrow B$. Figure 10.3b and c depict the result of TSA_1 and TSA_2, respectively. In more detail, we can observe that both TSA_1 and TSA_2 segmented trajectory $A \rightarrow D$ to subtrajectories $A \rightarrow O$ and $O \rightarrow D$, due to the fact that after O, both the density and the composition of the neighborhood change. The same holds for trajectories $A \rightarrow C$, $C \rightarrow B$, and $D \rightarrow B$, which are segmented to subtrajectories $A \rightarrow O$, $O \rightarrow C$, $C \rightarrow O$, $O \rightarrow B$, $D \rightarrow O$, and $O \rightarrow B$. However, when it comes to trajectory $A \rightarrow B$, we can observe that while TSA_2 segments it to subtrajectories $A \rightarrow O$ and $O \rightarrow B$, TSA_1 does not perform any segmentation. This is due to the fact that, after O, even though the density of the neighborhood remains the same (i.e., 3 moving objects), the composition of the neighborhood changes completely.

Both segmentation algorithms share a common methodology, which employs two consecutive sliding windows W_1 and W_2 of size w (i.e., w samples) to estimate the point $r_i \in CP$ (cutting point) where the "difference" between the two windows is maximized. This methodology has been successfully applied in the past on signal segmentation [24, 25]. To exemplify, let us consider trajectory $A \rightarrow D$ of Example 10.2. For simplicity, we assume that the voting of the specific trajectory from A to O is 3 and from O to D is 1. Figure 10.4 illustrates the two sliding windows W_1 and W_2 that traverse the voting signal of trajectory $A \rightarrow D$.

Similar Subtrajectories The next step is to calculate the similarity between all the pairs of subtrajectories, using Eq. (10.2). This cannot be done completely after the segmentation at the *Reducer* phase of Job 1, illustrated in Fig. 10.2, because at that point each reduce function has information only about the segmentation of the reference trajectory to subtrajectories. For this reason, at this point we cannot calculate the denominator of Eq. (10.2). However, for each subtrajectory $r' \in r$, where r is the reference trajectory, we can calculate the similarity between the matching points (enumerator of Eq. (10.2)).

At this point, each *Mapper* has now all the information needed to calculate the similarity between all the pairs of subtrajectories (Eq. (10.2)), for each temporal partition separately. The similarity between subtrajectories is output in a new relation,

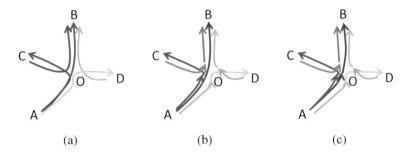

Fig. 10.3 (a) Five trajectories $A \rightarrow B$, $A \rightarrow C$, $A \rightarrow D$, $C \rightarrow B$, and $D \rightarrow B$, (b) TSA_1 segmentation, (c) TSA_2 segmentation

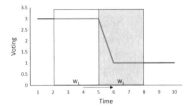

Fig. 10.4 The two consecutive sliding windows W_1 and W_2 used by the segmentation algorithms

called *SP*. Each tuple of this relation holds information about a subtrajectory r' and its similarity with all the other subtrajectories, whenever this similarity is larger than zero. More specifically, *SP* contains a set of key-value pairs where the key is the ID of the subtrajectory ($r'.ID$) and the value is a list $AdjLst$ containing elements of the form ($s'.ID$, Sim), where s' is a subtrajectory for which it holds that $Sim(r', s') > 0$.

10.3.2.4 Distributed Subtrajectory Clustering

Clustering After having calculated the similarity between all pairs of subtrajectories for each temporal partition, we can proceed to the actual clustering and outlier detection procedure. The intuition behind the proposed solution to the subtrajectory clustering and outlier detection problem (Problem 10.3) is to select as cluster representatives highly voted subtrajectories that have zero similarity with the already selected representatives $R_i \in R$, thus addressing the inter-cluster distance minimization. Then, we assign each subtrajectory to the cluster (i.e., Representative) with which it has the greatest similarity.

The input of the clustering algorithm is SP, ST, and parameters k and α and the output is the set of clusters C and the set of outliers O. More specifically, k is a threshold for setting a lower bound on the voting of a representative. This prevents the algorithm from identifying clusters with small support. Parameter α is a similarity threshold used to assign subtrajectories to cluster representatives.

It ensures that a subtrajectory assigned to a cluster has sufficient similarity with the representative of the cluster. This actually poses a lower bound to the average distance between the representatives and the cluster members and, consequently, guarantees a minimum quality in the identified clusters (intra-cluster distance).

Refinement of Results At this point we successfully accomplished to deal with Problem 10.3 for each temporal partition. However, this might result in having duplicates due to the fact that each subtrajectory that temporally intersects multiple partitions is replicated to each one of them. The actual problem that lies here is not the duplicate elimination problem itself but the fact that the result for such a subtrajectory might be contradicting in different partitions. In more detail, for each partition, the clustering procedure will decide whether a subtrajectory is a *Representative* (R), a *Cluster Member* (C), or an *Outlier* (O). Hence, for each intersecting subtrajectory q and for each pair of consecutive temporal partitions (i, j) with which q intersects, q can have the following pairs of states: (a) O–O, (b) R–R, (c) C–C, (d) R–C (C–R), (e) R–O (O–R), and (f) C–O (O–C).

For each of the above cases a decision has to be made, in order to eliminate duplicates and provide the correct result according to the problem definition. More specifically, in case of (a), q is marked as outlier in both partitions; hence, we only need to eliminate duplicates. In case of (b), the two clusters are "merged," since all of the subtrajectories that belong to them are similar "enough" with q, which is the representative of both clusters. In case of (c), let us assume that q belongs to cluster $C_i(R(q))$ in Partition i and $C_{i+1}(R(q))$ in Partition $i + 1$. Then, q is assigned to the cluster with which it has the largest similarity with its representative. In case of (d), q remains to be a cluster representative and is removed from the cluster C in which it is a member. Finally, in case of (e) and (f), q is removed from O. For more details concerning the *Distributed Subtrajectory Clustering* solution presented in this section, please refer to [35].

10.3.2.5 Experimental Results

In this section, we provide our experimental study on the solution that we proposed in order to address the *Distributed Subtrajectory Clustering* problem. The datasets employed for our experiments are:

- **IFS (April 2016)**—Flights between Madrid and Barcelona during April 2016 of size 43 MB and consisting of approximately 900K records.
- **NARI/Brest Area (6 months) Raw**—Vessels moving in Brest area, consisting of approximately 18 million records of size 697 MB.
- **FlightAware (April 2016)**—Trajectories of aircraft that consist of approximately 250 million records of 11.4 GB.
- **IMIS (3 years)**—consists of 699,031 trajectories of ships moving in the Eastern Mediterranean for a period of 3 years. This dataset contains approximately 1.5 billion records, 56 GB in total size.

Our experimental methodology is as follows: Initially, we illustrate that the subtrajectory clustering solution produces results of high quality as compared to the whole-trajectory clustering solution. Finally, we verify the scalability of our algorithms by varying the dataset size.

The experiments were conducted in a 49 node Hadoop 2.7.2 cluster. One node acted as the master and 48 nodes acted as slaves. The master node consists of 8 CPU cores, 8 GB of RAM, and 60 GB of hard disk, while each slave node is comprised of 4 CPU cores, 4 GB of RAM, and 60 GB of hard disk. Our configuration enables each slave node to launch 4 containers, thus resulting that at a given time the cluster can run up to 192 jobs (Map or Reduce). The operating system running on all the nodes is Debian 8.3.

Quality of Clustering Analysis To illustrate the quality of the results we employed the **IFS (April 2016)** (Fig. 10.5a). To assist with our analysis, we performed a preprocessing step where all trajectories were aligned to start at the same time. In order to be able to compare the two approaches (i.e., whole- and subtrajectory clustering), we deactivated the segmentation step of the subtrajectory clustering solution. The subtrajectory clustering algorithm identified 6 clusters, 3 clusters from Madrid to Barcelona and 3 clusters from Barcelona to Madrid. Moreover, an outlier was detected, which is not something common in aviation data. Both the cluster representatives and the outlier are depicted in Fig. 10.5b. However, if we consider only the spatial dimension, these 6 clusters and 1 outlier seem to be actually 2 clusters. But if we also take into consideration the temporal dimension, as presented in the space-time cube of Fig. 10.5d, we will actually see that there are clusters that follow the same path but have different behavior as far as it concerns the speed and/or the duration of the flight. These probably correspond to different types of aircrafts. In order to compare with the Spark MLlib-based solution, we vectorized the data and utilized the k-means algorithm on the same dataset with $k = 6$, which is the number of clusters that was previously identified. Figure 10.5c illustrates the result of k-means and Fig. 10.5d the corresponding space-time cube.

To conclude, the distributed subtrajectory clustering approach presents several advantages over the Spark MLlib-based. To begin with, an important issue is that the number of clusters is discovered by the algorithm and is not up to the user to give as input the correct number of clusters. In addition, another important functionality that a trajectory clustering algorithm should have is the outlier detection. Finally, due to the fact that the Spark MLlib-based solution with both vectorization methods does some kind of resampling, some movement patterns might be "lost," depending on the number of out samples of the resampling procedure.

Scalability We vary the size of our dataset and measure the execution time of our algorithms. To study the effect of dataset size, we created 4 datasets: 20%, 40%, 60%, and 80% of each of the original datasets. For the purpose of this analysis

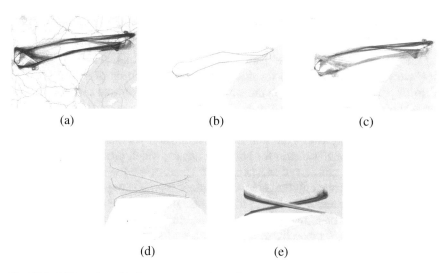

Fig. 10.5 (a) Raw data, (b) cluster representatives (6 clusters discovered), (c) k-means with $k = 6$, (d) cluster representatives—space-time cube, (e) k-means with $k = 6$—space-time cube

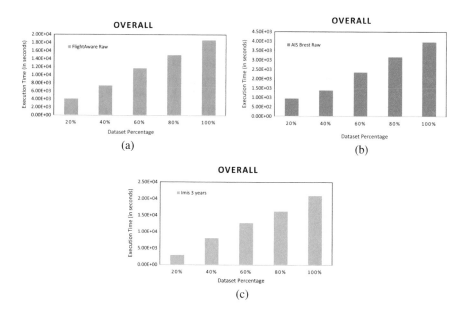

Fig. 10.6 Scalability analysis varying the size of the (a) FlightAware (April 2016), (b) NARI/Brest Area (6 months) Raw, and (c) IMIS (3 years) dataset

we are going to use the following datasets: **NARI/Brest Area (6 months) Raw**, **FlightAware (April 2016)**, and **IMIS (3 years)**.

As shown in Fig. 10.6, as the size of the dataset increases, the proposed clustering algorithm turns out to have linear behavior.

10.4 Distributed Hotspot Analysis

In this section, we present the THS (trajectory hotspot) and aTHS (approximate trajectory hotspot) algorithms for distributed hotspot analysis over big trajectory data, as presented in [20]. Our approach to Hotspot discovery and analysis is based on spatiotemporal partitioning of the 3D data space in cells. Accordingly, we try to identify cells that constitute hotspots, i.e., not only do they have high density, but also that the density values are statistically significant. We employ the Getis–Ord statistic [23], a popular metric for hotspot analysis, which produces z-scores and p-values. A cell is considered as a hotspot, if it is associated with high z-score and low p-value. Unfortunately, the Getis–Ord statistic is typically applicable in the case of 2D spatial data, and even though it can be extended to the 3D case, it has been designed for point data. In contrast, our application scenario concerns trajectories of moving objects, temporally sorted sequences of spatiotemporal positions, and the applicability of hotspot analysis based on a metric, such as the Getis–Ord statistic (but also any other metric), is far from straightforward. To this end, we formulate the problem of trajectory hotspot analysis, where our main intuition is that the contribution of a moving object to a cell's density is proportional to the time spent by the moving object in the cell. In particular, we adapt the Getis–Ord statistic in order to capture this intuition for the case of trajectory data. Then, we propose a parallel and scalable processing algorithm for computing hotspots in terms of spatiotemporal cells produced by grid-based partitioning of the data space under study. Our algorithm achieves scalability by parallel processing of z-scores for the different cells and returns the exact result set. Moreover, we couple our exact algorithm with a simple approximate algorithm that only considers neighboring cells at distance h (in number of cells), instead of all cells, thus achieving significant performance improvements. More importantly, we show how to quantify the error in z-score computation, thereby developing a method that can trade-off accuracy for performance in a controlled manner.

10.4.1 Definitions

Consider a database that contains trajectories of moving objects. A trajectory is a sequence of data points p described by 2D geospatial coordinates ($p.x$ for longitude and $p.y$ for latitude), a timestamp ($p.t$), and the moving object's id ($p.o$). Furthermore, consider a spatiotemporal partitioning $\mathscr{P}$ which partitions the 3D spatiotemporal domain into n 3D cells $\{c_1, \ldots, c_n\} \in \mathscr{P}$. Every data point p is mapped to one cell c_i, which is determined based on the spatiotemporal location of p. Individual objects moving inside single cells constitute a subset of data points p which form individual subtrajectories τ. The earliest (latest) point of a subtrajectory τ is denoted as $\tau.p_{start}$ ($\tau.p_{end}$). Also, we use $c_{i_{start}}$, $c_{i_{end}}$ to refer to temporal start and end of a cell c_i.

We also define the *attribute value* x_i of the cell c_i as: $x_i = \Sigma_{\tau \in c_i} \frac{\tau . p_{end} . t - \tau . p_{start} . t}{c_{i_{end}} - c_{i_{start}}}$.
Thus every object that moves in a spatiotemporal cell c_i contributes to the cell's attribute value by its temporal duration $\tau . p_{end} . t - \tau . p_{start} . t$ normalized by dividing with the cell's temporal lifespan $c_{i_{end}} - c_{i_{start}}$. This definition implies that the longer a moving object's subtrajectory stays in a spatiotemporal cell, the higher its contribution to the cell's attribute value.

Having calculated an attribute value for each cell, we opt to use the Getis–Ord statistic to calculate z-scores. The Getis–Ord statistic G_i^* is defined as [23]:

$$G_i^* = \frac{\Sigma_{j=1}^n w_{i,j} x_j - \overline{X} \Sigma_{j=1}^n w_{i,j}}{S \sqrt{\frac{[n \Sigma_{j=1}^n w_{i,j}^2 - (\Sigma_{j=1}^n w_{i,j})^2]}{n-1}}} \tag{10.6}$$

where x_j is the attribute value for cell j, $w_{i,j}$ is the spatiotemporal weight between cell i and j, n is equal to the total number of cells, and

$$\overline{X} = \frac{\Sigma_{j=1}^n x_j}{n} \tag{10.7}$$

$$S = \sqrt{\frac{\Sigma_{j=1}^n x_j^2}{n} - (\overline{X})^2} \tag{10.8}$$

The spatiotemporal weight $w_{i,j}$ used in the Getis–Ord statistic represents the score influence between neighboring cells; a cell needs to have a neighborhood of high attribute values to be considered as hotspot. Our goal is to have the influence of a neighboring cell c_i to a given cell c_j to be decreasing with increased spatiotemporal distance. Thus we employ a weight function that decreases exponentially with increasing distance; we define: $w_{i,j} = a^{1-\rho}$, where $a > 1$ is an application-dependent parameter, and ρ represents the distance between cell i and cell j measured in number of cells. For immediate neighboring cells, where $\rho = 1$, we have $w_{i,j} = 1$, while for the next neighbors we have, respectively: $1/a$, $1/a^2$, ...

Based on this, the problem of trajectory hotspot analysis is to identify the k most statistically significant cells according to the Getis–Ord statistic and can be formally stated as follows:

Problem 10.5 (Trajectory Hotspot Analysis) Given a trajectory dataset and a space partitioning $\mathscr{P}$, find the top-k cells $TOPK = \{c_1, \ldots, c_k\} \in \mathscr{P}$ based on the Getis–Ord statistic G_i^*, such that: $G_i^* \geq G_j^*, \forall c_i \in TOPK, c_j \in \mathscr{P} - TOPK$.

Our aim is to study the problem of trajectory hotspot analysis over massive spatiotemporal data by proposing a parallel and scalable solution. Thus, we assume that the trajectory dataset is stored in multiple nodes, without any more specific assumptions about the exact partitioning mechanism. Hence, here, we study a distributed version of the trajectory hotspot analysis problem.

10.4.2 Exact THS Algorithm

The proposed Exact THS algorithm is designed to be efficiently executed over a set of nodes in parallel and is implemented in Apache Spark. The input dataset D is assumed to be stored in a distributed file system, in particular HDFS. Intuitively, our solution consists of three main steps, which are depicted in Fig. 10.7. In the first step, the goal is to compute all the cells' attribute values of a user-defined spatiotemporal equi-width grid. To this end, the individual attribute values of trajectory data points are aggregated into cell attribute values, using Eq. (10.6). Then, during the second step, we calculate the cells' attribute mean value and standard deviation which will be provided to the Getis–Ord formula later. Furthermore, we compute the weighted sum of the values for each cell c_i : $\Sigma_{j=1}^n w_{i,j} x_j$. Upon successful completion of the second step, we have calculated all the individual variables included in the Getis–Ord formula, and we are now ready to commence the final step. The goal of the third step is to calculate the z-scores of the spatiotemporal grid cells by applying the Getis–Ord formula. The trajectory hotspots can then be trivially calculated by selecting either the top-k cells with the higher z-score values or the cells having a p-value below a specified threshold.

10.4.3 An Approximate Algorithm: aTHS

The afore-described algorithm (*THS*) is exact and computes the correct hotspots over widely distributed data. However, its computational cost is relatively high and can be intolerable when the number of cells n in $\mathscr{P}$ is large. This is because every cell attribute value must be sent to all other cells of the grid, thus leading to data exchange through the network of $O(n^2)$ as well as analogous computational cost, which is prohibitive for large values of n. Instead, in this section, we propose an approximate algorithm, denoted *aTHS*, for solving the problem. The rationale behind *aTHS* algorithm is to compute an approximation $\hat{G}_i^*$ of the value G_i^* of a cell c_i by taking into account only those cells at maximum distance h from c_i. The distance is measured in a number of cells. Intuitively, cells that are located far away from c_i will only have a small effect on the value G_i^* and should not affect its accuracy significantly when neglected. More interestingly, we show how to quantify the error $\Delta G_i^* = G_i^* - \hat{G}_i^*$ of the computed hotspot z-score of any cell c_i, when taking into account only neighboring cells at distance h. In turn, this yields an analytical method that can be used to trade-off accuracy with computational efficiency, having bounding error values.

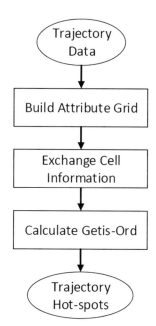

Fig. 10.7 Overview of THS algorithm

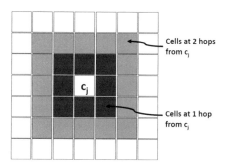

Fig. 10.8 Example of cells at distance from a reference cell c_j (the dark color indicates the weight of their contribution to c_j's value x_j)

The *aTHS* Algorithm Based on the problem definition, cells located far away from a reference cell c_i only have a limited contribution to the Getis–Ord value G_i^* of c_i. Our approximate algorithm (*aTHS*) exploits this concept and can be parameterized with a value h, which defines the subset of neighboring cells that contribute to the value of c_i. We express h in terms of cells, for instance, setting $h=2$ corresponds to the case depicted in Fig. 10.8, where only the colored cells will be taken into account by *aTHS* for the computation of $\hat{G}_i^*$ (an approximation of the value of G_i^*). In practice, the relationship between cell c_j and white cells can be expressed by setting their weight factor equal to zero. In algorithmic terms, *aTHS* is differentiated from *THS* in the second and third step.

10.4.4 Experimental Study

In this section, we evaluate the performance of our approach for trajectory hotspot analysis. We implemented all algorithms in Java, using Apache Spark 2.2.0 Core API.

Datasets We employed a real dataset containing surveillance information from the maritime domain. The data was collected over a period of 3 years, consisting of 83,735,633 individual trajectories for vessels moving in the Eastern Mediterranean Sea. This dataset is 89.4 GB in total size and contains approximately 1.9 billion records. Each record represents a point in the trajectory of a vessel and is made up by <*trajectoryID;timestamp;latitude;longitude*>. The dataset is stored in 720 HDFS blocks, in uncompressed text format.

Evaluation Methodology We picked four parameters to study their effect on the efficiency of our algorithm, namely (a) the spatial size of cells (in terms of both latitude and longitude), (b) the temporal size of cells, (c) the h distance which defines the number of neighboring cells contributing to the score of a reference cell c_i, and (d) the k number of hotspots to be reported in the final result set. In practice, the first two parameters affect the number n of cells of P, the third parameter h refers to the number of broadcasting messages which occur during the second step, and the fourth parameter k affects exclusively the last step of our algorithm. Also, we set a equal to 2 in all experiments.

Metrics Our main evaluation metric was the execution time needed for each individual step of our algorithm on the Spark cluster. In the following, the actual execution times will be presented, omitting any overhead caused by Spark and YARN initialization procedures. All execution times are depicted using the number of milliseconds elapsed for processing each step.

We deployed our code on a Hadoop YARN 2.7 cluster consisting of 10 computing nodes. Node 1 has 6 GB of RAM and 4 single-core CPUs running at 2.1 GHz. Nodes 2–10 have 8 GB of RAM, 4 single-core CPUs running at 2.1 GHz, and 100 GB hard disk each. Node 1 is configured to run the HDFS NameNode and YARN ResourceManager services, whereas all other nodes run the HDFS DataNode and YARN NodeManager services. In all our experiments, we use the YARN cluster deploy mode. We initiate 9 Spark executors, configured to use 5.5 GB of main memory and 2 executor cores each. We also configured HDFS with 128 MB block size and a replication factor of 2. On each node, Java 8 is installed on Ubuntu 16.04.

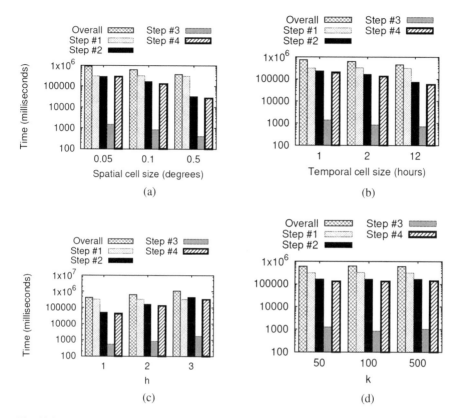

Fig. 10.9 Performance of our algorithm for various (**a**) spatial cell sizes of P, (**b**) temporal cell sizes of P, (**c**) values of h, and (**d**) values of k

10.4.4.1   Experimental Results

Varying the Spatial Cell Size   In Fig. 10.9a, we demonstrate the results of our experiments by varying the spatial cell size of P. Higher sized spatial cells decrease the total number of cells n used in the grid partitioning of the 3D space. In turn, this is expected to lead to reduced execution time, since fewer cells need to be computed and lower communication is required by the algorithm. Indeed, the overall execution time is reduced when the grid is of coarser granularity.

Varying the Temporal Cell Size   Figure 10.9b demonstrates the efficiency of our algorithm for various temporal cell sizes. The effect of larger cells in the temporal dimension to the overall execution time is similar to the previous experiment: Larger temporal cell size leads to fewer total cells in the grid, thus reducing the overall execution time. The experiment with the most coarse temporal partitioning (12 h) was measured to be twice more efficient than the experiment using the finest partitioning, in total execution time.

Varying h $aTHS$ can be parameterized with a user-defined variable h, which defines the set of neighboring cells contributing to the calculation of each cell's z-score. Figure 10.9c demonstrates the experimental results when using different values for variable h. The overall execution time is significantly reduced for lower values of h, since each cell broadcasts its attribute value to less neighboring cells, thus reducing the network overhead for exchanging such information between cells. By using a value of 3 for variable h, we measured three times higher overall execution time compared to the experiment having a value of 1. This significant reduction to the total execution time in $aTHS$, results in an approximate result $\hat{G}_i^*$. However, the deviation of $\hat{G}_i^*$ to G_i^* can be quantified as already explained.

Varying Top-k The value of top-k affects the size of the final result set. Figure 10.9d demonstrates the impact of this variable on dataset sizes throughout the execution of our algorithm and the individual steps' processing times. The overall execution time is not significantly affected by the value of this variable.

10.5 Distributed Data-Enriched Mobility Networks

Inference of the underlying network, given a large number of moving traces, both in aviation and maritime domain, is a challenging task that we try to address. The goal is to discover the directed graph of transitions, i.e., the set V of vertices and the set E of edges that form the routes network. Additionally, enriched information has to be taken into account in order to produce enriched graphs with contextual information. Domain experts may benefit a lot from such additional information. For example, one can then easily produce analytics of trajectories based on specific weather conditions and reveal how these conditions affecting or not the paths followed. Moreover, flight plans or predefined sea routes can be compared with real paths followed by ships or planes and the domain expert would be able to identify and explain the reason more easily.

10.5.1 Definitions

Definition 10.1 (Enriched Point) An enriched point r_i corresponds to a (raw point) p_i of a moving object and is defined as a tuple $< p_i, t_i, v_i >$, where v_i is a multi-dimensional vector consisting of categorical and/or numerical variables that annotate the raw point with associated context data.

Examples of v_i attribute values could be any user-defined tag or annotation valid regarding the specific domain application (e.g., consider annotations made by an event recognition module that detects the "top-of-climb" or "top-of-descent," "stop," "turn," etc.) or any numerical variable that can be attached to p_i, such as weather information (e.g., temperature, wind speed, humidity, etc.).

Definition 10.2 (Semantic Trajectory)   An enriched trajectory R corresponds to a (raw) trajectory T of a moving object, which is defined as the sequence of the enriched points of T.

Definition 10.3 (Data-Enriched Mobility Network)   A data-enriched mobility network N is a graph denoted by $N = (V, E)$, where V is a set of vertices and $E \subseteq V \times V$ is a set of edges.

The set V of vertices corresponds to the union of sets of enriched points where each set of enriched points is of the same enriched category, while the set E of edges corresponds to the union of paths in between vertices found. Given the above definitions the problem can now be formally expressed.

Problem 10.6 (Data-Enriched Network Inference)   Given a database of enriched trajectories, infer the underlying data-enriched network.

The data-enriched network that is to be found should meet the following requirements, in order to provide added value to domain experts:

1. **Consistency**. Network vertices inferred ideally should belong to one connected component, or the network must be of a number of connected components.
2. **Node Validity**. Vertices of the network correspond to enriched categories and are valid if derived from an aggregation or an assemblage (depending on the corresponding method that is used) of a number of more than m enriched points, thus having real value for the domain experts. For example, if a network vertex is derived from less than m points, then this vertex might be excluded from the final vertices set.
3. **Edge Validity**. Edges must correspond to frequent paths followed by objects. The frequency of paths is determined as defined by an application-dependent threshold σ.

10.5.2   Discovering Data-Enriched Mobility Networks

We follow an approach where first the vertices are discovered and then edges connecting these vertices are inferred from the trajectories. The input to the process comprises of a set of enriched trajectories. An enriched trajectory is modeled as a sequence of timestamped enriched points. The output of the process is a semantic-aware mobility network modeled as a directed graph $G = (V, E)$, where the vertices V correspond to semantic nodes and the edges E correspond to the discovered paths between semantic nodes. The process comprises two main steps, described subsequently in detail and illustrated in Fig. 10.10a–d.

The network discovery method described above is applicable to both domains, maritime and aviation, although the set of enriched trajectories which is the input to the process is formed differently for each domain. In the maritime domain the input is comprised of trajectories made by one vessel and covers a large temporal time frame. Thus, one maritime network is discovered for each vessel. Similarly, in the aviation domain the input is comprised of trajectories of aircraft that flew between two given airports. This way, one network is computed for every pair of airports. Note that the network is a directed graph, meaning that, e.g., Madrid–Barcelona is treated differently than the Barcelona–Madrid, implying that flights of only one direction at a time are taken for each pair of airports.

10.5.2.1 Step 1: Enriched Nodes Extraction

The input dataset of enriched trajectories is efficiently managed according to the enriched points and the enriched information each trajectory carries. By exploiting that each enriched point belongs to a semantic category (i.e., points carry an application depended tag, e.g., HOME, SPORT, etc., and these tags then make up the categories), we can split the input into enriched points of the same category. For every such input, nodes are formed based on a spatial-only clustering algorithm. Then, each cluster becomes a network node, carrying also semantic information from the corresponding category (Fig. 10.11). Moreover, the membership of a node in its corresponding cluster is above a given threshold m, thus having real value for the domain experts (i.e., suppress outliers).

Additionally, for each cluster/node several statistics are calculated and used later to refine and enhance the prediction accuracy. Statistics are modeled as Normal distributions (mean and standard deviation). Timing distributions describe when a ship or aircraft traverses the corresponding cluster/node within a day (24-h time frame). Elapsed time statistics describe the duration of staying in each cluster/node and speed statistics provide the mean speed passing through the corresponding cluster/node. Note that statistics may well be calculated after the discovery of the network as a post-processing step.

10.5.2.2 Step 2: Enriched Paths Discovery

Trajectories are processed separately of each other to identify enriched nodes and edges of the network, resulting in the discovery of semantic paths. More specifically, each cluster/node (i.e., corresponding enriched points of a moving object that clustered together) is processed separately and sequentially, ordered in the temporal dimension. If any two consecutive trajectory points correspond to two different network nodes/clusters, then a transition is recorded from one network node to the other. For each such transition, the beginning (from) and end (to) nodes of the network are identified and marked as additional information of the transition. These transitions form the set of candidate network edges.

Moreover, multiple transitions between pairs of network nodes are recorded by increasing (+1) the corresponding weight of the edge. In the end of the process,

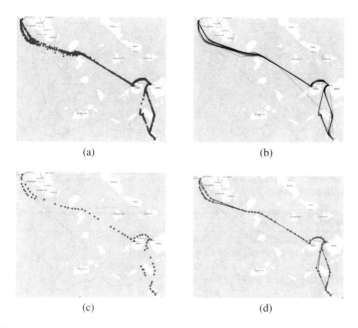

(a)   (b)

(c)   (d)

Fig. 10.10 Overview of semantic-aware network inference solution in maritime domain: (**a**) all enriched points from input, (**b**) enriched trajectories formed from enriched points, (**c**) enriched nodes extraction, (**d**) enriched path discovery

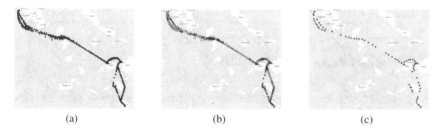

(a)   (b)   (c)

Fig. 10.11 Overview of network nodes extraction step in maritime domain: (**a**) all enriched points from input, (**b**) enriched points clustered spatially to candidate nodes, (**c**) semantic nodes extraction

all the edges/transitions are identified, along with their weights, which are simply the cardinality (absolute number) of all trajectories from the same edge. As an optional post-processing step, a threshold filtering can be applied by the domain expert if needed, in order to keep routes only above a specific weight or support. Finally, semantic nodes and edges are assembled to produce the complete map of the network. The weighted-edge network allows us to create a hierarchy of networks based on filtering edges with weight less than an application defined threshold σ. An example of paths discovery for maritime is shown in Fig. 10.12.

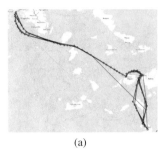 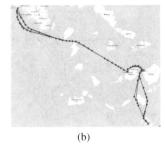

        (a)   (b)

Fig. 10.12 Overview of network paths discovery step in maritime domain: (**a**) all paths found, (**b**) only edges with more than σ weight are kept

---

Algorithm 10.4 SeaAirNet

Input: A semantic trajectory database STD, node membership m
Output: A semantic-aware spatial network N

1: **for** $e \in STD$ **do**
2:     /* e is the category of points (CP) in STD */
3:     $ClP \leftarrow SpatialClustering(CPe)$
4:     /* ClP are tuples of the form $< clid, oid, p_i, t_i, e > $ */
5:     **for** $ClP_i \in ClP$ **do**
6:         **if** $ClP_i \neq Noise \wedge |ClP_i| > m$ **then**
7:             $N.V \leftarrow N.V \cup createNode(ClP_i)$
8:             $ClP_s \leftarrow ClP_s \cup ClP_i$
9:     $N.E \leftarrow discoverEdges(ClPs, N.V)$
10: **return** N

10.5.3   The SeaAirNet Algorithm

We presented the steps of our methodology from an abstract point of view, accompanied with a running example and respective figures. Next, we present these steps in an algorithmic view. The main algorithm, named *SeaAirNet*, is the starting point.

Algorithm *SeaAirNet* takes as input a semantic trajectory database (STD) and a node membership limit m and outputs a semantic-aware mobility network. In line 1 all enriched points of the STD are partitioned based on the category each one belongs (semantic category) and then each partition is processed separately. In line 3 a spatial clustering algorithm is utilized to cluster enriched points of the specific category. For each cluster found (line 5), if the cluster is not considered as "noise" and its membership is over the application-dependent limit m (line 6), then it is considered valid and a network node is created (line 7) along with its statistics. It must be noted that the clustering algorithm might identify some points as noise: These do not belong in any cluster, or they do belong to a non-valid cluster depending on the clustering algorithm. In line 5 the newly created node is added to the set of network nodes. In line 8 each valid cluster is added to the set of valid

---

Algorithm 10.5 discoverEdges

Input: Clusters of enriched points CIP_s, network nodes V
Output: Network edges E

1: $nodesAndPoints \leftarrow CIP_s$.join V on $(clid, e)$
2: $P \leftarrow partition(nodesAndPoints$ on $oid).orderBy(t)$
3: **for** each $P_i \in P$ **do**
4:    $E_i \leftarrow edge(getNode, getNextNode, 1)$
5:    /* each E_i is of form $< fromNode, toNode, weight > $*/
6:    **if** $E_i.node <> E_i.nextNode \land E_i.nextNode$ isValid **then**
7:       $E \leftarrow E \cup E_i$
8: $E \leftarrow aggregate(E$ on $fromNode, toNode).sum(weight)$
9: **return** E

---

clusters that will later be used for the discovery of network edges. After extracting the semantic nodes (step 1), the paths discovery (step 2) task follows, where (in line 9) the *discoverEdges* algorithm is called by passing the set of valid clusters found and the set of network nodes extracted. Lastly, the returned nodes and edges (line 10) form the desired mobility network.

The algorithm *discoverEdges* discovers network-weighted edges from the clusters and network nodes found in algorithm *SeaAirNet*. Initially (line 1), variable *nodesAndPoints* holds a join of network nodes found (V) and corresponding enriched points that formed the node. In line 2, a partitioning technique is applied. In detail, the join set is partitioned by the object identifier (i.e., the id of vessel or aircraft), while the data of each partition is ordered based on time (t). Now, every partition holds one semantic trajectory with its enriched points ordered in time and also, each enriched point holds the cluster (network) node it belongs. For each such partition (line 3), enriched points are scanned sequentially in line 4 and every two consecutive points form an edge with weight of 1. Validity of the edge is checked in line 6: To be valid, the two network nodes it connects must be different and also the last node must be valid too, due to the sequential scanning of enriched points. If the edge is a valid network edge, then it is considered a candidate edge (line 7) and is added to the set of edges. When all partitions are processed, then results are aggregated on weight w.r.t. the beginning and ending nodes of each edge.

10.5.4 Experimental Results

In this section, we evaluate our approach of computing semantic-aware mobility networks both on aviation and on maritime domains. Our approach in this study is qualitative, meaning that we evaluate the produced networks by visually inspecting whether the network provides an accurate representation of the data used to extract it. We implemented all algorithms in Scala programming language, using Apache Spark 2.2.0 API. We implement a grid-density-based clustering algorithm in Spark to avoid the need of predefining the number of clusters in the beginning.

All algorithms are packed in one Scala project and when necessary, intermediate results are stored and retrieved using a database to avoid long lineages. Parameters of the algorithms are:

- *epsilon*: the radius of the disk that defines the neighborhood of a point;
- *minPts*: minimum number of members to consider a neighborhood as a valid cluster;
- *m*: node membership to consider a cluster as a valid network node;
- σ: threshold to keep edges of a certain weight and above in post-processing.

10.5.4.1 Datasets

In the maritime domain we selected to explore the movement of the DELOS vessel. DELOS has an mmsi of 241087000 and the spatiotemporal extend of its data is set to: $extend[x, y, t] = [23.61 \rightarrow 25.43, 36.39 \rightarrow 37.95, 2016-01-0100:11:34 \rightarrow 2016-01-3121:46:28]$. The above extent consists of 5761 raw points from IMIS Global AIS messages concerning specific vessel. These raw points are passed from synopses generator (SG) (cf. Chap. 3 of this book), which produced 2195 synopses whose enriched points are used for discovering the network.

In the aviation domain the MADRID (MAD)–BARCELONA (BCN) airports pair is selected using IFS radar surveillance data provided by CRIDA. The temporal extent is set to 2016-04-01 05:16:54 until 2016-04-30 20:06:08. The dataset consists of 997,450 raw points (both directions) which are derived from 1396 flights. The number of corresponding synopses produced by the SG are 254,330.

The methodology of discovering semantic-aware networks can be applied to raw trajectories as well as to semantic trajectories. Due to space limitations in the following we show an application of the methodology to raw trajectories in the aviation domain and an application to enriched trajectories in the maritime domain.

10.5.4.2 Qualitative Results in Aviation

We apply algorithm to raw 447,234 points of MAD to BCN airport pair (one direction), with parameters: $epsilon = [10000, 10000, 1000]$, $minPts = 100$, $m = 200$. We get 54 clusters (Fig. 10.13).

(a)   (b)

Fig. 10.13 All points colored by cluster; (**a**) with noise (19,882) and (**b**) without noise

These clusters are transformed to nodes by keeping only the medoid of each cluster. We consider only clusters with more than 200 members; thus we get 39 nodes (Fig. 10.14).

Then edges are discovered between above nodes. In total we found 182 edges with weights from 1 to 1274 (average 89.51). Based on domain expert, edges can be filtered out using their weight (Fig. 10.15).

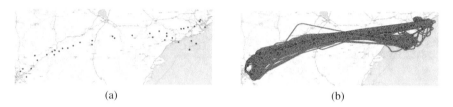

(a) (b)

Fig. 10.14 Medoids of the 39 clusters found (**a**) and how medoids cover the whole dataset (**b**)

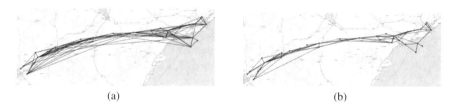

(a) (b)

Fig. 10.15 Discovering paths (edges) of the network, (**a**) all edges found with weight over 0 (182), (**b**) keep only edges with weight over 2 (130)

10.5.4.3 Qualitative Results in Maritime

We applied our algorithm to the 2195 points of trajectories' synopses of DELOS in the Aegean Sea, with parameters $epsilon = [1000, 1000, 100]$, $minPts = 1$, $m = 5$. Synopses are grouped in two groups: either as stops and/or as changes (Fig. 10.16).

These clusters are transformed to nodes, where each node represents one cluster. In total we get 128 nodes with membership from 1 to 224 (average 19.84). We consider only clusters with more than 5 members and thus we get 73 nodes (Fig. 10.17).

Then edges are discovered between these nodes. The algorithm finds 359 edges with weights from 1 to 91 (average 4.98). Figure 10.18 shows network edges for various values of σ.

From the above qualitative evaluation, it can be concluded that the higher the weight of the edges, the higher the compactness in the representation of the

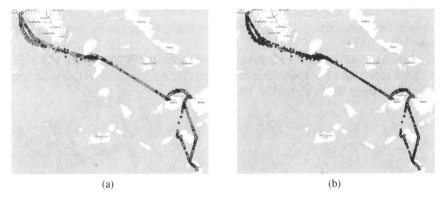

(a)   (b)

Fig. 10.16 All points colored by cluster above (**a**) and grouped based on their semantics in (**b**)

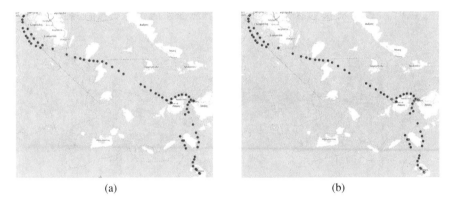

(a)   (b)

Fig. 10.17 Medoids of the 73 clusters found (**a**) and colored by type in (**b**)

dataset with this data-enriched network structure. Also, the network extracted from synopses is more or less the same as the one produced by the raw data. The advantage of this is that not only we may extract the network by processing much less data, but more importantly, we exploit the synopses to attach semantics to the vertices of the network.

10.6 Related Work

In recent years, an increased research interest has been observed in knowledge discovery out of mobility data. Towards this direction, several methods, which are directly related to our work, have been proposed.

Co-movement Patterns One of the first approaches for identifying such collective mobility behavior is the so-called flock pattern [14]. Inspired by this, a less "strict"

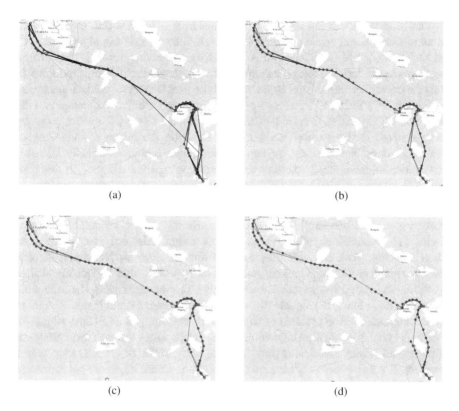

(a)   (b)

(c)   (d)

Fig. 10.18 Discovering paths (edges) of the network, (**a**) all edges found with weight over 0 (359), (**b**) edges with weight over 1 (226), (**c**) edges with weight over 2 (176), (**d**) edges with weight over 3 (136)

definition of flocks was proposed in [12] where the notion of a moving cluster was introduced. There are several related works that emerged from the above ideas, like the approaches of convoys, swarms, platoons, traveling companion, and gathering pattern [39]. However, all of the aforementioned approaches are centralized and cannot scale to massive datasets. In this direction, the problem of efficient convoy discovery was studied both in centralized [22] and distributed environment by employing the MapReduce programming model [21]. An approach that defines a new generalized mobility pattern is presented in [9]. In more detail, the general co-movement pattern (GCMP) is proposed, which models various co-movement patterns in a unified way and is deployed on a modern distributed platform (i.e., Apache Spark) to tackle the scalability issue.

Trajectory Clustering Most of the aforementioned approaches operate at specific predefined temporal "snapshots" of the dataset, thus ignoring the route of each moving object between these "snapshots." Another line of research tries to discover groups of either entire or portions of trajectories considering their routes. A typical strategy in dealing with trajectory clustering is to transform trajectories to a multi-dimensional space and then apply well-known clustering algorithms such as OPTICS [2] and DBSCAN [8]. Alternatively, another approach is to define an appropriate similarity function and embed it to an extensible clustering algorithm. In this direction, there are several approaches whose goal is to group whole trajectories, including T-OPTICS [19], that incorporates a trajectory similarity function into the OPTICS [2] algorithm. CenTR-I-FCM [26], a variant of Fuzzy C-means, proposes a specialized similarity function that aims to tackle the inherent uncertainty of trajectory data. Nevertheless, trajectory clustering is a computationally intensive operation and centralized solutions cannot scale to massive datasets. In this context, [6] introduces a scalable GPU-based trajectory clustering approach which is based on OPTICS [2].

Subtrajectory Clustering Nonetheless, discovering clusters of complete trajectories can overlook significant patterns that might exist only for portions of their lifespan. To deal with this, another line of research has emerged, that of *Subtrajectory Clustering*. The predominant approach here is TraClus [15], a partition-and-group framework for clustering 2D moving objects (i.e., the time dimension is ignored) that enables the discovery of common subtrajectories. A more recent approach to the problem of subtrajectory clustering is S^2T-Clustering [28], where the goal is to partition trajectories into subtrajectories and then form groups of similar ones, while, at the same time, separate the ones that fit into no group, called outliers. It consists of two phases: a neighborhood-aware trajectory segmentation (*NaTS*) phase and a sampling, clustering, and outlier (*SaCO*) detection phase. A slightly different approach is presented in QuT-Clustering [27] and [33], where the goal is, given a temporal period of interest W, to efficiently retrieve the clusters and outliers at subtrajectory level that temporally intersect W. In order to achieve this, a hierarchical structure, called ReTraTree (for Representative Trajectory Tree) that effectively indexes a dataset for subtrajectory clustering purposes, is built and utilized. An alternative viewpoint to the problem of subtrajectory clustering is presented in [1], where the goal is to identify "common" portions between trajectories, w.r.t. some constraints and/or objectives, cluster these "common" subtrajectories, and represent each cluster as a pathlet, which is a point sequence that is not necessarily a subsequence of an actual trajectory. A pathlet can be viewed as a portion of a path that is traversed by many trajectories. Similarly, in [40] the goal is to identify corridors, which are frequent routes traversed by a significant number of moving objects. As already mentioned, all of the above subtrajectory clustering approaches are centralized and cannot scale to the size of today's trajectory data.

Hotspot Analysis The problem of trajectory hotspot analysis is related to the spatial and spatiotemporal hotspot analysis. Several studies exist for conducting hotspot spatiotemporal analysis, such as [11, 17]. Spatiotemporal event data are analyzed and visualized in [17]. It consists of two steps: first, it uses multivariate kernel density estimation in space and time to estimate the density of the input data. Interestingly, different kernels in spatial and temporal domains can be used. In the second step it identifies hotspots from the most dense kernels and proposes a new visualization technique, based on Reeb graphs to illustrate the identified spatiotemporal hotspots. Hong et al. [11] studied the case of human mobility data such as taxi trips, bike rides, and subway trips. This data can be modeled by using spatiotemporal directed graphs (STG). The goal is to find subgraphs of the STG that have interesting flows (e.g., a black hole has the overall inflow greater than the overall outflaw). The user needs to input a threshold in order for the algorithm to successfully identify the interesting flows. A similar study is presented in [13] where human mobile traffic data are analyzed to discover hotspots on graphs. This study identifies spatial locations where the data volumes from wireless network transmissions are unusually high, based on user-defined thresholds. After identifying these locations, the algorithm detects the distribution of mobile data traffic hotspots to propose an efficient cell deployment strategy.

The trajectory hotspot analysis problem is also related to the trajectory mining domain [39]. Such trajectory and subtrajectory clustering techniques have been presented previously, in this section.

Data-Enriched Network Discovery Several methods rely on k-means clustering of raw tracking data using distance and direction as criteria to introduce cluster seeds at fixed locations along a trajectory. Edelkamp and Schroedl [7] use various heuristics for segmentation, map matching, and lane clustering from GPS traces. Schroedl et al. [30] use k-means clustering to refine an existing network map rather than construct the entire network starting from a blank terrain. Other methods transform GPS traces to density-based discretized images and are based on kernel density estimation (KDE). Most of these algorithms function well either when the data are frequently sampled (e.g., once per second) or when there is a lot of data redundancy. Biagioni and Eriksson [3] use a dataset which is being sampled very frequently (from 2 to 6 s). Steiner and Leonhardt [32] present an approach which uses tracking data of lower frequency, but still with intervals not exceeding 15 s. The limitation of KDE-based algorithms is that they are quite sensitive with respect to noisy data and outliers.

Another category, to which the present work relates, involves trace clustering approaches. These methods either adopt map matching or heuristics by aggregating GPS traces into an incrementally built transportation network. Moving object's heading and distance measures are also used to perform additions and deletions onto the incremental construction of the map. Rogers et al. [29] use trace clustering to refine an existing network rather than extracting it from scratch. Cao and Krumm [4] eliminate noise in GPS traces, while Fathi and Krumm [10] provide an approach that discovers intersections by using a prototypical detector trained on ground truth data from an existing map. This approach works best for well-aligned transportation networks (e.g., vertically aligned road networks) and with frequently sampled data of up to 5 s. Liu et al. [16] efficiently build a map but require accurate data and high sampling rates (i.e., 1 s). Zhang et al. [38] use a method similar to GPS trace clustering to continuously refine existing maps.

In general, although the problems of map construction, update, and enhancement are complementary, typically each individual work focuses on a single one of them. For example, a recent work by Wang et al. [37] applies trace clustering techniques to introduce a new KDE-based road fitting algorithm. The authors achieve an important contribution in terms of map entries in terms of data records on the OpenStreetMap collection, but their application mainly focuses on updating a map rather than constructing it. Similarly, Shan et al. [31] extend by proposing an automatic map update system which focuses on the identification of missing segments and is robust w.r.t. low sampling rates (on average of 120 s). Wang et al. [37] efficiently tackle the hard time performance of current approaches, deal with tracking data of low sampling rate but they mainly focus on inferring a map attributed with topological characteristics.

10.7 Discussion: Lessons Learnt

In this chapter, we reported on offline data analytic methods over moving object trajectories. The overall objective was to develop advanced, beyond current state-of-the-art data analytics methods and tools over a repository of trajectories of moving objects. In detail, we initially studied the problem of trajectory clustering by utilizing a methodology, which transforms trajectories to vectors, so as existing big-data-ready, point-based clustering algorithms (such as those provided by the Spark MLlib machine learning library) can be used. The goal of this approach is on the one hand to first provide solutions for the whole-trajectory clustering problem and to study the limitations of using off-the-shelf clustering algorithms for big trajectory data. Subsequently, the problem of distributed (sub)trajectory clustering over massive mobility data [35] has been addressed. In order to provide a solution to this we build upon a scalable distributed trajectory join method [34], which utilizes the popular MapReduce distributed programming model. Successively, we presented a scalable distributed trajectory-based hotspot analysis [20]. In this line of research, we followed a different clustering approach that provides statistical guarantees for the identified clusters. Interestingly, as a proof of concept, with this approach, we are able to discover hotspots that in the maritime domain can be used to measure the fishing pressure at sea, while in the aviation domain it identifies air-blocks that present demand-capacity problems. Finally, we studied the problem of distributed data-enriched mobility network discovery. The algorithms proposed provide contextually enhanced spatial graphs, which can successfully be utilized to support online location and trajectory prediction/forecasting scenarios (cf. Chap. 8).

References

1. Agarwal, P.K., Fox, K., Munagala, K., Nath, A., Pan, J., Taylor, E.: Subtrajectory clustering: models and algorithms. In: PODS, pp. 75–87 (2018)
2. Ankerst, M., Breunig, M.M., Kriegel, H., Sander, J.: OPTICS: ordering points to identify the clustering structure. In: SIGMOD, pp. 49–60 (1999)
3. Biagioni, J., Eriksson, J.: Map inference in the face of noise and disparity. In: Proceedings of the 20th International Conference on Advances in Geographic Information Systems, pp. 79–88 (2012)
4. Cao, L., Krumm, J.: From GPS traces to a routable road map. In: Proceedings of the 17th ACM SIGSPATIAL International Conference on Advances in Geographic Information Systems, pp. 3–12 (2009)
5. Claramunt, C., Ray, C., Camossi, E., Jousselme, A., Hadzagic, M., Andrienko, G.L., Andrienko, N.V., Theodoridis, Y., Vouros, G.A., Salmon, L.: Maritime data integration and analysis: recent progress and research challenges. In: Proceedings of the 20th International Conference on Extending Database Technology, EDBT, pp. 192–197 (2017)
6. Deng, Z., Hu, Y., Zhu, M., Huang, X., Du, B.: A scalable and fast OPTICS for clustering trajectory big data. Clust. Comput. **18**(2), 549–562 (2015)
7. Edelkamp, S., Schrödl, S.: Route Planning and Map Inference with Global Positioning Traces, pp. 128–151. Springer, Berlin (2003)
8. Ester, M., Kriegel, H., Sander, J., Xu, X.: A density-based algorithm for discovering clusters in large spatial databases with noise. In: KDD, pp. 226–231 (1996)
9. Fan, Q., Zhang, D., Wu, H., Tan, K.: A general and parallel platform for mining co-movement patterns over large-scale trajectories. Proc. VLDB Endowment **10**(4), 313–324 (2016)
10. Fathi, A., Krumm, J.: Detecting road intersections from GPS traces. In: Geographic Information Science, pp. 56–69 (2010)
11. Hong, L., Zheng, Y., Yung, D., Shang, J., Zou, L.: Detecting urban black holes based on human mobility data. In: Proceedings of the 23rd International Conference on Advances in Geographic Information Systems SIGSPATIAL, pp. 35:1–35:10 (2015)
12. Kalnis, P., Mamoulis, N., Bakiras, S.: On discovering moving clusters in spatio-temporal data. In: SSTD, pp. 364–381 (2005)
13. Klessig, H., Suryaprakash, V., Blume, O., Fehske, A.J., Fettweis, G.: A framework enabling spatial analysis of mobile traffic hot spots. IEEE Wirel. Commun. Lett. **3**(5), 537–540 (2014). https://doi.org/10.1109/LWC.2014.2349520
14. Laube, P., Imfeld, S., Weibel, R.: Discovering relative motion patterns in groups of moving point objects. Int. J. Geogr. Inf. Sci. **19**(6), 639–668 (2005)
15. Lee, J., Han, J., Whang, K.: Trajectory clustering: a partition-and-group framework. In: SIGMOD, pp. 593–604 (2007)
16. Liu, X., Biagioni, J., Eriksson, J., Wang, Y., Forman, G., Zhu, Y.: Mining large-scale, sparse GPS traces for map inference: comparison of approaches. In: Proceedings of the 18th ACM SIGKDD International Conference on Knowledge Discovery and Data Mining, pp. 669–677 (2012)
17. Lukasczyk, J., Maciejewski, R., Garth, C., Hagen, H.: Understanding hotspots: a topological visual analytics approach. In: Proceedings of the 23rd International Conference on Advances in Geographic Information Systems SIGSPATIAL, pp. 36:1–36:10 (2015)
18. Moran, P.: Notes on continuous stochastic phenomena. Biometrika **37**(1), 17–23 (1950)
19. Nanni, M., Pedreschi, D.: Time-focused clustering of trajectories of moving objects. J. Intell. Inf. Syst. **27**(3), 267–289 (2006)
20. Nikitopoulos, P., Paraskevopoulos, A., Doulkeridis, C., Pelekis, N., Theodoridis, Y.: Hot spot analysis over big trajectory data. In: IEEE International Conference on Big Data, Big Data 2018, Seattle, WA, 10–13 December 2018, pp. 761–770 (2018). https://doi.org/10.1109/BigData.2018.8622376
21. Orakzai, F., Calders, T., Pedersen, T.B.: Distributed convoy pattern mining. In: IEEE MDM, pp. 122–131 (2016)

22. Orakzai, F., Calders, T., Pedersen, T.B.: k/2-hop: fast mining of convoy patterns with effective pruning. Proc. VLDB Endowment **12**(9), 948–960 (2019)
23. Ord, J.K., Getis, A.: Local spatial autocorrelation statistics: distributional issues and an application. Geogr. Anal. **27**(4), 286–306 (1995)
24. Panagiotakis, C., Tziritas, G.: A speech/music discriminator based on RMS and zero-crossings. IEEE Trans. Multimedia **7**(1), 155–166 (2005)
25. Panagiotakis, C., Kokinou, E., Vallianatos, F.: Automatic p-phase picking based on local-maxima distribution. IEEE Trans. Geosci. Remote Sens. **46**(8), 2280–2287 (2008)
26. Pelekis, N., Kopanakis, I., Kotsifakos, E.E., Frentzos, E., Theodoridis, Y.: Clustering uncertain trajectories. Knowl. Inf. Syst. **28**(1), 117–147 (2011)
27. Pelekis, N., Tampakis, P., Vodas, M., Doulkeridis, C., Theodoridis, Y.: On temporal-constrained sub-trajectory cluster analysis. Data Min. Knowl. Discov. **31**(5), 1294–1330 (2017)
28. Pelekis, N., Tampakis, P., Vodas, M., Panagiotakis, C., Theodoridis, Y.: In-DBMS sampling-based sub-trajectory clustering. In: EDBT, pp. 632–643 (2017)
29. Rogers, S., Langley, P., Wilson, C.: Mining GPS data to augment road models. In: Proceedings of the Fifth ACM SIGKDD International Conference on Knowledge Discovery and Data Mining, pp. 104–113 (1999)
30. Schroedl, S., Wagstaff, K., Rogers, S., Langley, P., Wilson, C.: Mining GPS traces for map refinement. Data Min. Knowl. Discov. **9**, 59–87 (2004)
31. Shan, Z., Wu, H., Sun, W., Zheng, B.: Cobweb: a robust map update system using GPS trajectories. In: Proceedings of the 2015 ACM International Joint Conference on Pervasive and Ubiquitous Computing, pp. 927–937 (2015)
32. Steiner, A., Leonhardt, A.: A map generation algorithm using low frequency vehicle position data contents. In: 90th Annual Meeting of the Transportation Research Board (2011)
33. Tampakis, P., Pelekis, N., Andrienko, N.V., Andrienko, G.L., Fuchs, G., Theodoridis, Y.: Time-aware sub-trajectory clustering in hermes@postgresql. In: ICDE, pp. 1581–1584 (2018)
34. Tampakis, P., Doulkeridis, C., Pelekis, N., Theodoridis, Y.: Distributed subtrajectory join on massive datasets. ACM Trans. Spatial Algorithms Syst. **6**(2) (2019). https://doi.org/10.1145/3373642
35. Tampakis, P., Pelekis, N., Doulkeridis, C., Theodoridis, Y.: Scalable distributed subtrajectory clustering. In: IEEE BigData 2019, pp. 950–959 (2019)
36. Vlachos, M., Gunopulos, D., Kollios, G.: Discovering similar multidimensional trajectories. In: ICDE, pp. 673–684 (2002)
37. Wang, S., Wang, Y., Li, Y.: Efficient map reconstruction and augmentation via topological methods. In: Proceedings of the 23rd SIGSPATIAL International Conference on Advances in Geographic Information Systems, pp. 25:1–25:10 (2015)
38. Zhang, L., Thiemann, F., Sester, M.: Integration of GPS traces with road map. In: Proceedings of the Third International Workshop on Computational Transportation Science, pp. 17–22 (2010)
39. Zheng, Y.: Trajectory data mining: an overview. ACM Trans. Intell. Syst. Technol. **6**(3), 29:1–29:41 (2015)
40. Zygouras, N., Gunopulos, D.: Corridor learning using individual trajectories. In: IEEE MDM, pp. 155–160 (2018)

Part V
Big Data Architectures for Time Critical Mobility Forecasting

This part presents how methods addressing data management, data processing, and mobility analytics are integrated in a big data architecture for time critical mobility analytics. We propose an architecture paradigm that has distinctive characteristics compared to well-known big data paradigmatic architectures. The chapter included in this part presents the δ architecture and its instantiation in the datAcron integrated system. The software stack of the datAcron architecture is provided in detail, together with technical solutions concerning individual, online, and offline components integration.

Chapter 11
The δ Big Data Architecture for Mobility Analytics

George A. Vouros, Apostolis Glenis, and Christos Doulkeridis

Abstract Motivated by needs in mobility analytics that require joint exploitation of streamed and voluminous archival data from diverse and heterogeneous data sources, this chapter presents the δ architecture: Denoting "difference," δ emphasizes on the different processing requirements from loosely coupled components, which serve intertwined processing purposes, forming processing pipelines. The δ architecture, being a generic architectural paradigm for realizing big data analytics systems, contributes principles for realizing such systems, focusing on the requirements from the system as whole, as well as from individual components and pipelines. The chapter presents the datAcron integrated system as a specific instantiation of the δ architecture, aiming to satisfy requirements for big data mobility analytics, exploiting real-world mobility data for performing real-time and batch analysis tasks.

11.1 Introduction

The technical challenges associated with big data analysis are manifold, and perhaps better illustrated in [1] using a Big Data Analysis Pipeline, such as the one depicted in Fig. 11.1. Similar pipelines have been proposed elsewhere, e.g. the Ingest-Enrich-Store-Train-Query pipeline of IBM Watson discovery generic process.

The different snapshots of the big data analysis process result from the need to "ingest and digest," i.e. gather, process, integrate, and analyze the increasing amounts of data coming from different data sources. These sources may include sensors providing data-in-motion (streamed data) or stores providing data-at-rest (archival or historical data). The aim is to satisfy requirements for realizing internet of things and cyber-physical systems exploiting big data, as well as for advancing the human monitoring, awareness, prediction, and decision-making abilities in

G. A. Vouros (✉) · A. Glenis · C. Doulkeridis
University of Piraeus, Piraeus, Greece
e-mail: georgev@unipi.gr; aglenis@unipi.gr; cdoulk@unipi.gr

© Springer Nature Switzerland AG 2020
G. A. Vouros et al. (eds.), *Big Data Analytics for Time-Critical Mobility Forecasting*, https://doi.org/10.1007/978-3-030-45164-6_11

Fig. 11.1 The big data analysis process

critical domains, such as in transportation and traffic. As discussed in the first two chapters of this book, the goal is to support data analytics in order to reveal models that provide sufficient abilities towards identifying and predicting important happenings, occurring trends, important situations, as well as prescribing appropriate actions and plans.

The major question to be answered is *how such an abstract process should be realized as a big data architecture*, orchestrating the functionality and satisfying domain-specific requirements related (a) to the variety of data coming from different, isolated data sources designed for different purposes, and including data with different semantics and formats; (b) to the volume, velocity, and veracity of data to be processed and/or managed by the overall system.

Proposals for paradigmatic architectures or architectural patterns for big data systems emphasizing on "analytics" include the λ and κ architectures. The λ architecture separates the batch from the real-time processing needs, incorporating components that realize the same data processing tasks in separate layers. However, to resolve a query function [2], one has to merge results from the batch view and the real-time view. This blending of results from different layers may hinder the satisfaction of real-time requirements. Aiming to prove results from a real-time layer without using a batch layer, the κ architecture [3] considers that any data source can be treated as a provider of streamed data. This view impacts the way we do batch processing: For instance, we may need to fetch a well-selected subset of pre-processed historical data from a store, not retained in any log, and do any batch task that inherently requires voluminous data, computationally demanding processing steps, human involvement, application of iterative exploration-filtering-subset selection, data-transformation steps in visual analytics workflows, etc. Thus, we need a type of architecture where both batch and real-time layers co-exist, following the λ prescriptions, but enabling different functionality at different layers of processing, allowing components to run quite independently from those in other layer(s), much like the κ paradigm. Hence, to realize the overall big data pipeline, we need to decide on the arrangement of individual components in architecture layers according to functionality and performance requirements. This may result in multiple pipelines, each one serving specific purposes and adhering to specific performance requirements. Different pipelines may share components or incorporate components that realize the same data processing tasks, much like the λ paradigm.

The work reported in this chapter is motivated by requirements on mobility analytics in time-critical domains, specifically, in aviation and maritime (as these have been presented in the first part of this book), and contributes a paradigmatic big data architecture that emphasizes on the need for different layers of processing, targeting to different, albeit interacting, real-time and batch analytics tasks, aiming to satisfy their performance requirements. We opt for denoting the differences on layers' components requirements and call the architecture "delta," i.e. δ.

The δ architecture, going beyond mobility analytics tasks, contributes generic principles for incorporating components into a layered architecture, realizing multiple facets of the generic big data analysis pipeline, where each component may function both as a consumer and as a producer. In doing so, we clearly separate the functionality and performance requirements from each of the components and provide rules on performance constraints that should be satisfied by pipelines. The article describes an implemented instantiation of the δ architecture: The datAcron integrated system for real-world big mobility data analytics.

The chapter is organized as follows: Sect. 11.2 presents background concepts and requirements for exploiting mobility data in time-critical domains. Section 11.3 presents the proposed architecture paradigm and snapshots of the δ architecture. Section 11.4 presents a specific realization of δ: The datAcron implemented prototype for mobility analytics. Section 11.6 presents specific architectural choices to support performance requirements in time-critical domains. The chapter concludes with discussion on the strengths and limitations of δ and presents plans for future work in Sect. 11.7.

11.2 Background, Motivating Points, and Requirements

11.2.1 Motivating Points and Background

Transportation, either at sea, in air, or in urban areas, has a significant role and impact on the global economy and our everyday lives. The improvements along the last decades of these transportation means in terms of traffic management, planning of operations, security, information to operators and end-users have been driven by location-based information, while the current traffic control approaches are based on feedback loops or model-based predictive methods. Increased traffic introduces new challenges that these "classical" approaches cannot address effectively: This presents challenges for data-driven traffic management approaches. Thus, we need to process, manage, and exploit systematically mobility data [4], so as to provide insights on happenings, reveal the rationale driving these happenings, gain insight into situations occurring, build valid models for long-term predictions, driving decision-making and effective planning of operations. These are issues that have been presented and discussed in the other chapters of this book. We have seen that there are requirements that span from real-time to batch analytics.

To achieve these goals, beyond historical data, we need to exploit multiple data sources, including surveillance data, contextual data about entities' environment (e.g., weather, traffic, regulations imposed, infrastructure, and constraints on using the infrastructure), and data regarding the entities themselves (e.g., attributes of aircraft/vessels): The integrated exploitation of these sources presents emerging challenges for analytics that are as big as the data itself, in all four dimensions: Volume, Velocity, Variety, and Variety [4]: Regarding volume, the ever-increasing amount of data emphasizes the need for advanced, computationally efficient, scalable data processing, query-answering, and analysis. Also, regarding velocity, it is true that we need to process data—in most of the cases—under tight performance constraints (e.g., with low latency and high-throughput) addressing operational needs for large fleets of moving objects, covering large geographic areas. Batch processing is also important to process the bulk of historical data, serving multiple analysis needs. Variety is an important aspect, since data come from disparate sources providing either data-at-rest (e.g., past trajectories of moving entities, attributes of entities, data regarding infrastructure) or data-in-motion (e.g., surveillance data, weather forecasts, timely regulations, and dynamically adjusted infrastructure): While integrating data from these sources under a common model is a challenge, the variety of data imposes the need to apply transformations to data according to analysis goals, which on their own turn, may require changes to the way data are aggregated, projected, or extracted. Finally, the veracity of data is important to the analysis tasks. However, it is very difficult to achieve the ideal quality of data, and thus, analysis methods have to deal with noisy and fluctuating data.

11.2.2 Requirements

Given the above-mentioned motivating points, data processing, management, and analysis components need to perform under various performance requirements. For the purposes of this chapter, focusing mostly on throughput and latency requirements, we consider a continuum from high-throughput and real-time processing, to low-throughput and batch processing. For instance, in mobility analytics, the latency requirements for real-time tasks are in the order of milliseconds, and these tasks need to process data coming in high velocity, covering large fleets and large geographical areas. We characterize any task that is not real time, as a batch task. Therefore, for the purposes of this chapter we consider that the batch mode ranges from a few seconds to some minutes.

Driven by processing/exploitation and analytics requirements on mobility data, specific issues and requirements from a big data architecture include:

– *Detecting errors* in raw data and signifying errors' existence, indicating also the type of error, as discussed in Chaps. 3 and 4. Errors are frequent in raw surveillance data and may be due either to human error or to data provision system quality or malfunctions. Depending on the modality of the data source, there may co-exist different error detection components, each with different performance requirements.

- *Detecting low-level events in raw data*, i.e. detecting "primitive" types of events coming from a single source and concerning each single moving entity: These events may further trigger patterns of high-level (or complex) events. Examples of low-level events, as discussed in Chap. 4, include "moving in high speed," "abrupt turn," "lack of communication." These events should be detected by exploiting streamed surveillance data in real time, as required by monitoring and situation awareness applications.
- *Aggregating and compressing data* so as to reduce data volume, as presented in Chap. 4. These tasks usually need to be done under tight performance requirements. For instance, tasks aggregating and compressing surveillance data for large numbers of moving objects must adhere to the pace and volume of data.
- *"Ingesting and digesting" data* from any source, as presented in Chaps. 5 and 6: This includes converting data to the appropriate format and form, as needed by system components. There may be more than one "ingesting and digesting" component, depending on the types of data sources and on the requirements imposed by the components exploiting this data.
- *Answering spatiotemporal queries and transforming data* to appropriate forms, providing subsets of data to other components, as discussed in Chap. 7. This is a typical batch task, although transformation of data in appropriate forms— discussed in Chap. 5—(e.g., transforming time series of events into moving entities trajectories), if chosen to be done in query time, should be computed with low latency.
- *Integrating data from multiple sources*: Regarding data integration for supporting mobility analytics, as discussed in Chap. 6, special emphasis is given in identifying spatiotemporal relations among entities, as well as in the enrichment of surveillance data with additional features. This task should be performed in real-time given that typically this data is exploited by real-time analytics components. On the other hand, some data integration tasks may be done in batch mode (e.g., integrating historical trajectories with weather and moving entities attributes or detecting relations between archived trajectories and areas of interest [5]).
- *Analysis tasks* for predicting mobility and events (cf. Chaps. 8–10), for prescribing actions to be taken to resolve problems, or for data exploration via visual means (cf. Chap. 3) can have low latency performance requirements (e.g., for monitoring and prediction/prescription), or they may target to batch processing with increased latency, towards providing analysis results over voluminous archived data. In particular, visual analytics [6], aiming to provide tools and methods that combine human and computational data processing, are inherently batch tasks.

Given these specific issues and processing requirements,

- ranging in a continuum from strict real time (high-throughput) to "pure" batch (resp. low-throughput) processing, and
- taking into account the intertwined functionality of different components realizing these requirements' (i.e., the fact that components act as consumers and as producers); in conjunction with the fact that

- the same functionality may be realized by multiple components adhering to different processing requirements,

it is evident that the λ and the κ architectural patterns do not suffice themselves to realize an architecture that incorporates all necessary components implementing multiple processing and storage tasks adhering to various processing requirements.

11.2.3 Related Work

This section describes efforts to fill the gap between the generic prescriptions for a Big Data Analysis Pipeline and the intertwined requirements of data processing and analysis components for domain-specific real-world purposes.

Purpose-based approaches, aiming to satisfy the performance requirements of analysis goals for specific purposes, support analysis components to have direct access to data sources or allow them to store and process data as needed. These solutions have major drawbacks: (a) Each component needs its own access point to any of the data sources available, while (b) components need their own data processing, transformation, data integration sub-components, increasing the software management costs, and the required computational resources. High performance is not guaranteed in this case, as complex pipelines may be implemented. (c) Components cannot in principle access results provided by other components, but only in cases where special connectors are provided; and (d) components should handle access to own copies of data, while consistency should be guaranteed among replicas.

In a more principled way, two paradigmatic architectures have been proposed. The λ architecture [2] separates the batch from the real-time processing needs, incorporating components that realize the same data processing tasks, with different performance requirements, in separate layers. In doing so, input data is sent to both a batch and a real-time processing system. Both systems execute the same processing logic and output results to a service layer. Queries from back-end systems are executed based on the data in the service layer, reconciling the results produced by the batch and real-time processing systems. Beyond the software management cost incurred and increasing the computational resources required, reconciling the results from the real-time and batch layers incurs performance cost beyond the one introduced by the batch layer, while in many cases this is not necessary: Having different components operating in different layers independently, each with specific latency requirements, is sufficient, given that these components have access to the data sources, via a common point. It must be pointed out that although the λ architecture does not cater for components implementing different functionality, a λ pattern in an architecture may be necessary. Approaches following the λ paradigm include SOLID [7] which is a 3-tier architecture, Alljoyn Lambda [8], and Spark Structured Streaming [9].

The κ architecture [3] introduces a stream processing pattern. Layers of processing do exist here as well, but for executing on the same data in parallel. Results finally provided are those produced by one of the processing layers. Although this pattern resolves the software management issues implied by the λ architecture, it provides a paradigm where the "mere" batch layer (e.g., the advanced data management operations) is missing: Such a layer is necessary in cases where (a) batch analytics in large amounts of data need to be performed (e.g., to feed a machine learning algorithm with meticulously filtered integrated views of training data), (b) analysis is done interactively with humans (e.g., in visual analytics tasks), or in any case where (c) data needs to be transformed according to analysis needs in low latency. The κ pattern is necessary in cases where different processing components operating in parallel need to access the same data from a single log.

An architecture that is very close to the κ architecture and our proposal is Liquid [10]. Liquid accentuates the need for satisfying different latency requirements, focusing on data integration tasks. Latency requirements, incremental processing (i.e., avoiding re-processing all data after updates), cost effectiveness (low operational cost), and resource isolation executing many components concurrently are major issues addressed. The proposed architecture is separated into two independent layers: The processing layer providing low latency results and incremental processing, and the messaging layer for storing high-volume data with high availability and ability to access data through metadata annotations. As noted in [10], this decouples producers and consumers, without affecting each other's performance. In addition, the separation improves the operational characteristics of the data integration stack in a large organization, allowing for management flexibility. Independency of layers is what distinguishes Liquid from a κ architecture.

Beyond the in-principle architectural approaches, domain-specific architectures for intelligent transportation systems related to our proposal either focus in implementation-specific issues using big data analytics technologies, e.g. [11], or in cloud-based architectures, e.g. [12, 13], mostly focusing on batch processing tasks. An exception to that is the approach specified in [11], built on using Kafka as the main communication medium between the traffic system and the analysis module. While the architectural solution targets at real-time analytics, it is an instantiation of the κ pattern, at a single layer of processing.

11.3 The δ Big Data Architecture

11.3.1 The δ Architecture: An Overview

The overall pattern of the δ architecture is depicted in Fig. 11.2. This comprises multiple processing layers, each comprising multiple processing components that typically do not interact. Components, connected in a loosely coupled way, act as consumers or producers, or in both roles, subscribing to logs, or publishing data in logs. By "logs" we refer to append-only messaging systems (such as Kafka) with high-throughput and low latency reads, independent of log size, able to serve multiple consumers. Logs can retain (where possible) a huge volume of data for a time period using a natural sequential order. Subsequently, we use the term *communication medium* to refer to a specific implementation of a log.

In addition to logs, data management components, supported by high-volume store(s), may be used to provide query-answering services: While data logs may also be used as data stores, we use the term "stores" to refer to facilities for storing voluminous, integrated archival data from disparate sources, according to a meticulously crafted model, supporting efficient query-answering and data transformations services, serving specific analysis tasks. Although query-answering services for data stores can be efficient, we consider that data from stores cannot serve real-time analysis needs, as logs could do.

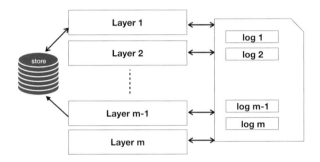

Fig. 11.2 The δ architecture

Therefore, architecture components interact by means of exchanging data either via data logs or via data stores.

This distinction between data stores and data logs has certain implications:

- By retaining multiple logs shared among components, a system may implement multiple, interacting processing pipelines. Each pipeline realizes a specific flow of data between components and has own processing requirements. Pipelines may share components and/or logs, interacting with other pipelines via constituent components' interactions.
- Components requiring a fast communication medium for satisfying demanding end-to-end performance requirements (e.g., high-throughput, low latency) can communicate by publishing/subscribing data to logs, while others that have batch processing requirements on large amounts of archival data, demanding the retrieval and transformation of archival data, may, in addition to accessing logs, fetch data by querying one of the architecture's data stores.

11.3.2 Principles and Rules

To be more specific on the δ architecture, we will provide background information, definitions of important terms, and introduce the denotation of important aspects.

A *pipeline P* is any sequence of processing components and communication media realized as a series of data processing, publishing, and fetching steps, as follows:

$$P = (C_1, CM_1, C_2, CM_2, \ldots .CM_k, C_k), \tag{11.1}$$

where any processing component C_m, $1 \leq m < k$, publishes results to a communication medium CM_m and the next component C_{m+1} fetches and exploits the specific results published by C_m. The subscripts in components are not associated to the placements of components into layers.

We denote a *pipeline P* in which a component C_i participates, by $(C_i \cdots P)$, while a *component C_i* in a pipeline P is denoted $(P \cdots C_i)$.

It must be noted that the main aspect here is not the "layered" configuration of δ to which other architectural paradigms adhere to, as well, but the satisfaction of performance requirements imposed on individual components and analysis tasks. However, architecture layers (or levels) serve as a means to make an architecture concrete, making explicit the processing pipelines and the ways requirements should be satisfied. So, starting from performance requirements, we may clarify what is the role of the δ architecture layers, and how one should decide the placement of processing components into these layers.

Given a set of *performance indicators* **I** (e.g., throughput and latency), the *performance requirements* on any component C are expressed as sets of constraints on a subset of indicators in **I**. Such constraints take the form of equalities or inequalities among performance indicators and specific values (e.g., *Latency* ≤ 10 ms).

Performance requirements from a specific, isolated component C are denoted by $constr(C)$, and performance requirements of the component through a pipeline P are denoted by $constr(P \cdots C)$. The performance constraints on a pipeline P are denoted by $constr(P)$.

The placement of components in architecture layers depends on the flow of data and on the processing requirements imposed on components: Considering performance requirements, typically, components in $layer_1$ have more strict performance requirements w.r.t. performance indicators (e.g., they must provide results in real-time) than components in a higher layer $layer_i$, $i > 1$.[1] We assume data flows from lower to higher layers' components. Thus, performance requirements on higher-layer components in any pipeline affect the requirements of components in lower layers in this pipeline. We will elaborate on this issue in subsequent paragraphs.

This ordering of components to layers implies that, typically, only processing components in higher layers should have batch processing requirements, requiring demanding (in terms of computational resources) access on system's data stores. Components with strict processing requirements typically should fetch data from logs. However, all components, at any layer, may provide/subscribe to logs and provide data to data stores.

---

[1] As the index of a layer becomes larger, the layer is considered to be higher in the architecture.

To ease presentation, we consider a single indicator $I \in \mathbf{I}$, assuming, without loss of generality, a discretization of values that this indicator may be assigned. For instance, regarding latency, we may define an ordered set of discrete values in which latency can range: *real time, very low latency, nearline, batch with considerable latency, batch with high latency.* Of course, in a real system we may impose very specific constraints that processing components must satisfy, using quantitative constraints, but discrete values for indicators may suffice as well, corresponding to humans' understanding of indicator values.

In general we assume that the following values can be assigned to I, ordered by magnitude: $\{V_0, V_1, V_2, V_3, \ldots\}$. We further assume a summation operation on these values, satisfying the well-known properties of summation on integers, such that $V_i + V_j = V_k$, where $0 \le i, j$ and $max(i, j) \le k$. Let V_0 be the zero element.

We denote by I_C the performance indicator's I value regarding a component C, including the tasks of fetching and publishing data: Similarly, we denote by $I_{(P \cdots C)}$ and I_P the performance indicator value regarding the performance of C through a pipeline P, and regarding the performance of a pipeline P as a whole, respectively. For example, $I_{(C \cdots P)}$ is the performance indicator regarding a pipeline P in which the component C participates.

To see the differences between these indicators, let us assume three components C_1, C_2, and C_3 in a pipeline $P = (C_1, CM_1, C_2, CM_2, C_3)$, and let $I_{C_1} = V_i$ and $I_{C_2} = V_j$. Then, the performance of C_2 through the pipeline P is $I_{P \cdots C_2} = V_i + V_j$, which may be greater than V_i and V_j. Also, the performance of the pipeline P is $I_P = I_{C_i \cdots P} = I_{P \cdots C_2} + I_{C_3}$, for $i = 1, 2, 3$.

Regarding performance requirements, and thus, constraints on performance indicators, it holds that, given a requirement

$$constr(P \cdots C_k) = (I_{(P \cdots C_k)} < V_j), j = 0, 1, 2, \ldots \tag{11.2}$$

on any component C_k, and a pipeline

$$P = (C_1, CM_1, C_2, CM_2, \ldots CM_k, C_k \ldots), \tag{11.3}$$

it must hold that

$$I_{C_1} + I_{C_2} + \cdots + I_{C_k} \le I_{P \cdots C_k} \le V_j, \tag{11.4}$$

or equivalently, that

$$I_{P \cdots C_{k-1}} + I_{C_k} \le I_{P \cdots C_k} \le V_j. \tag{11.5}$$

It must be pointed out that the equality in the first part of relations (11.4) and (11.5) holds when $I_{P \cdots C_{k-1}}$ or I_{C_k} is equal to V_0.

Therefore, given any pipeline

$$(C \cdots P) = (C_1, CM_1, C_2, CM_2, \ldots CM_k, C_k = C \ldots) \tag{11.6}$$

up to C, which may be part of a larger pipeline, then each constraint $I_{C\cdots P} \leq V_x$ in $constr(C \cdots P)$ implies further constraints in $constr(C_i)$, for any other component C_i, $1 \leq i \leq k$, participating in $(C \cdots P)$. For example, requiring that $C \cdots P$ should provide results with latency less than 1 s, then the sum of each component's latency in that pipeline must be less than 1 s.

More specifically, given the constraint

$$I_{C\cdots P} \leq V_x, \tag{11.7}$$

then it holds that

$$(I_{P\cdots C_i} < V_{P\cdots C_i}) \in constr(P \cdots C_i), \tag{11.8}$$

where $V_{P\cdots C_i} = V_x - \sum_{j\neq i} I_{C_j}$,[2] and I_{C_j} is the performance indicator of component C_j, $i \leq j \leq k$.

Furthermore, given the participation of a component C in different pipelines, the performance requirement for C regarding I is as follows:

$$constr(C_i) = (I_{C_i} \leq min_P\{V_{P\cdots C_i}\}). \tag{11.9}$$

Although the main objective is not on the placement of components to layers, this is necessary towards realizing a Big Data Analysis Pipeline (comprising multiple processing pipelines), as a whole. Generally, we should expect components having "nearly the same performance" requirements to be assigned to the same layer, if there is not any data flow between them. Therefore, given a pipeline $P=(\ldots C_i, CM_i, \ldots C_j, \ldots)$, components C_i and C_j will be placed in different layers, with C_j in a higher layer than C_i, given that there is a data flow from C_i to C_j. The exact layer of C_i depends not only of the performance constraints on C_j, and thus C_i, but on the requirements on any pipeline $C_i \cdots P$ in which C_i participates.

To formulate the term *"nearly the same performance,"* we define that two components C_x and C_y have *negligible performance difference* w.r.t. the indicator I, when for any $i \in \{x, y\}$ it holds that $I_{C_i} = V_m$, for a specific m in $\{0, 1, 2, \ldots\}$. More generally, it must hold that $I_{C_x} - I_{C_y} < \epsilon$, where ϵ is a threshold value specifying the minimum performance difference in order two components to be considered to have negligible performance difference.

Denoting the layer of a component C_i by $layer(C_i)$, ranging in $\{1, 2, 3 \ldots\}$, the following rules determine the placement of components in layers in a δ architecture:

R1: Components C_x and C_y in a pipeline $(\ldots C_x, \ldots C_y \ldots)$ are placed at different layers, such that $layer(C_x) < layer(C_y)$. The placement of each component is with respect to $constr(C_i)$, i.e. with respect to all the pipelines in which it participates. An exception to this rule may concern components in cyclic pipelines (see R4, below).

---

[2]The result of $V_i - V_j$ is V_k s.t. $V_i = V_k + V_j$.

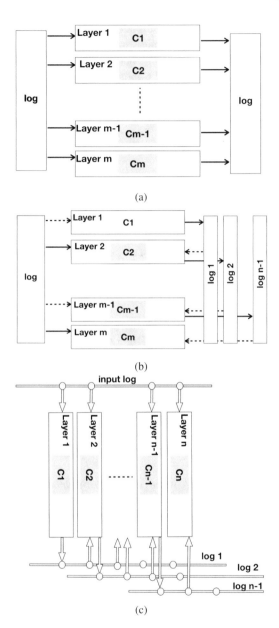

Fig. 11.3 The 1 log snapshot of δ (**a**), the n logs snapshot of δ (multiple pipelines) (**b**), alternative view of the n logs snapshot of δ (**c**)

R2: Components with negligible performance difference are placed at the same layer, given that no flow of data exists between them.

R3: Components C_x and C_y with non-negligible performance difference are placed at different layers: If $I_{C_x} < I_{C_y}$, then $layer(C_x) < layer(C_y)$.

R4: Components C_x and C_y that participate in a cyclic pipeline have negligible performance difference through that pipeline and should be placed in the same layer. That is, $layer(C_x) = layer(C_y)$, given that both components do not participate in other non-cyclic pipelines.

It must be noted that the exact layer of a component may be a more complicated decision, given that performance constraints imposed to pipelines (thus, to components) may concern multiple performance indicators, while processing components may inherit constraints on different indicators from multiple pipelines. In this general case where multiple indicators are considered, we need to use multiple-criteria decision-making algorithms, given the specification of performance requirements in terms of constraints. In addition to this, we may also decide to take advantage of the existing flexibility to having multiple components for the same functionality, each with different performance requirements.

In these cases, more interestingly, in order to maximize the performance of multiple (interacting) pipelines w.r.t. multiple performance indicators (which may present conflicting decisions as to which component to be used) we need to use multi-objective search algorithms [14].

11.3.3 Snapshots of the δ Architecture

This subsection provides specific configurations of the δ architecture, in generic forms, showing its potential to realize different paradigms of data processing architectures. Given our interest in mobility analytics tasks that exploit big surveillance data in conjunction with other sources, we consider that there is at least one log (the input log) that provides real-time data to all components and to which all components subscribe.

11.3.3.1 A Single Log, No Data Management Configuration

This is the most simple and trivial configuration of the architecture shown in Fig. 11.3a: Each layer comprises one component, and each component fetches data from the input log and publishes results to a common log. Here, components process input data in parallel, requiring no interaction with others.

The single log where results are being published implies that results are ordered according to the time published, but may require these to be tagged by the producer component. This is useful in cases where two components having different performance requirements target similar analysis results.

11.3.3.2 Multiple Logs, No Data Management Configuration

In this configuration of the δ architecture, shown in Fig. 11.3b, components do interact and form pipelines. The interaction happens via publishing results and subscribing to common logs. The final architecture may comprise several—interacting or not—pipelines. An alternative view of it, showing clearly the pipelines and flows of data, is shown in Fig. 11.3c.

In the configurations already presented, each component may have access to archival data according to own processing needs, although there is not any provision for a data store in the overall architecture.

11.3.3.3 The λ Architecture in Multiple Layers

This snapshot, shown in Fig. 11.4, follows the same pattern of the "one log, no data management" snapshot in Fig. 11.3a, but all components do publish results in a common data store.

The component at the highest layer may reconcile results published by other components and provides a service layer where queries from back-end systems are executed. In case processing components realize the same processing tasks with different processing requirements, this snapshot of δ resembles the λ architecture at multiple layers.

11.3.3.4 The κ Architecture in Multiple Layers Configuration

This configuration, shown in Fig. 11.5, follows a similar pattern as the "multiple logs, no data management" snapshot, but the component at the highest layer fetches results from all logs and provides final results. In case processing components realize the same processing tasks, this snapshot of δ resembles the κ architecture at multiple layers, where (a) each component produces results in parallel to other components, executing on the same input data and (b) the component at the higher layer may switch between logs.

Concluding this section, it must be noted that these snapshots of the δ architecture are the most simple and primitive ones. Of course, mixtures of these generic patterns may result in more sophisticated, albeit more complex, architectures.

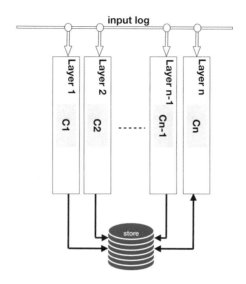

Fig. 11.4 The λ snapshot of δ architecture

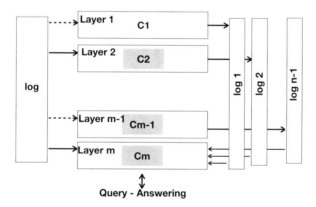

Fig. 11.5 The κ snapshot of δ architecture

11.4   Instantiating δ in the datAcron Big Data System for Mobility Analytics

This section presents an implemented instantiation of the δ architecture, shown in Fig. 11.6, realizing the datAcron integrated prototype, which supports real-time and batch analytics over big mobility data.

Given that datAcron aims at mobility analytics for large number of moving objects in the aviation and maritime domains, emphasis is put on the latency and throughput performance indicators, both for the individual components, as well as for the implemented pipelines. While throughput is measured by the number of messages per second, translated to the number of moving objects whose position is processed, latency ranges in the following set of values ordered by orders of magnitude: $\{operational, tactical, strategic\}$.[3] Thus, the datAcron system must operate so as to serve large numbers of moving objects, covering large geographical areas, with various latency requirements.

As shown in Fig. 11.6, the datAcron architecture consists of the following components, addressing the requirements of big data mobility analytics specified in Sect. 11.2.2.

The raw surveillance data feeders provide maritime data from a range of terrestrial AIS receivers and orbit satellites and aviation data from European Surveillance data feeds. The input data stream is collected, tested for veracity using a streaming analytics module, cleaned and annotated with various error flags.

These in situ (i.e., close to the data sources) components detect errors and detect and compress trajectories from raw surveillance data. Trajectories are enriched with low-level events. These events can be provided by the Low-Level Event Detector (LED) and the Synopses Generator (SG) [15]: For the purposes of this presentation we consider that both functionalities are implemented in a single component in a seamless way, the SG component, as also described in Chap. 4 in this book.

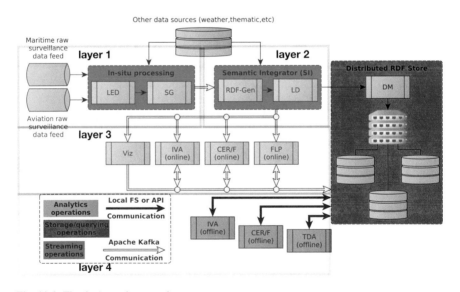

Fig. 11.6 The datAcron integrated prototype

The Semantic Integrator (SI) components transform data from any source into a common model [24] (RDF-Gen component [23]), addressing also data integration

---

[3]Operational latency: milliseconds, Tactical latency: few seconds, Strategic latency: tens of seconds or minutes.

requirements (Link Discovery (LD) component) [16], as also presented in Chaps. 5 and 6. The requirements for providing data loading and storage, query-answering, and data transformation services are addressed by the Data Management component [17], also presented in Chap. 7.

All components at level 2 and at level 1 must operate in operational latency, consuming the stream fed into the system in low latency and contributing to large throughput of the system pipelines.

The components at the next level perform data analysis tasks, with varying performance requirements. Specifically, the FLP module of the Trajectory/Future Location Predictor (TP/FLP) [18] predicts the short-term future location of a moving object in operational latency [19]. TP presents a similar functionality, targeting at predicting the future trajectory of a moving object as far in time horizon as possible, with tactical latency [20]. These components are presented in Chap. 8 in this book. The CER module of the Complex Event Recognition/Forecasting (CER/F) components aims at detection of complex events [15], whereas CEF aims to forecasting and prediction of complex events [21]. These components (cf. Chap. 9) perform at operational (online components) or strategic (offline components). The online components are working on synopses of trajectories and exploit spatial relations provided by the link discovery tasks performed by the SI. The output of CER is a stream with detected events. On the other hand, CEF enriches the input stream with a probability of events occurrence, per event monitored. Real-time visualizations of moving entities and their semantic trajectories are also supported by the Real-time visualization (Viz) component. This provides a map-based visualization of the stream of enriched spatiotemporal events generated by the TP/FLP and CER/F components operating with operational latency and large throughput. It is able to display different event types simultaneously, with an individual visual encoding for each type. Finally, the Interactive Visual Analytics (IVA) components build on top of the visualization module to provide limited analytical capacity on streaming data, either with operational or with strategic latency. The primary use is to allow analysts, and possibly maritime and aviation operators, to fine-tune and observe the impact of parameter adjustments to the TP/FLP and CER/F modules compared to actual data. Functionality and paradigmatic workflows for visual analytics components are presented in Chap. 3.

Finally, the Trajectory Data Analytics (TDA) modules (cf. Chap. 10) provides advanced trajectory analytics (e.g., data-driven discovery of the networks/routes upon which the movement of the vessels/aircrafts take place), while on the other hand it provides global patterns that represent meta-models (e.g., clusters and sequential patterns of trajectories, as in [22]). It operates with strategic latency.

The flows of data between components in the integrated datAcron system result in three different pipelines. In particular:

- *Information processing and management* pipeline: This comprises (a) the in situ components in the first layer, as well as (b) the SI components in the second system layer.

- *Real-time analytics* pipelines: Each of these pipelines comprises an analytics component in the third architecture layer, operating with operational latency. Each of these pipelines must be able to accommodate a sufficient fleet of moving objects. TP/FLP, CER/F components operate in parallel, with no flow of data between them.
- *Batch analytics* pipelines consume the available stored information by querying the RDF store [17] and store selected results back to the RDF store for future use.

11.5 Implementation of datAcron Integrated Prototype

In this section, we provide more technical details on the datAcron pipelines and flows, thus presenting implementation of system pipelines, describing also specific interactions between the individual components.

11.5.1 Information Processing and Data Management

In this pipeline, the prototype consumes as input raw surveillance data, both for maritime and aviation surveillance sources. This input is provided as a Kafka stream. The components that this pipeline comprises are as follows:

- LED is a Flink component consuming streaming data. It processes a stream of raw messages and enriches it with derived attributes such as min/max, average, and variance of original fields (e.g., of speed). The output is provided as a Kafka stream that contains the original data, enriched with the afore-described fields.
- SG accesses the output of LED and performs two major operations. First, it performs data cleansing, thus eliminating noisy data. Second, it identifies "critical points" on per trajectory basis. The output is provided in Avro[4] format as a Kafka stream.
- SI receives the Kafka stream produced by SG and performs transformation to RDF (using the RDFgen component) as well as data integration (i.e., link discovery, using the Link Discovery (LD) component), by enriching trajectory positions with information about weather as well as other contextual information. The output is provided in RDF, encoded in Terse RDF Triple Language (TTL)[5] and serialized in binary format as Java objects, also provided as a Kafka stream. The implementation of RDF data generators (RDFgen) is in Java, whereas the link discovery (LD) framework—which is the most processing-intensive operation in SI—is implemented in Apache Flink.

---

[4]https://avro.apache.org/.

[5]https://www.w3.org/TR/turtle/.

- Viz receives this enriched stream of information and provides real-time visualizations that can be used by operational users for improved situational monitoring and awareness.

As an extension of the previous pipeline, we can consider the pipeline ending at the Distributed RDF Store, which receives the RDF data provided by SI in a Kafka stream and stores it in the distributed RDF store, offering query-answering services.

11.5.2   *Online Future Location Prediction and Trajectory Prediction*

This pipeline addresses the trajectory and future location prediction (TP/FLP) in an online fashion.

The pipeline in the datAcron architectural diagram ends at the TP/FLP components. Normally, this involves data from full-resolution data, synopses and enriched data integrated and then published in the "enhanced surveillance data stream" provided to the TP/FLP (online) components.

The TP/FLP components operate in two different modes: (a) normal mode, in which they consume the output of the information processing and management pipeline, or (b) streaming simulation mode, in which they consume CSV files in a streaming fashion, mainly for testing their functionality.

The output from TP/FLP components includes forecasts of variables and they are (a) stored as JSON Arrays in HDFS, in order to further enrich synopses results (e.g., by filling communication gaps) and give extended input to the batch analytics components; (b) published in the "enhanced surveillance data stream" according to a stream specification that is used as template for the final integration with IVA. The stream specification is Kafka in JSON format.

11.5.3   *Complex Event Recognition/Forecasting Online*

The modules Complex Event Recognition (CER) and Complex Event Forecasting (CEF) consume the data provided by the SI module and either read in the data from files or from a Kafka topic. Outputs are produced as streams on Kafka.

The complex events forecasting (CEF) component is implemented in Flink with Java and forecasts complex patterns of events that are sent to Kafka.

Regarding event recognition, the CER module consumes serialized ST_RDF java objects and produces serialized ComplexEvent java objects that contain a JSON string with the recognized complex events. Several complex events patterns have been implemented.

11.5.4 Visualization and Interactive Visual Analytics: Online

Viz digests the enriched stream produced by fusing output streams from SI, TP/FLP, and CER/F components into the single enhanced surveillance data stream. This stream is automatically integrated into trajectory objects, displayed jointly as points and lines, respectively, on a (2D) map display.

The IVA module builds on top of Viz to provide limited analytical capacity on streaming data. Therefore, the principal input to the IVA is identical to that of the Viz. As already specified above, the primary use of IVA is to allow analysts, and possibly operators, to fine-tune analytics processes and observe impact of parameter adjustments. For this purpose, there is a flow of parameter settings from the TP/FLP, CER/F, and SG components to the IVA module and a flow of analysis results from these modules to IVA: Given the cyclic pattern of interaction between IVA and the analysis components, and the fact that CER/F, TP/FLP, and IVA have negligible performance differences, these components are shown to be at the same architecture layer.

In terms of actual integration of these modules, the networked connection between modules Viz, CER/F, and TP/FLP utilizes the same Kafka with JSON-encoded payload.

The implementation of the real-time visualization VA module follows the client-server architecture. On the server side, Kafka streams are consumed. All the applications on the server side are developed in Java. The client side is browser-based and written in JavaScript. The incoming messages from the Kafka streams are processed one by one and from every incoming Kafka message a JSON object is generated. Each JSON object is forwarded to the client using WebSockets. In fact, any input source providing enriched positional information of moving objects can be consumed and fed to the visualization front end, as long as it is provided as a Kafka stream. Essentially, this facilitates the integration with the Kafka streams provided by the components TP/FLP and CER/F, as well as with the enhanced surveillance stream provided by SI.

11.5.5 Trajectory Data Analytics Offline

There is a cyclic interaction pattern between the offline Trajectory Data Analytics (TDA) and the Data Management (DM) components. More specifically, the input from the DM is provided to the offline Trajectory Data Analytics component, which writes back the output to the DM.

Actually, the analyst selects the desired process of the offline TDA component (shown in Fig. 11.7) and poses a SPARQL query to the DM so as to retrieve the transformed RDF triples into the appropriate format (one record per point, including the moving object id, the timestamp, and the spatial position attributes).

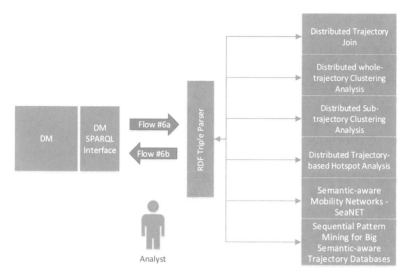

Fig. 11.7   The offline Trajectory Data Analytics (TDA) module

The output of this component is again converted to RDF, stored to the DM, and is available to the analyst via the SPARQL interface provided by DM. This output can comprise one trajectory cluster per record, specifying the id of the cluster and a list of the sub-trajectories that belong to the cluster.

11.5.6   Complex Event Recognition/Forecasting Offline

This flow realizes the offline complex event recognition and forecasting functionality. This is an example of a datAcron pipeline that is technically possible in the architecture, but actually it was not required from datAcron: However, it has been kept as a placeholder for future developments, given that currently, the functionality of the online counter-component suffices.

11.5.7   Interactive Visual Analytics Offline

The IVA component provides facilities for the visual exploration of data and visual-interactive support for building and refining models and their parameter settings. It therefore targets analysis experts working on the strategic level using data-at-rest, but may in suitable cases also provide visualizations with limited interactivity on the tactical or even operational level. Therefore, the IVA component operating in batch mode (offline) can be fed with data from a wide variety of data sources, including data from the DM.

The purpose of the Visual Analysis approach is to combine algorithmic analysis with the human analyst's insight and tacit knowledge in the face of incomplete or informal problem specifications, and noisy, incomplete, or conflicting data.

11.6 Experimental Evaluation

To evaluate the datAcron instantiation of the δ architecture we performed a "per-component" evaluation (Fig. 11.8), and a "per pipeline" evaluation focusing on the pipelines required to operate in real time and with large enough throughput to cover a sufficient number of moving objects.

To show the scalability of the prototype, we measure throughput and latency of individual components and of the system pipelines, increasing the number of cores available to the system (parallelism). Our computational platform consists of 10

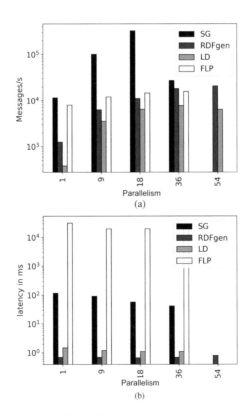

Fig. 11.8 Throughput and latency for each component

nodes (2*XEON e5-2603v4 6-core 1.7 GHz, 128 GB of RAM, 256 GB SSD). Three nodes are configured as Kafka brokers. One of the nodes was designated as the driver node; hence, all the parallelism configurations are a multiple of 9.

Per-Component Evaluation In terms of throughput, as shown in Fig. 11.8a, the Synopses Generator (SG) scales effectively achieving 100K messages/s using 9 workers and 360K messages/s for 18 workers. The throughput of SG decreases with 36 workers, because the YARN cluster could not allocate sufficient resources for the execution. Therefore, for this component, it does not make sense to test higher level of parallelism in the given infrastructure. As mentioned, the Semantic Integrator incorporated RDF generators implemented with RDFgen and Link Discovery (LD) component. The RDFgen and FLP components scale well up to 36 workers reaching 17K and 15K messages/s, respectively. For the LD component we observe that the maximum speedup is detected in the configuration of 36 workers, reaching 7.6K messages/s. Given an average sampling rate of 1 message per 5 s for each moving object, this means that LD is able to accommodate 38K moving objects with 5 s update.

Figure 11.8b depicts the latency for all components. The latency of the Synopses Generator (SG) is approx. 60 ms given 18 workers. This is an improvement over the centralized case, where the average latency is around 120 ms. The latency remains under 2 ms for all executions of the RDFgen and LD components. This proves that the RDFgen scales gracefully as the parallelism increases. Thus, all the components in the information processing and data management pipeline manage to meet the requirement of operational latency. The FLP component has the highest latency of all the components, reaching 32 s for the single worker case and 20 s for the 18 and 36 workers case. The batch interval for the FLP component was set at 10 s.

Per Pipeline Evaluation For this type of experiments each component was given 9 or 18 workers, so that all the components can achieve maximum performance.

The results obtained for the Semantic Integrator (SI) pipeline, which is the most crucial part of the architecture given the high-throughput and operational latency of the SG component, show that for parallelism equal to 9 it achieves 3.5K messages/s for the slowest component (which is also the last component in the pipeline). For parallelism equal to 18 the SI pipeline achieves 4.8K messages/s for the slowest component. For the FLP pipeline (one of the real-time analytics pipelines) and for parallelism equal to 9 the throughput achieved is approx. 10K messages/s, while for parallelism 18 is approx. 15K messages/s. This means that the FLP pipeline can accommodate a very large fleet of moving objects with quite frequent update of positioning information (covering large parts or even the whole Europe in sea or air), thus achieving the requirements of the overall datAcron prototype.

11.7 Concluding Remarks

In this chapter, we introduced the δ architecture for big data processing, and an instantiation of this architecture for the domain of mobility analytics. The δ architecture offers an in-principle architecture that can be instantiated to other big data architectural patterns—such as κ or λ—focusing on loosely coupled components that act both as producers and consumers. One major advantage is its flexibility; components may act independently or be placed in pipelines whose performance may be fine-tuned according to application requirements. Last, but not least, we demonstrate and evaluate an implemented instantiation of the generic architecture, realizing the datAcron system for mobility analytics.

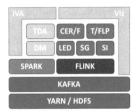

Fig. 11.9 The datAcron software stack and related big data technologies

In summary, the datAcron integrated prototype demonstrates its functionality and operation in real-time processing and online scenarios, most notably with low-level event detection, detection and generation of trajectory synopses, semantic enrichment of positional data with contextual and weather data, as well as with higher data analysis tasks, including future location prediction, complex event recognition and forecasting and real-time visualizations. Also, the prototype is able to demonstrate batch processing and offline analytics over integrated RDF data, using Spark as the big data platform for development. Figure 11.9 provides an illustration of the datAcron software stack, showing the individual modules and the big data technologies on which they rely.

Acknowledgments This work has been supported by the datAcron project, which has received funding from the European Union's Horizon 2020 research and innovation program under grant agreement No. 687591.

References

1. Jagadish, H.V., et al.: Big data and its technical challenges. Commun. ACM **57**(7), 86–94 (2014)
2. Marz, N.: How to beat the CAP theorem. nathanmartz.com/blog, October 13, 2011. Retrieved 10 May 2018
3. Kreps, J.: Questioning the lambda architecture. radar.oreilly.com. O'reilly, July 2, 2014. Retrieved 10 May 2018

4. Vouros, G.A., et al.: Big data analytics for time critical mobility forecasting: recent progress and research challenge. In: 21st International Conference on Extending Database Technology (EDBT/ICDT 2018), Vienna
5. Koutroumanis, N., et al.: Integration of mobility data with weather. In: Proceedings of BDMA@EDBT2019, CEUR, vol. 2322 (2019). https://doi.org/10.5281/zenodo.2563133
6. Andrienko, G., et al.: Visual analytics of mobility and transportation: state of the art and further research directions. IEEE Trans. Intell. Transp. Syst. **18**(8), 2232–2249 (2017)
7. Martinez-Prieto, M.A., et al.: The solid architecture for real-time management of big semantic data. Future Gener. Comput. Syst. **47**, 62–79 (2015)
8. Villari, M., et al.: AllJoyn lambda: an architecture for the management of smart environments in IoT. In: 2014 International Conference on Smart Computing Workshops, Hong Kong, 2014, pp. 9–14. https://doi.org/10.1109/SMARTCOMP-W.2014.7046676
9. Armbrust, M., et al.: Structured streaming: a declarative API for real-time applications in Apache Spark. In: Proceedings of the 2018 International Conference on Management of Data (SIGMOD '18), pp. 601–613. ACM, New York (2018). https://doi.org/10.1145/3183713.3190664
10. Fernandez, R.C., et al.: Liquid: unifying nearline and offline big data integration. In: 7th Biennial Conference on Innovative Data Systems Research (CIDR'15), Asilomar, 4–7 January 2015
11. Amini, S., Gerostathopoulos, I., Prehofer, C.: Big data analytics architecture for real-time traffic control. In: 5th IEEE International Conference on Models and Technologies for Intelligent Transportation Systems (MT-ITS) (2017). https://ieeexplore.ieee.org/document/8005605
12. Fiosina, J., Fiosins, M., Mueller, J.P.: Big data processing and mining for next generation intelligent transportation systems. J. Teknol. **63**(3), 23–38 (2013). https://doi.org/10.11113/jt.v63.1949
13. Kemp, G., et al.: Towards Cloud big data services for intelligent transport systems. In: Concurrent Engineering, Delft, Jul 2015
14. Sastry, V.N., Janakiraman, T.N, Mohideen, S.I.: New algorithms for multi objective shortest path problem. Opsearch **40**(4), 278–298 (2003). https://doi.org/10.1007/BF03398701
15. Patroumpas, K., et al.: Online event recognition from moving vessel trajectories. Geoinformatica **21**(2), 389–427 (2017)
16. Santipantakis, G.M., et al.: SPARTAN: semantic integration of big spatio-temporal data from streaming and archival sources. Future Comput. Gener. Syst. Available online, https://doi.org/10.1016/j.future.2018.07.007
17. Nikitopoulos, P., et al.: Parallel and scalable processing of spatio-temporal RDF queries using Spark. Geoinformatica. https://doi.org/10.1007/s10707-019-00371-0
18. Petrou, P., et al.: Online long-term trajectory prediction based on mined route patterns (2019). http://www.master-project-h2020.eu/wp-content/uploads/2019/07/MASTER2019_paper_5.pdf
19. Georgiou, H.V., et al.: Moving objects analytics: survey on future location & trajectory prediction methods. https://zenodo.org/record/1309181#.XToHg5MzZTY
20. Georgiou, H.V., et al.: Semantic-aware aircraft trajectory prediction using flight plans. Int. J. Data Sci. Anal. Available online, https://doi.org/10.1007/s41060-019-00182-4
21. Alevizos, E., Artikis, A., Paliouras, G.: Wayeb: a tool for complex event forecasting. In: Artificial Intelligence and Reasoning (22nd LPAR), Awassa, 2018
22. Pelekis, N., et al.: In-DBMS sampling-based sub-trajectory clustering. In: Proceedings of EDBT 2017, 21–24 March, pp. 632–643 (2017)
23. Santipantakis, G.M., et al.: RDF-Gen: generating RDF from streaming and archival data. In: WIMS'18, Novi Sad
24. Vouros, G.A., et al.: The datAcron ontology for the specification of semantic trajectories. J. Data Semant. **8**, 235–262 (2019). https://doi.org/10.1007/s13740-019-00108-0. The datAcron ontology: http://ai-group.ds.unipi.gr/datacron_ontology/

Part VI
Ethical Issues for Time Critical Mobility Analytics

This last part focuses on important ethical issues that research on mobility analytics should address: This is deemed to be crucial, given the growth of interest in that topic in computer science and operational stakeholders, necessitating the sharing of data and distributing the processing among stakeholders.

Chapter 12
Ethical Issues in Big Data Analytics for Time Critical Mobility Forecasting

Gemma Galdon Clavell and Victoria Peuvrelle

Abstract Big data analytics for time critical mobility forecasting involves the use of large amounts of data from different sources, which are combined to gather new insights and foresee potential needs and developments. Due to this intensive use of data and the sensitivity of projects involving critical infrastructures and technologies, the use of big data analytics for time critical mobility forecasting has legal and ethical impacts and risks that need to be addressed and mitigated. While this may not be obvious at first glance in some cases, as the presence of personal data and direct impacts on individuals may be minimal, ethics issues are still relevant. These are related to the possible privacy- and security-related consequences, as well as potential misuse and "function creep", both during product development and testing and in the actual use of the final product. This chapter thus seeks to tackle those ethical challenges as they were addressed by an expert ethics team in the EC-funded datAcron project. We start by explaining the efforts made at the EU level to ensure the ethical development of publicly funded technologies and the framework and risks all projects need take into account, and then we briefly go over the different insights and actions regarding the datAcron project to ensure the ethics compliance of the project. While the process and advice developed is specific to datAcron, we believe it holds lessons for other similar initiatives seeking to be aware of and comply with their ethics obligations.

12.1 Introduction

There is an increasing awareness of the need to combine investment in new, innovative technologies with a deeper understanding of their legal, social, and ethical impacts. Recent controversies surrounding the legality, acceptability, and unexpected impact of technological developments in the field of security have led

G. G. Clavell · V. Peuvrelle (✉)
Eticas Research and Innovation, Barcelona, Spain
e-mail: gemma@eticasconsulting.com; victoria@eticasfoundation.com;
victoria@eticasfoundation.org

© Springer Nature Switzerland AG 2020
G. A. Vouros et al. (eds.), *Big Data Analytics for Time-Critical Mobility Forecasting*, https://doi.org/10.1007/978-3-030-45164-6_12

343

to an increased interest and concern in addressing societal issues in the field of EU research and innovation, with a specific emphasis on legal compliance, longer-term phenomena, and the need to better govern and exploit, at an early/conception stage, the negative and positive externalities of technological innovation processes and products in the short, medium, and long term.

Big data is defined as "high-volume, high-velocity, and/or high-variety information assets that demand cost-effective, innovative forms of information processing that enable enhanced insight, decision-making, and process automation" [6]. In projects dealing with big data, responsible research is a particularly sensitive issue, as data from different sources and of different times is gathered/generated, mined, combined, shared, and used to forecast future events in ways that are often unaccountable or that render the information vulnerable. The positive impact of these innovations can nevertheless be maximized if the necessary legal, social, and ethical precautions and mechanisms are put in place. In this context, our involvement in the datAcron mobility analytics project aimed at addressing these issues in big data analytics for time critical mobility forecasting. While this was considered a low-risk project, the main ethics issues were related to the possible consequences of event recognition and trajectory analytics for privacy and the potential misuse or "function creep" (defined below) of the technologies developed. The identification and monitoring of these issues involved a continued relationship with the research team and the review of several key documents to anticipate and tackle potentially sensitive activities. In the process, the awareness of the team around these issues was raised and several mitigation strategies were incorporated in the project.

The following sections of this chapter thus go over the necessary steps to address the aforementioned ethical issues in a way that is consistent with the principles of responsible research and innovation, privacy and data protection rules and guidelines, and ethical safeguards, specifically in the handling of data, development of experiments, and potential future use of the technology. First, existing methodologies to carry out ethical research at the European level are presented. Secondly, ethics issues in mobility forecasting are explored, focusing on the mobility of moving objects. Recommendations are made to tackle those ethics issues in the third and final part of this chapter.

12.2   Ethics in Research at the EU Level

Several methodologies and definitions have been provided at the EU level to ensure responsible research and innovation, linking the specific needs of projects with broader principles. In 2010, for instance, researchers from several EU-funded projects got together to compile a policy brief on "Responsible Research and Innovation in the Information and Communication Technologies and Security Technologies Fields." The group agreed on the following shared definition of responsible research and innovation:

Responsible Research and Innovation is a transparent, interactive process by which societal actors and innovators become mutually responsive to each other with a view on the (ethical) acceptability, sustainability and societal desirability of the innovation process and its marketable products (in order to allow a proper embedding of scientific and technological advances in our society). [8]

At around the same time, the European Commission's Directorate-General for Enterprise and Industry (DG ENTR) commissioned a report from an Expert Working Group on Societal Impact, mentioning how the Security Research Programme was at a "key moment," as "the agenda for security research and development in Horizon 2020 is gradually taking shape" [5]. The group comprised of experts from the security industry, academia, the non-governmental organizations, and policy communities concluded, among other things, that:

- Citizen rights should be a fundamental requirement which could and should lead to drawing boundaries of what is and what is not acceptable in EC funded security research.
- The principles of research ethics should include accountability for scientific procedures, clarification of criteria, and choice of research objects, disinterestedness, regard for conflicts of interest, consent of participants in research, confidentiality, transparency of methods and results, respect for data protection, and ownership.
- EC-funded research should lead to enhancing the security of European citizens and show how it will affect the lives of citizens in doing so.
- Societal impact should be addressed in the following phases: work program and annual calls, proposals, negotiation, project execution, and implementation of a completed product, system, or techniques in different contexts.

These concerns have since then been embedded in the ethics monitoring scheme of the European Commission,[1] which identifies, tracks, and monitors the ethical compliance of all EC-funded projects dealing with sensitive issues and, most notably, using personal data. Thus, consortia wishing to be funded by the European Commission must fulfill a number of ethics requirements in order to receive funding. This entails explaining to the Commission the project's commitment to ethics in the proposal and completing an ethics table, for which the Commission has issued a Guidance.[2] Throughout any project, the consortium must hand in a number of documents that guarantee the project's compliance with ethical guidelines and regulations, such as:

- A data management plan explaining how the data processed within the project will be handled;
- A legal document explaining the regulations to be followed and how the project complies with them;

---

[1]https://ec.europa.eu/programmes/horizon2020/en/h2020-section/ethics.
[2]https://ec.europa.eu/research/participants/data/ref/h2020/grants_manual/hi/ethics/h2020_hi_ethics-self-assess_en.pdf.

- A document explaining how the technologies developed follow the principle of privacy by design;
- A guidance brief on how the consortium is carrying out research ethically.

Other documents are added according to the specificities of each project, and thus ensure that any ethical issues are tackled before and throughout the project's duration. Following this type of rigorous ethical process is mandatory for H2020 projects, but can also be of use for any projects developing new technologies, especially those such as datAcron, where ethical issues may not be clear at first glance. The following section goes over the ethics issues that are linked to new technologies.

12.3   Ethical Issues

12.3.1   Consequences of Outputs for Privacy

Data protection and privacy risks depend on the specific features of the developed system; it depends on the types of data involved, whether it is personal, and on the uses of that data. Indeed, privacy risks are particularly heightened when the personal data being processed is of a sensitive nature, such as biometric data, or data concerning religious, sexual, or political beliefs.

Mobility forecasting of humans can be problematic as it can divulge sensitive information about individuals. Indeed, the grand majority of people have routines and patterns of movements. Hence for many, if their movements are divulged, their workplace or home address can easily be found by correlating the location data with maps. This applies to more sensitive and personal data as well, such as an individual's religious activity, political meetings, or other types of personal hobbies which such individual would rather keep personal. If a person with the wrong intentions has access to this data, the physical security of the person can be jeopardized. Not only this, national security can be put at stake, for instance, in the case of the fitness apps Strada and Polar, which released flow maps of their users' movements, thus divulging the location of military bases and soldiers' training patterns in at-risk areas [7, 9].

These privacy issues are greatly reduced when exploited data concern moving objects rather than individuals, as was the case in datAcron scenarios. The privacy issues linked to the processing of the personal data themselves were low in that case, for the following reasons:

- Only the movement of vessels and airplanes was subject to monitoring. Most large moving entities do not belong to individuals and therefore do not constitute personal data nor can they be linked to individuals. However, smaller moving entities that do belong to individuals can be more problematic as the analysis of their moment might lead to the identification of their owner. In order to minimize this risk, datAcron only considered the movement of very large numbers of entities; hence, the potential issues due to the singling out of individual vessels and airplanes was highly reduced;

- datAcron only used publicly available information concerning the movement of vessels and airplanes. This decision also lowered the ethics risks of the project.

Nevertheless, the following privacy and security concerns continued to be relevant concerning the monitoring of the movement of entities;

- For individually owned vessels and planes, the technologies developed could lead to the unintended revelation of the owner's identity;
- The tracking of moving entities linked to governmental activities or actors could compromise national security, as well as the security of those on board of such vessels.

These privacy risks, however, are understood in terms of probabilities, which means measuring the likelihood of an undesired event and weighing the consequences (e.g., the potential identification of vessels/aircraft that should remain anonymous and the potential outcomes of this situation).

12.3.2 Function Creep and Dual Use

The concept of "function creep" refers to the "gradual widening of the use of a technology or system beyond the purpose for which it was originally intended, especially when this leads to potential invasion of privacy" [3]. Different aspects of a project or system can vary with time—functions, scope and objectives, geographical reach, end-users, etc. and so risks that were addressed at one point or deemed non-problematic from an ethical perspective may later on have legal or ethical implications, cause unforeseen or unintended privacy harms, or even constitute criminal offenses. Therefore, while not each and every future practice or use of technology can be anticipated, it is important to assess what the possibilities for function or mission creep for any given project may be.

Function/mission creep is particularly relevant when the technologies or procedures being developed can be used in different contexts. In the context of datAcron, the need to specify users and actors was a clear priority, as the contexts of utilization the technology include:

- Military/Security
- ATM and maritime management

  - Transportation of passengers
  - Transportation of goods
  - Private use

- Government—Management of infrastructures
- Industrial/commercial

Furthermore, the capacity of these technologies to be used for military purposes opens the door to dual-use risks which is something that should also be considered, as the restrictions on the funding, usage, and selling of dual-use items have important limitations.

When technologies can be used for military purposes, they are called dual-use items and are subject to the EU dual use Regulation (EC, No 428/2009) [4]. This regulation provides a list of dual-use items and sets up a Community regime for the control of their exports, transfer, brokering, and transit. Any project seeking funding from an EU public body should ensure that none of the products or components is part of that list and provides specific guarantees to this effect.

12.3.3 Other Ethics Issues

Other ethical issues in the context of big data analytics for time critical mobility forecasting can arise. These relate mainly to unexpected or controversial uses, as was the case with Strada and Polar flow maps, and data management issues, such as lack of transparency and accountability, which can later lead to bigger issues such as data breaches. In the context of datAcron, for instance, we identified the following issues and discarded them as risks after going over their likelihood in the specific context of the project and in the potential use of the technologies developed later on.

12.3.3.1 Subjective Thresholds

Thex technologies developed within datAcron included among their aims the detection of "threatening" or "abnormal" activity and "important" events. There are culturally charged definitions that require that specific protocols to determine what is "threatening," "abnormal" and "important" are defined to ensure the consistency of the decisions made by the relevant actors or automatic processes. Projects faced with similar challenges therefore need to define clearly the terms they employ when they are setting their goals and see whether these might be subject to a range of interpretations.

Clearly defining terms helps ensure that certain negative values are not inadvertently built into systems. Additionally, thinking through the terms in light of equality, fairness, and non-discrimination promotes accountability and social acceptability.

12.3.3.2 Algorithms and Automatic Decision-Making

Forecasting tools use algorithms to make decisions, e.g. on the potential future location and trajectory of a moving object, or on events that may occur. If these algorithms only make simple decisions (identify speed gating and displacement of vessels, for instance), they may not be problematic.

However, if at any point algorithms are expected to make complex decisions (beyond database matching, for instance), the potential societal impact and risk of these should be assessed and established. It is crucial in such cases that the teams developing the algorithms be trained on these issues, as well as on the topic of algorithmic bias. Algorithmic bias can occur at various stages of the algorithm's development. It may be due to the data exploited by the algorithm, which might be biased or outdated or wrong; or it may be due to the way the design of the algorithm translates the developers' societal biases, or because the algorithm's output is wrongly used or interpreted. Assessing these biases is important in ensuring that the algorithm fulfills its purpose properly.

12.3.3.3   Accuracy of Data

Data can be contradictory or imprecise. Making decisions on the basis of bad and low-quality data (e.g., noisy, imprecise, inadequate, biased, outdated, uncleaned). In essence, data unfit for its purpose can have important consequences for data-driven technology developments and the decision-making processes. The adequacy and reliability of data sources in conjunction with our abilities to identify "corner cases" (i.e., cases where our algorithms are not trained at all, or not trained properly) are therefore crucial to the development of successful technological outcomes.

12.3.3.4   Undesirable Reuse and Threat Modeling

Once data is created, it cannot be protected against all risks, as the possibility of a threat gaining access to it is never zero. Therefore, it is important to define who or what would constitute the threat model of the forecasting tool—could other entities want unwarranted access to its input data and/or forecasting? Could this data benefit any category of entities (insurance companies, for instance)? Would the data developed by the tool have any commercial value that would make it attractive to third parties? These are questions that need to be discussed in all projects.

12.3.3.5   Technological Divide and Discrimination

The use of technology is not evenly distributed geographically or socially. Therefore, any data-driven initiative needs to take steps to ensure that it does not reproduce or reinforce existing discrimination. In the case of datAcron, for instance, only 60% of all passenger aircraft around the world are equipped with an Automatic Dependent Surveillance-Broadcast (ADS-B) transponders. Hence, whether those entities using less or different technologies can be left out of the benefits of the tools developed should be envisaged. Furthermore, whether this is due to geographical differences that can limit the scope of the solutions developed, and if this will translate into a differential access to its benefits for different actors, and to what extent this can be minimized, should be considered.

All of these ethical issues are to be kept in mind and delved into when developing tools and systems for big data analysis in time critical mobility forecasting. The following section goes over the actions to take in order to avoid ethics issues in big data analytics for time critical mobility forecasting, taking as a model the European Commission's monitoring scheme.

12.4 Tackling Ethics Issues

12.4.1 Identification of Factors for Privacy Concerns

In order to evaluate the privacy risks in big data analytics for time critical mobility forecasting, it is essential to first understand the data ecosystem surrounding the technology. Therefore, three factors should be taken into account: the types of moving entities covered, the provenance and type of data exploited, and the actors and end-users involved.

12.4.1.1 Types of Moving Entities Involved

A key factor to determine the potential privacy risks and issues depends directly on the kind of moving entities that are monitored, as their nature might render their monitoring problematic from a privacy perspective, or even illegal. If only very large fleets of moving entities are included, the privacy risks decrease immensely; however, privacy issues may arise when datasets include data on smaller individually owned moving entities, or governmentally owned moving entities. During the design of the system, it is necessary to consider the potential (intended or unintended) effects of the identification of the moving entities:

- If the identification of a certain model of vessel or aircraft could provide information on its use beyond what the owner/s disclosed;
- If the combination of model and geolocation data could facilitate the identification of a specific model or type of vessel or aircraft—information that, in its turn, could reveal information on use or other types of information that one may want to keep private.

When developing a technology which tracks movement, these risks should be taken into account, and specific security, privacy, and data protection processes should be created to tackle them.

12.4.1.2 Provenance and Type of Data

Another factor to determine the risks related to privacy and personal data is the type of data processed and its provenance. Since big data exploitation implies the processing of several data sources, it is necessary to evaluate the potential consequences of the correlation of various data sources (archival or streams). Hence, the following concerns arise:

- The existence of unique identifiers for mobile entities (e.g., aircraft and vessels), and the potential development of "reverse search" mechanisms.
- The consequences of "in situ" data processing (i.e., processing of data close to the sources that generate this).
- The use of open data and the possibilities of re-identification (both for the retrieved data as for the resulting datasets).
- Exploitation of messaging systems (e.g., Controller Pilot Data Link): content, meta-data, etc.
- Data revealed by monitoring and information provision systems like the Automatic Identification System and others.

A possible solution to ensure good practices is the development of a specific protocol for the selection of data sources. This includes a comprehensive list of all the data sources and of interesting links between them, and their assessment in terms of whether they include or allow the inference of personal data, among other relevant aspects.

12.4.1.3 Actors/End-Users

Identifying the actors involved in the development of technology and identifying the foreseen end-users can also help identify issues. Depending on the characteristics of the collected, stored, and shared data, confidentiality concerns could also arise. For this reason, once the data sources and resulting datasets are defined, it is necessary to define who will be able to access them and how this information will be recorded (logs to monitor the activities in the system).

It is important to foresee who will have access not only to the technology, but also to the resulting information, and under which conditions and protocols, i.e. who will be able to monitor the moving entities, what technical and managerial capabilities will they have, which is the responsibility chain in case of an unforeseen event, and which processes need to be in place, for instance, in case criminal activities are discovered. Setting out the interfaces used and the context in which the technology is being used also helps in assessing possible vulnerabilities.

12.4.2   Identification of Factors to Avoid Function Creep

The identification of these three factors will help foresee possible privacy issues. In the same vein, the steps to be taken to avoid function creeps include:

- Describing the initial definition of the expected application of the project at hand:

  - Geographical areas/countries included:

    Areas covered,
    Open sea/coastline, also indicating potential function differences,
    Airspace volumes included;

  - Entities and end-users that are expected to have access to the information and make use of the system;
  - Utilization guidelines and protocols (especially in case of abnormal events).

- Considering the later incorporation of data sources through the scalability options.

  - Broadening of the scope of the system;
  - Moving entities involved;
  - Utilization possibilities (e.g., for law enforcement purposes).

- Updating and reviewing the system and its function/mission creep risks.
- Avoiding the use of the resulting systems and tools for other contexts/other moving entities (e.g., vehicle traffic management), as different moving entities might pose different privacy issues.
- Identifying unexpected actors that may make use of the technologies developed (e.g., criminals attempting to act in open seas).

Once these identification exercises have been carried out, it is much easier to determine the likelihood of privacy concerns, function creeps, or the occurrence of an undesired event.

12.4.3   The Integration of Privacy by Design Within the Process

"Privacy by Design" (PbD) is a concept developed back in the 1990s by Ann Cavoukian, Information and Privacy Commissioner for Ontario (Canada) [2]. It aims at establishing privacy assurance as the default mode of operation of an organization. Principles of Privacy by Design should apply to all types of processing which might involve personal information. The objectives are to ensure privacy to achieve control over one's information (even indirectly), and gaining a sustainable competitive advantage for organizations.

In the case of projects like datAcron, the rather low probability of ethics issues due to the low occurrence of personal data did not mean Privacy by Design should not be followed. It rather meant that applying the principles is easier, as less areas and processes need to be developed. Even in such projects, following a PbD approach is important to avoid ethics issues linked to privacy, but also other ethics issues such as unforeseen disclosure of information.

The Privacy by Design approach is based on 7 foundational principles, which are the following:

1. **Proactive Not Reactive; Preventative Not Remedial**

   The Privacy by Design (PbD) approach is characterized by proactive rather than reactive measures. It anticipates and prevents privacy invasive events before they happen. PbD does not wait for privacy risks to materialize, nor does it offer remedies for resolving privacy infractions once they have occurred—it aims to prevent them from occurring. In short, Privacy by Design comes before-the-fact, not after.

2. **Privacy as the Default Setting**

   We can all be certain of one thing—the default rules! Privacy by Design seeks to deliver the maximum degree of privacy by ensuring that personal data are automatically protected in any given IT system or business practice. If an individual does nothing, their privacy still remains intact. No action is required on the part of the individual to protect their privacy—it is built into the system, by default.

3. **Privacy Embedded into Design**

   Privacy by Design is embedded into the design and architecture of IT systems and business practices. It is not bolted on as an add-on, after the fact. The result is that privacy becomes an essential component of the core functionality being delivered. Privacy is integral to the system, without diminishing functionality.

4. **Full Functionality—Positive-Sum, Not Zero-Sum**

   Privacy by Design seeks to accommodate all legitimate interests and objectives in a positive-sum "win-win" manner, not through a dated, zero-sum approach, where unnecessary trade-offs are made. Privacy by Design avoids the pretense of false dichotomies, such as privacy vs. security, demonstrating that it is possible to have both.

5. **End-to-End Security—Full Lifecycle Protection**

   Privacy by Design, having been embedded into the system prior to the first element of information being collected, extends securely throughout the entire lifecycle of the data involved—strong security measures are essential to privacy, from start to finish. This ensures that all data are securely retained, and then securely destroyed at the end of the process, in a timely fashion. Thus, Privacy by Design ensures cradle to grave, secure lifecycle management of information, end-to-end.

6. **Visibility and Transparency—Keep It Open**

   Privacy by Design seeks to assure all stakeholders that whatever the business practice or technology involved, it is in fact, operating according to the stated promises and objectives, subject to independent verification. Its component parts and operations remain visible and transparent, to users and providers alike. Remember, trust but verify.

7. Respect for User Privacy—Keep It User-Centric

> Above all, Privacy by Design requires architects and operators to keep the interests of the individual uppermost by offering such measures as strong privacy defaults, appropriate notice, and empowering user-friendly options. Keep it user-centric. [2]

Privacy by Design includes both operational and organizational measures. Since the use of personal data in the big data analytics for time critical mobility forecasting of moving objects is minimal, there are perhaps less measures to put in place to ensure Privacy by Design, though these might be harder to identify.

In the case of big data analytics for time critical mobility forecasting, the following table provides recommendations for the application of Privacy by Design for operational issues:

Principle	Application to big data analytics for time critical mobility forecasting
Proactive not Reactive	• Consider the risks of the processed data by listing all the expected data sources and its characteristics. • Consider the actors/end-users involved.
Privacy as the Default Setting	• Consider the possibility of not tracking if it is not compulsory by the law. • Put measures in place to avoid the identification of individuals.
Privacy Embedded into Design	• Assess the privacy issues that might arise.
Full Functionality	• Consider privacy as important as other factors.
End-to-End Security	• Taking into account the risks vs. benefits, consider implementing end-to-end encryption.
Visibility and Transparency	• Allow the inspection and verification of the resulting system through external evaluation audits.
Respect for User Privacy	• Inform the affected identifiable users about the monitoring activities.

These recommendations overlap with the identification of factors developed in the two previous sections and are meant to give the developers a better idea of how to develop their project in a manner that respects the privacy of those whose personal data might be involved. It is worth mentioning that the application of Privacy by Design principles is especially relevant in case of unexpected situations. In those cases, the pertinent staff responsible for the system should have event management procedures to ensure good practice.

12.4.4   Drafting a Data Management Plan

The identification of the previously-listed items is the first step to drafting a Data Management Plan (DMP), a necessary step to ensure compliance with Privacy by Design principles. DMPs set out the whole ecosystem surrounding data and especially personal data within a project. They are divided into three sections (and are to be adapted according to the specificities of every project):

- A data summary: This gives the overall outline of the characteristics of the data being processed, and the purposes which justify its collection. This is especially relevant to any personal data being processed.
- Allocation of resources: This identifies the roles and responsibilities for each actor involved in the development of a technology.
- Data security: This part gives information on data protection, storage, recovery, transfer, and retention.

The adequate measures to put in place to protect data are dependent on the type of data being processed, its use, and who will have access to it. Therefore, the Data Management Plan will depend on these factors. Having an adequate DMP is essential in order to tackle potential privacy issues as they come along.

12.4.5   Identification and Documentation of the Licensing Options

Lastly, the identification of the appropriate licensing options is an important step in every project, especially those involving datasets, as it helps mitigate potential misuses by controlling the dissemination and reuse of the technologies and their outputs.

There are currently several projects and commercial products aimed at making available datasets reflecting the activities of aircraft and vessels. These projects give an interesting overview of data sharing procedures, as well as the related questions to be considered (privacy policies, licensing, provenance of datasets, etc.). These are some of the relevant initiatives that can be used as benchmarking for big data analytics for time critical mobility forecasting:

OpenFlights.org   This website shares with the public airport, airline, and route data since 2009. The databases are made available under the Open Database License and any rights in individual contents of the database are licensed under the Database Contents License. http://openflights.org/data.html

FlightRadar24   Flightradar24 is a flight tracking service that provides users with real-time info about thousands of aircraft around the world. Under the terms and conditions section, the group refers to the "non-transferable right to access and use the services." This right is granted for personal, non-commercial use only. There are also references to frame the data collection and transmission procedures and possible unauthorized uses. https://www.flightradar24.com/data/

Sea-Web Sea-web combines comprehensive ships, owners, shipbuilders, fixtures, casualties, port state control, and real-time ship movements and ports information into a single application. Sea-web is an enhanced service that includes all information from the Internet Ships Register but provides many additional benefits. It covers over 180,000 ships down to 100 GT. http://www.ships-register.com/

FleetMon This database allows searches by name, identification numbers, flag state, length, and vessel type; it also allows tracking specific vessels to follow their activity. The website grants a limited, revocable, non-exclusive, non-transferable, and non-sublicensable license for beneficial use of the content accessible on the site. https://www.fleetmon.com/vessels/

MarineTraffic It is a vessel tracking service that covers monthly up to 800 million vessel positions and 18 million vessel and port related events; it also provides details of over 650 thousand marine assets available (vessels, ports, lights). Regarding utilization rights, it states that "the user shall use the information and data for his/her own internal use only and shall have no other rights with respect to the data, including without limitation, any right otherwise to use, distribute, furnish or resell the data or any portion or derivative thereof." http://www.marinetraffic.com

In the case of big data analytics for time critical mobility forecasting, the following licensing issues are to be tackled:

Licensing of the Source Data Licensing issues do not only concern the technology or output data, but also the rights in terms of the use of data sources. Once the provenance of the data is detailed, it is necessary to ensure that all utilization permissions are granted (including all types of data, like ship pictures). Mapping systems may be based on different alternatives like OpenStreetMap (OSM) or Google Maps. Product and company names may be the trademarks of their respective owners. Open data sources also have conditions and limitations for their use, which depend on the type of license upon which they were published.

Licensing of the Output Data The output data should be subject to clear licensing options. It is necessary to foresee the use scenarios and the scope of end-users and look at the terms regarding storage, property rights, reuse, reutilization, and commercialization.

Licensing of the Resulting Tools The resulting tools (interfaces, analysis software, etc.) will be subject to specific distribution, utilization, and, in certain cases, modification rights.

Transparency and Open Data Issues In case of publicly accessible data, transparency regulations are subject to limitations related to privacy and security issues.

Protocols It is recommendable to designate a contact person and define protocols in case of misuse or conflict.

A useful solution to keep control of the licensing issues is to create a table where source data and output data are listed, and adding a column for the licensing concerns as they arise.

Taking these issues into account, the following standard licensing options for output data and tools allow enough flexibility and control to be used for projects such as datAcron [1]:

- Creative Commons:

  Creative Commons is a non-profit corporation set up in 2001 for the purpose of producing simple yet robust licenses for creative works. While originally aimed at works such as music, images and video, Creative Commons licenses have been used widely for most forms of original content, including data. There are six main Creative Commons licenses. While the spirit behind them has remained constant, the wording of their legal deeds has been revised over time, resulting in different versions, and adapted to different legal jurisdictions, resulting in different ports.

  All versions of the licenses treat datasets and databases as a whole: they do not treat the individual data themselves differently from the collection/database. This might be considered an advantage in terms of simplicity, but means they cannot be used without difficulty in certain complex cases such as collections of variously copyrighted works. Similarly, the licenses do not distinguish using data as part of a new collection/database from using them to generate content (graphs, models, maps, etc.).

- Open Data Commons:

  The Open Data Commons Project was set up in 2007 to develop a successor to the Talis Community License (TCL). The first license to be produced was a public domain dedication for databases. The project transferred to the Open Knowledge Foundation in 2009 and has produced two further licenses having some of the character of the Creative Commons licenses, but designed specifically for databases. The Open Data Commons Attribution License (ODC-By) allows licensees to copy, distribute and use the database, to produce works from it and to modify, transform and build upon it for any purpose. If content is generated from the data, that content should include or accompany a notice explaining that the database was used in its creation. The Open Data Commons Open Database Licence (ODC-ODbL) is the same as ODC-By but for a couple of additional conditions.

Deciding on the licensing option once again depends on the nature of the technology or dataset, its use, and the desired openness assigned to it.

12.5   Conclusion

This chapter has gone over the ethics issues that projects such as datAcron might encounter and how to solve such issues. The efforts carried out at the European level to tackle ethics issues in research projects developing new technologies were first explained. These led to the ethics monitoring scheme developed by the European Commission in order to ensure that the projects it funds follow international standards in terms of ethics.

The most relevant ethics issues were then delved into. These are the possible consequences of event recognition, trajectory analytics, and misuse by end-users for privacy and security, which can arise when individually owned vessels and planes are included in the dataset, and potential misuse of the technology by end-users. Other potential issues were also discussed, such as subjective thresholds in defining terms, automated decision-making, accuracy of data, undesirable reuses of data, and technological divide and discrimination. These, however, are more high-level and linked to data management.

Lastly, recommendations were made to tackle ethics. To this effect, the usefulness of the identification of certain factors which can lead to issues was explained, followed by an explanation of the concept of Privacy by Design and of a data management plan. Finally, licensing options and their relevance were discussed. The implementation of these measures leads to robustly tackling ethics issues both before and during the development of projects developing new data-driven technologies such as datAcron.

References

1. Ball, A.: How to license research data. How-to guides, Edinburgh, Digital Curation Centre, 2014. Available at: https://alexball.me.uk/docs/ball2011hlr?style=plain
2. Cavoukian, A.: Privacy by Design. The 7 Foundational Principles. Information and Privacy Commissioner of Ontario, vol. 5, 2009, Available at: https://www.ipc.on.ca/wp-content/uploads/resources/7foundationalprinciples.pdf
3. Collins English Dictionary. Collins English Dictionary - Complete & Unabridged 2012 Digital Edition. [Online] Available at: http://www.dictionary.com/browse/function-creep
4. Council Regulation (EC) No 428/2009 of 5 May 2009 setting up a Community regime for the control of exports, transfer, brokering and transit of dual-use items, 2009
5. European Commission, Report of the Societal Impact Expert Working Group EC DG ENTR Report February 2012. , s.l.: s.n., 2012
6. Gartner. [Online], IT Glossary. 2013. Available at: http://www.gartner.com/it-glossary/big-data/
7. Hern, A.: Fitness tracking app Strava gives away location of secret US army bases. The Guardian, 18 January 2018
8. Schomberg, V.: Towards responsible research and innovation in the information and communication technologies and security technologies fields. , s.l.: s.n., 2011
9. Tan, R.: Fitness app Polar revealed not only where U.S. military personnel worked, but where they lived. The Washington Post, 18 July 2018

Index

Printed in the United States
by Baker & Taylor Publisher Services